# BIOELECTRIC PHENOMENA

# BIOELECTRIC PHENOMENA

**Robert Plonsey**
*Professor of Biomedical Engineering*
*Case Western Reserve University*

*with the Introduction and first chapter by*

**David G. Fleming**
*Professor of Biomedical Engineering*
*Case Western Reserve University*

McGraw-Hill Book Company

New York    St. Louis    San Francisco    London
Sydney    Toronto    Mexico    Panama

**BIOELECTRIC PHENOMENA**

*Library of Congress Catalog Card Number* 69-17189

07-050342-7

456789 KPKP 79876

*For my parents:*
*Louis and Betty*

*my wife:*
*Vivian*

*and my son:*
*Daniel*

# Preface

This book is the outgrowth of the course in bioelectric phenomena developed as part of the bioengineering program at Case Institute of Technology (now Case Western Reserve University). The objective of the course has been to acquaint those graduate students whose backgrounds are in the physical sciences and engineering with the electrical behavior of nerve and muscle and, in particular, where possible, to utilize mathematical techniques to provide a quantitative basis for observed phenomena. In this task it has been necessary to be selective, and the material presented covers membrane phenomena, the propagated action potential, the volume conductor, and electrocardiography.

Although an effort was made to be timely in all topics, the rapid advance of science was clearly evident in the several revisions of class notes that have led to the present text. This has underscored, for the author, the importance of emphasizing the "fundamentals." Unfortunately, it is not always possible to determine what priority to assign the fundamentals, nor which they are.

Since the author's interest in the life sciences is relatively recent, this book represents, in a sense, his own education. In particular, it follows the pattern of thinking characteristic of the engineer (bioengineer) and physicist (biophysicist). Hopefully, this presentation will be of interest to the life scientist, too.

The author wishes to acknowledge the debt owed to many who assisted, in one way or another, his developing interest in and understanding of bioengineering and the realization of this book. To:

The late Dr. Donald P. Eckman, who confidently "staked" the author to participation in the Systems Research Center Life Sciences Group.

Dr. David G. Fleming, friend and colleague, who first suggested that we jointly teach a course on bioelectric phenomena, who continually encouraged the development of the course and of this text, and who has given freely much constructive criticism.

Dr. Nick Sperelakis, who, when he was at Western Reserve University, helped the author see how much more he needed to learn in the area of electrophysiology.

The many graduate students, past and present, and, in particular, Drs. John Clark, Ronald Cechner, Fred Terry, and Messrs. Roy Shubert and Dennis Heppner, for their penetrating questions and comments, and special thanks to Mr. Paul Palatt and Mr. Maurice Klee for a host of corrections and suggestions.

Several hard-working secretaries undaunted by a hieroglyphic-like handwriting—in particular, Elaine Brunkala, Janet Leonard, Kris Grimes, and Donna Baznik—who became so skilled they found numerous technical, as well as grammatical, errors. Thanks also to Donna Baznik for her gallant efforts in behalf of the Name and Subject Indexes.

My wife for accepting my, often, inaccessibility while buried in the manuscript.

The National Institutes of Health, through whose support much of the time to develop this material, and the course on which it is based, was made available. In particular, credit goes to N.I.H. grants GM 12203, GM ITI-1090, and HE 10417, from the Public Health Service.

McGraw-Hill Book Company and Editing Supervisor, Molly Scully, for the helpful editorial assistance provided.

To all the above, my heartfelt gratitude.

*Robert Plonsey*

# Contents

# Introduction

The phenomenon of communication in living organisms may be examined in a variety of ways. There is a panoramic view from the outside, filled with a plethora of detail, awesome in complexity, and at least for the present somewhat unknowable. This same phenomenon when viewed from the inside, as if through a pinhole, is in sharp focus, but dim and of a restricted field. This metaphor, albeit trite, illustrates the conflict facing the physical scientist who wishes to apply his training and talents to the study of this problem in some meaningful way. For example, I was aware of the conflict in Dr. Plonsey's mind when, in preparing the outline for this book, he wished to produce a volume based on physical laws, yet nontrivial in its scope. His decision was to limit the coverage to bioelectric phenomena describable by physical law. For this reason the material in the volume is largely devoted to topics considered axon and membrane biophysics by neurophysiologists.

Many of the details and almost all of the principles relating electrical activity in the nervous system to behavior are still unknown to us.

The squid, the limulus, and the frog have provided their share of information.   The cat, the monkey, and, to a lesser extent, man have all contributed to the pool of data.   The myriad of interactions between electrical and biochemical mechanisms are slowly being identified.   Systems theoretic approaches to higher-order events are coming into existence. These methods will be useful in ordering and mapping the panoramic scene.   But for the present, however, it does not matter how we may eventually unravel the complexity of the nervous system, if we keep in mind the principle that any complete description must be consistent and compatible with physical laws.   The primary contribution of this book is that it brings together much of the pertinent physical theory.   For this reason the volume will serve as a general reference for the experienced investigator and as an introduction for the novice graduate student.

*David G. Fleming*

# BIOELECTRIC PHENOMENA

# 1
# Physiology of Nerve and Muscle
## by David G. Fleming

## 1.1 INTRODUCTION

This book deals with the general phenomenon of bioelectricity and has been written with the physical scientist and engineer in mind. As this may be the reader's first exposure to this material, an introductory chapter on the elements of nerve and muscle physiology is included. This chapter presents a rather cursory description of the structure of nerve and muscle along with a set of observations on the electrical behavior of each. This treatment, however brief, is intended to provide background for the remainder of the book. The reader who is already familiar with these subjects should feel free to proceed immediately to Chap. 2.

## 1.2 THE NERVE CELL (NEURON)

The nervous systems of multicellular organisms are typically composed of ensembles of *neurons* or nerve cells. These may be arranged in a simple net as seen in coelenterates, or in complex arrays as observed in

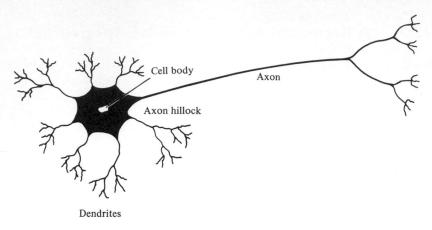

**Fig. 1.1**   Parts of a typical neuron.

advanced invertebrates and vertebrates.   In the latter group of animals, individual nerve cells may vary both in anatomical configurations and in functional properties.   While these differences must be considered critically in assessing the overall operation of the nervous system in any given organism, there are both structural and functional features common to a wide variety of neuronal cell types.

Current concepts of idealized neurons stem primarily from studies on mammalian spinal motoneurons, crustacean sensory neurons, the squid giant axon, and the nerve-muscle junction in the frog.[1,2]   A general conclusion is that all neurons, regardless of individual differences in shape, size, or location, have four functional components: (1) one or more input elements; (2) an integrative element; (3) an active transmission line; (4) one or more output elements or terminals.

The anatomical counterpart of the idealized neuron is illustrated in Fig. 1.1.   Classical studies with the light microscope, augmented by contemporary investigations using the electron microscope, suggest that the neuron is divisible into *dendrites*, the cell body or *soma*, the *axon*, and the terminal region.

The neuron, in common with other cells, is surrounded by a complex plasma membrane whose thickness is estimated to be between 50 and 150 Å.   Inputs to the neuron may occur at many points on its surface.   The majority of the inputs, however, enter through the dendrites, which consist of single- to many-branched structures extending

[1] J. C. Eccles, "The Physiology of Synapses," Springer-Verlag OHG, Berlin, 1964.
[2] S. W. Kuffler, Excitation and Inhibition in Single Nerve Cells, *Harvey Lectures*, **54**:176–218 (1958–1959).

from the cell body.   At their terminal ends the twigs of dendritic *trees* interface with extensions from other neurons or sensory cells.   The specialized structures called *synapses*, seen at the junction between neurons, are of prime importance, for they contain the mechanisms for information flow from one cell to another.   The integrative process takes place either in the dendritic structure or in the soma.   If the algebraic sum of input excitations exceeds a threshold level, the cell "fires," generating a signal which is actively transmitted down the length of the axon to the terminal regions.   These axon endings constitute the presynaptic portions of the following set of junctions.

As noted, there is a considerable variation in the size of nerve cells.   The diameters of the somata of vertebrate neurons range downward in size from approximately 30 microns to just a few microns.   It is difficult to estimate the length of dendrites; however, they probably reach a maximum of 2 mm in the outer layers of the cerebral cortex. Axon length varies from a minimum in the order of 50 microns to a maximum of several meters in large mammals.   Axonal cross-sectional area also spans a broad range.   In vertebrates, large sensory and motor fibers may have diameters of 20 to 25 microns.   Axons with diameters less than 0.5 micron are found in sensory ganglia.   In the squid there are about 20 giant nerve fibers with diameters ranging up to 1 mm. The largest two of these fibers are more than 20 cm long.

Large nerve fibers are characterized by the fact that they are surrounded by a *myelin sheath* of mainly lipoid material, as shown in Fig. 1.2.   The myelin sheath in turn is surrounded by a specialized type of cell called the *Schwann cell*.   According to present concepts, the myelin sheath actually consists of many layers of Schwann cell membrane which were left behind as the cell body rotated around the axon during growth.   The sheath is regularly interrupted at axial intervals of 1 mm to give pronounced indentations known as *nodes of Ranvier* (Fig. 1.3).   The nodal region is characterized by its low electrical resistance.   This geometry imparts certain constraints on conduction in myelinated axons, and this will be discussed in detail later in the book.

The plasma membrane surrounding a neuron has been the subject of intensive investigation using electron microscopy and x-ray diffraction methods.   Additional information on the properties of the membrane has been deduced from permeability, electrical conductivity, and surface-tension measurements.[1]

Robertson[2] has proposed a *unit-membrane theory*.   According to

[1] E. D. Korn, Structure of Biological Membranes, *Science*, **153**:1491–1498, 23 September 1966.

[2] J. D. Robertson, Unit Membranes, in M. Locke (ed.), "Cellular Membranes in Development," Academic Press Inc., New York, 1964.

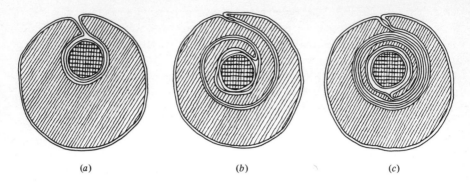

$(a)$                          $(b)$                          $(c)$

**Fig. 1.2**  Formation of myelin sheath.  $(a)$ The axon is embedded in a Schwann cell.
$(b)$ Myelin sheath is produced when layers of membrane separating the overhanging
lips of Schwann cell cytoplasm come together and elongate.  $(c)$ Myelin sheath is
formed into a tightly packed spiral around the axon.  [*After J. D. Robertson, New
Unit Membrane Organelle of Schwann Cells, in A. M. Shanes (ed.) "Biophysics of
Physiological and Pharmacological Actions," American Association for the Advancement
of Science, Washington, D.C., 1961.*]

this proposition there is a basic structure common to all membranes,
or major portions of all membranes, in essentially every type of cell.
This structure is composed of a bimolecular leaflet of phospholipid,
whole nonpolar portions are inwardly oriented perpendicular to the
plane of the membrane.  The polar moieties of the phospholipids are
covered with a layer of protein.  An appreciable amount of cholesterol,
carbohydrates, and some neutral fats may also be present.  There is
some reason to believe that the inner and outer protein layers serve
both structural and enzymatic functions.  The nature of the binding
forces within the membrane is not clearly understood.  The unit membrane is shown in Fig. 1.4.

The existence of potential gradients across cell membranes has been

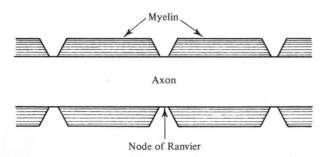

**Fig. 1.3**  Myelinated nerve fiber showing nodes of Ranvier.

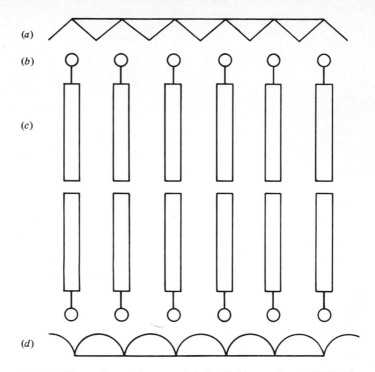

**Fig. 1.4** The unit membrane. (*a*) Nonlipid monolayer; (*b*) lipid molecules—polar end of the molecule; (*c*) hydrophobic carbon chains; (*d*) nonlipid monolayer, different from (*a*).

verified repeatedly for a wide variety of cell types from single large plant cells through animal nerve, muscle, and secretory cells. The manner of the establishment and maintenance of these potentials, although by no means completely elucidated, is the subject of extensive treatment in the later chapters of this book. The magnitude of these potentials in resting cells is in the order of 100 mv with the inside negative.

A unique feature of excitable cells is the brief transition in membrane properties which takes place when the transmembrane potential is reduced below a threshold value. In normal neural activity this is brought about by an appropriate superposition of inputs. When the threshold is surpassed, a complex limit cycle oscillation takes place. Since the response is essentially independent of the magnitude of the transthreshold stimulus, it has been classically termed an *all-or-none* phenomenon. During this response, the transmembrane potential rapidly switches from negative to positive and then, more slowly, recovers. The total event is known as the *action potential*. As a conse-

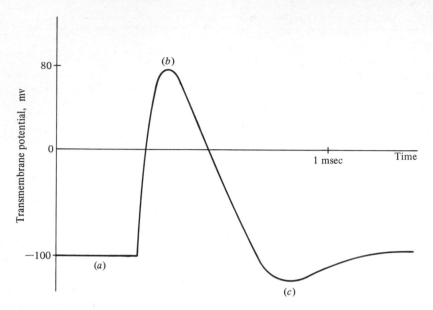

**Fig. 1.5**  Transmembrane potential vs. time for an action potential.   (*a*) Resting potential; (*b*) spike; (*c*) overshoot.

quence of the changing polarity, which characterizes the action potential at a point on the membrane, an ionic current flow takes place through the internal and external media.   This *local current* pattern serves to reduce the resting-membrane potential beyond threshold in the regions adjacent to the active area.   This sequential triggering spreads the action potential contiguously to the remaining membrane.   In this manner *propagation* of the action potential takes place.   A representative transmembrane action potential is illustrated in Fig. 1.5.

The presence of the myelin sheath around large axons tends to alter the mode of propagation in this type of fiber.   The sheath, which may be composed of a considerable number of tight turns of 120-Å-thick lamina, has low electrical conductivity and functions much as insulating material on a metallic wire.   The node of Ranvier is covered only by Schwann cell cytoplasm, and from a functional point of view, the axon plasma membrane is bare at the node.   Activation in myelinated axons, therefore, occurs at a nodal region and produces a local circuit current which completes the closed loop via adjacent nodes.   As current flow is constrained primarily to nodal areas, current density is relatively high at these circumscribed sites.   For this reason activation "jumps" from node to node, and this phenomenon is called *saltatory* conduction.

A connective-tissue system binds individual peripheral nerve fibers

**Table 1.1  Properties of mammalian nerve fibers***

| Property | A | B | sC | drC |
|---|---|---|---|---|
| Fiber diameter, microns | 1–22 | ≤3 | 0.3–1.3 | 0.4–1.2 |
| Conduction velocity, m/sec | 5–120 | 3–15 | 0.7–2.3 | 0.6–2.0 |
| Spike duration, m/sec | 0.4–0.5 | 1.2 | 2.0 | 2.0 |
| Absolute refractor period, m/sec | 0.4–1.0 | 1.2 | 2.0 | 2.0 |

* T. Ruch and H. Patton (eds.), "Physiology and Biophysics," W. B. Saunders Company, Philadelphia, 1965.

into a *nerve trunk*. Individual axons are covered by a connective-tissue tube called the *endoneurium*. Bundles of nerve fibers are bound together by a laminated capsule, the *perineurium*, which has alternating layers of connective tissue and endothelial cells in mammals. The entire nerve trunk is enveloped by a system of loose connective tissue, the *epineurium*. The sheaths appear to act as a diffusion barrier between the fibers within the nerve trunk and the extracellular fluid space.

The individual axons which make up the nerve trunk may vary in terms of diameter, myelin-sheath thickness, and other electrical properties. There are four separable classes of axon types known as A, B, sC, and drC. Table 1.1 summarizes some of the electrical and physical differences among them. Type A fibers are myelinated and have the largest diameter. Type B fibers are somewhat narrower and are more thinly myelinated. The C fibers are small and not myelinated.

A schematized cross-sectional view of a typical nerve trunk is shown in Fig. 1.6. Since the nerve trunk consists of a collection of axons

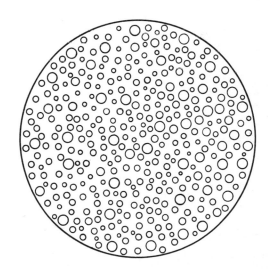

**Fig. 1.6**  Cross section of a nerve trunk.

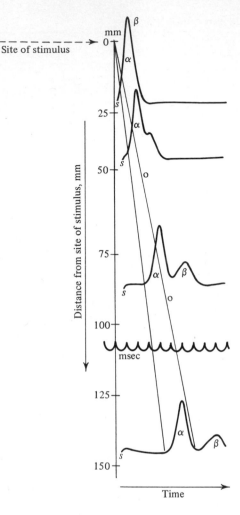

**Fig. 1.7** Compound action potential recorded at four sites along a nerve trunk. Straight lines connect the onset of $\alpha$ and $\beta$ deflections; their slopes are inversely proportional to the conduction velocity. (*After J. Erlanger, and H. S. Gasser, "Electrical Signs of Nervous Activity," University of Pennsylvania Press, Philadelphia, 1937.*)

differing in both size and type and, therefore, with different conduction velocities, propagation proceeds dispersively. As a result, the shape of an action potential initiated at one end varies as a function of axial distance. This phenomenon is demonstrated in Fig. 1.7, which shows the recorded *compound action potential* at different distances from the origin at which the stimulating electrode is located. This illustrates that the sciatic nerve from the frog contains two main fiber groups, $\alpha$ and $\beta$,† with the $\alpha$ group exceeding the velocity of the $\beta$ group. Each

---

† The type A fibers are subdivided into groups $\alpha$ to $\delta$ according to decreasing fiber diameter, with the $\alpha$ and $\beta$ group comprising fibers of approximately 16 to 6 microns diameter.

straight line, which connects points at the onset of activity of a group, has a slope that is inversely proportional to the velocity of that group. Measurements made on isolated A fibers show an approximate proportionality between conduction velocity and diameter. This empirical result is expressed mathematically as[1]

$$V = 2.50D \qquad (1.1)$$

where $V$ is the velocity in meters per second and $D$ is the diameter in microns. If the velocity, as determined from compound-action-potential measurement, is compared with the largest myelinated fiber, as determined subsequently by histological measurement, the coefficient in (1.1) comes out between 6 and 9.[†]

The threshold for activation is lower for the larger-diameter fibers. Consequently, as the stimulus intensity of a nerve trunk is raised, the magnitude of the compound action potential increases. This apparently "graded" response is actually a consequence of the change in numbers of fibers which are being excited. Ultimately, a stimulus strength that "fires" all fibers is reached, and higher stimulating strengths produce no further increase in compound-action-potential magnitude. For each component fiber, the all-or-none law is obeyed, as discussed earlier.

## 1.3 THE ELECTRICAL PROPERTIES OF SKELETAL MUSCLE

From a superficial point of view, the electrical properties of voluntary or skeletal muscle are comparable to those of nonmyelinated axons. Fast *twitch* muscle fibers resemble nerve axons in that they can produce action potentials at any point on their surface. In these twitch muscle fibers a mechanism which is similar to that in nerve enables the excitation to spread quickly over the whole cell at a velocity of several meters per second. Muscle-nerve differences lie mainly in the special anatomical features of muscle cells and in the role played by the $Ca^{++}$ ion in muscle excitation.

A skeletal muscle is built up from a set of individual 10- to 100-micron-thick fibers which are functionally separable. The resting length of a fiber ranges from a few millimeters to several centimeters, depending on the muscle in which it is found. Each fiber is enveloped by a fine double membrane, the *sarcolemma*. The outer membrane consists of a

[1] I. Tasaki, I. Ishii, and H. Ito, On the Relation between the Conduction Rate, the Fiber-Diameter, and the Internodal Distance of the Medulated Nerve Fiber, *Jap. J. Med. Sci., III Biophysics*, **9**:189 (1943).

[†] H. S. Gasser and H. Grundfest, Axon Diameter in Relation to the Spike Dimension and Conduction Velocity in Mammalian A Fibers, *Am. J. Physiol.*, **127**:393 (1939).

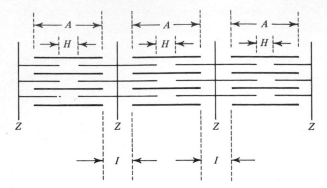

**Fig. 1.8**  Arrangement of thick and thin filaments within a
myofibril.   Creation of striation bands $A$, $I$, $H$ due to overlap is indicated.   Thin filaments attach at the $Z$ line.

fine network of slender collagen fibers which encloses a clear layer 200 Å
thick.   The inner membrane has the morphological characteristics of
a plasma membrane with the type of structure already described for
nerve cells.   The internal structure of each fiber resembles a grid in
that there are striations in both axial and transverse directions.   The
longitudinal components responsible for this appearance are *myofibrils*,
1 to 2 microns in diameter, lying in parallel.   The myofibrils are enclosed
in a clear medium called the *sarcoplasm*.   The *cross striations* arise from
alternating isotropic ($I$) and anisotropic ($A$) bands along the length of
each myofibril.   The alignment of these $A$ and $I$ bands within a fiber
accounts for its cross-ribbed appearance when viewed optically, particularly when a polarizing light microscope is used.

Each myofibril is composed of two contractile protein filaments,
thick ones containing myosin and slender ones of actin, as shown in
Fig. 1.8.   Recent evidence provided by electron-microscope studies have
led to the formulation of a *sliding interaction mechanism* for muscle
contraction.   According to this hypothesis, shortening results when the
stacked arrays of thick myosin filaments are caused to slide over the
interdigitated actin filaments.

Areas within a myofibril where the actin and myosin filaments
overlap correspond to the $A$ band and account for its optical density.
The regions of only actin filaments are identical with the $I$ bands.   At
the center of each $A$ band there is a somewhat lighter area, the $H$ band,
which is established because of the absence of actin filaments.   One of
the criteria for the correctness of this model is the requirement that
upon contraction of the muscle fiber the $I$ and $H$ bands narrow while
the $A$ band remains unchanged, a conclusion that should be apparent

on examination of Fig. 1.8. Experimental observations confirm that this is indeed the case. An additional feature of the muscle fiber is the *Z line* which transects each *I* band. The region between two adjacent Z lines is called the *sarcomere*, the classical repeating unit of structure. According to the sliding-filament model, the contractile unit is actually centered on the *Z* line. In resting unstretched muscle, the sarcomere has an approximate length of 2.5 microns.

Large muscle fibers have two distinct tubular networks constituting a pair of internal membrane systems.[1] Although there is wide variation in the form of these structures in different kinds of muscle cells, they all have certain features in common.

One of these, the sarcoplasmic reticulum (SR), is longitudinal, truly intracellular, and has never been observed to be continuous with the plasma membrane, although it may come into contact with it. In the frog sartorius muscle, the SR of each fiber forms a hollow three-dimensional lacework between and surrounding each of the myofibrils. The system consists of a perforated collar centered in the *A* band. Longitudinal tubules extend from the collar toward the adjacent *I* bands, where they fuse.

The second membrane system is an extension of the fiber surface into the interior and is called the transverse tubular system (T system). Its structure suggests a possible role in transporting the excitation signal to all internal parts of the muscle fiber. In operation, the SR system is closed and bounded, separating part of the internal cell contents from the rest. The transverse tubular system, on the other hand, is open to the external medium and may have an ionic content similar to that of the external fluid. The sarcoplasmic reticular and transverse tubular systems constitute a significant fraction of the total volume of a muscle fiber, as well as markedly increasing its effective surface area.

As noted, the electrical behavior of a muscle fiber is, in many respects, similar to that of a nerve fiber. An action potential initiated at one end of a muscle fiber spreads by virtue of local current effects to the other end. Although the action potential is carried by the surface membrane, its effect is transmitted through the transverse tubules, causing the reticulum to release $Ca^{++}$. This is thought to diffuse to the region of thin and thick filament overlap, causing a reaction that results in the production of mechanical contraction.

The role of the transverse tubules as noted above is illustrated in experiments by Huxley and Taylor.[2] They applied local depolarizing

[1] L. D. Peachey, The Role of Transverse Tubules in Excitation Contraction Coupling in Striated Muscles, *Ann. Acad. Sci.*, **137**:1025 (1966).

[2] A. F. Huxley and R. E. Taylor, Local Activation of Striated Muscle Fibers, *J. Physiol.*, **144**:426 (1958).

currents along the surface of a muscle fiber and found that a local contraction (involving half of or an entire sarcomere) was produced only when the site of stimulation was at the transverse tubule. Since little delay between excitation and contraction was noted, it was assumed that the tubules convey the stimulus directly to the fibrils and that $Ca^{++}$ diffusion for a path length of only 1 micron or less is required.

Increasing the stimulus strength to a muscle produces an increasing response until a maximal level is reached, at which point no further effect can be produced. The increasing response results from the activation of larger and larger numbers of muscle fibers; saturation arises when all fibers are responding. Individual muscle fibers, like nerve fibers, follow the all-or-none law. However, this remark must be qualified because it is possible, as noted in the Huxley and Taylor experiment, to produce a local response by artificial stimulation. While such responses are graded, they are not associated with an action potential. When a stimulus initiates a propagated action potential on a fiber, a maximal contraction results throughout the fiber. In this sense, the all-or-none contraction depends on the all-or-none action potential. When a stimulus does not initiate propagation, the contraction is confined to the activated region alone.

*Efferent*, or *motor nerve*, fibers convey electrical activity to the appropriate muscle, causing that muscle to contract. The number of motor axons innervating a muscle is a small fraction of the total number of muscle fibers since each axon normally activates from 3 to 150 muscle fibers. All muscle fibers innervated by a single *motoneuron* constitute a *motor unit;* this is the smallest unit of normal muscular activity. Its contraction is known as a *twitch response*. The contractile tension of the muscle as a whole is the sum of each fiber's contributions. A graded smooth response can be achieved by increasing the number of units activated (recruitment) and/or their frequency of stimulation.

## 1.4  THE SYNAPSE

As noted in Sec. 1.2, many nerves are linked together in a chain which starts with a specialized transducer sensitive to some enviromental property and ends with a (motor) nerve that activates a motor unit. The connection between each nerve cell and the one following is a *synapse*, and a single axon may form synaptic connections to many nerve cells. Since the synapse affects the transmission of excitation between cells, a study of its properties is important for an understanding of the behavior of a neural network. For our present purposes, it is sufficient to keep in mind the simplest of neural organizations, namely, the reflex arc.   In this case there is a sensory nerve with dendrites that ramify at the skin and possess specialized sensitivity to stimuli of various kinds, e.g., touch or heat.   The cell body is located in the dorsal-root ganglion, and the axon connects to other neurons in the spinal cord. The simplest example involves a single synapse with a motoneuron.

Synapses vary considerably among different nerves and species. As noted earlier, the axon of a cell, in general, ramifies at its ending; its many branches make contact with dendrites and somata of adjoining cells.   The axonal side of the synapse is referred to as the *presynaptic* region; the corresponding portion of the succeeding cell is termed *postsynaptic*.   Contact with the postsynaptic neuron is often made by *terminal boutons* which consist of bulbous endings.   Such boutons may cover much of the dendritic and somatic surface of the postjunctional neuron.

The pre- and postsynaptic cells are physically separated at the synapse by a membrane called the *synaptolemma*.   This consists, essentially, of the plasma membrane of the pre- and postsynaptic cells. The *synaptic gap*, which separates the two, is in the order of 200 Å. Because both separation and close coupling exist at the synapse, the latter behaves somewhat like a switch in a computer network.

The bouton, or knob, contains many small, round structures called *synaptic vesicles*.   These contain chemical substances (transmitter) which are important in transmission of excitation from pre- to postsynaptic cells.   Although the identity of the transmitter is not known, it is presumably released by the arriving action potential and diffuses across the gap to the postsynaptic membrane.   For one class of junction, the chemical causes an increased ionic permeability, and consequently a membrane depolarization is produced.   If a sufficiently large number of synaptic knobs are simultaneously activated so that a depolarization of 20-30 mv results, then an action potential will be produced in the postjunctional neuron.   These synapses are classified as excitatory, and the voltage change resulting from their stimulation is called an *excitatory postsynaptic potential* (EPSP).   At other junctions, an inhibitory transmitter is released, resulting in the hyperpolarization of the membrane and a reduction of excitability.   This is called an *inhibitory postsynaptic potential* (IPSP).

While a nerve axon can carry an impulse in either direction, the synapse is a unidirectional device.   It is capable of exciting only the postsynaptic cell.   Although a single axon is unable, alone, to release sufficient transmitter to cause a postsynaptic discharge, because of its arborization and contact with many cells, it plays a part in the excitation of many cells.

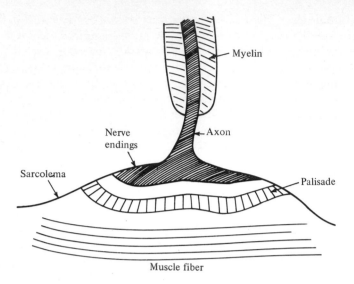

**Fig. 1.9**   The neuromuscular junction.

## 1.5 NEUROMUSCULAR JUNCTION

The events subsequent to the activation of a vertebrate motor nerve
fiber lead to the excitation of a small group of muscle fibers.   The
interface between a motor nerve ending and a muscle fiber, known as
the *neuromuscular junction* or *end plate*, is a specialized region; it is
illustrated in Fig. 1.9.   As in the neural-neural synapse, a definite gap
or cleft appears in the nerve-muscle junction.   The terminal axon at
the junction is partially enveloped by a trough formed by a depression
of the muscle-fiber surface and running parallel to it.   As revealed by
electron microscopy, the groove contains a series of postjunctional folds
0.5 to 1 micron in length and 500 to 1000 Å wide.   These folds are called
*palisades* by virtue of their appearance in light-microscope studies.
The synaptic space is essentially free of Schwann cytoplasm, which
otherwise separates the axon from overlying connective tissue.   The
synaptic space itself has a width of several hundred angstroms.   The
cytoplasm of the terminal axon contains numerous synaptic vesicles
which are 300 to 400 Å in diameter.

In a manner similar to the neural synapse, the arrival of an action
potential at the terminal end of a motor axon results in the release of a
chemical transmitter.   For skeletal muscle this substance has been
identified as acetylcholine (ACh).   The release of acetylcholine appears
to result from the breakdown of a number of synaptic vesicles in some

manner as yet unknown.  The actual quantity liberated per action potential is quite small (about $10^{-17}$ mole).  The ACh diffuses across the gap between the nerve axon and muscle end plate and reacts with a *receptor* at the end plate.  This ACh–receptor complex causes an increase in membrane permeability to cations (mainly $Na^+$ and $K^+$), and this, in turn, drives the membrane potential toward activation threshold. If the potential change is sufficiently large, threshold is reached and an impulse is propagated away from the end plate in all directions.  This process, called the *neuromuscular delay*, requires about 0.65 to 0.7 msec of time in the cat.  Neuromuscular delay in the frog sartorius is in the order of 1.1 to 1.5 msec.

A second substance, the enzyme acetylcholinesterase (AChE), is also found in the vicinity of the junction.  It serves to limit the action of ACh by splitting it into its two component molecules, acetate and choline.  As the concentration of AChE is sufficiently high to hydrolyze all free ACh in a matter of milliseconds, ACh has but a brief opportunity to act at the junction.  This mechanism serves to prevent sustained electrotonic depolarization of the muscle membrane.

The importance of localized potential changes in the process of excitation transmission was studied by Fatt and Katz,[1] using intracellular recording techniques.  Their work revealed that as the motor nerve was stimulated, the *end-plate potential* recorded on the muscle-fiber side of the junction rose in a stepwise fashion until the threshold of excitation was exceeded and an action potential appeared.  That the end-plate potential is distinct from the action potential it initiates is illustrated in Fig. 1.10.

[1] P. Fatt and B. Katz, An Analysis of the End-plate Potential Recorded with an Intra-cellular Electrode, *J. Physiol.*, **115**:320 (1951).

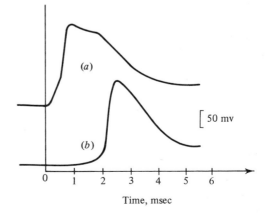

Fig. 1.10  Action potential.  (*a*) Inside muscle fiber at the neuro-muscular junction; (*b*) recorded 2.5 mm from the site in (*a*). [*P. Fatt and B. Katz, An Analysis of the End-plate Potential Recorded with an Intra-cellular Electrode, J. Physiol.*, **115**:320 (1951).]

in particular, by an increase in sodium permeability. Since the membrane resting potential is critically dependent on the relative permeability to each ion species, the result is a change to a new steady-state value closer to threshold.

Dynamic studies on the Pacinian corpuscle reveal some additional phenomena. For example, if a steady stimulus is applied, the resulting generator potential will ordinarily diminish with time. This decrease in amplitude of potential (and consequent decrease in spike frequency) is called *adaptation*. For the Pacinian corpuscle, the potential decreases to zero following a sustained pressure; while for a stretch receptor, a decrease from the initial value to a lower steady-state value normally occurs. When the steady stimulus is removed, a transient generator potential, known as the *off effect*, is produced. Depending on the type of receptor, this may be hyperpolarizing or depolarizing.

## 1.7  THE ACTION POTENTIAL

Detailed discussions of the electrical behavior of nerve and muscle fibers undergoing an action potential are found in succeeding chapters of this book. This chapter concludes with a descriptive review of simple observations that can be made readily in the laboratory.

Figure 1.13 illustrates the nerve chamber. This consists of several parallel silver wires on which a nerve fiber may be placed and at one end of which a stimulating current can be introduced. The remaining wire electrodes record the electrical activity which ensues. In addition, the electrode spacing allows the properties of the propagated action potential, including its velocity, to be measured.

The passive electrical properties of the nerve make it possible to elicit an electrical response even when a subthreshold stimulus is applied. The magnitude of this response decreases with distance from the stimulation site. The phenomenon is known as *electrotonus*, and it is of importance when measuring passive nerve and muscle properties. The distance from the site of a step current stimulation to the point at which the steady-state amplitude has diminished to $1/e$ is known as the *space*

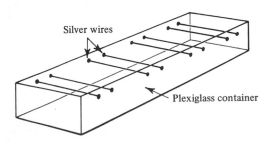

Silver wires

Plexiglass container

**Fig. 1.13** Nerve chamber. The nerve is supported by silver wires, one pair of which is used for stimulation; the propagated action potential is recorded at each pair.

*constant.* The time it takes for this response to reach steady state depends on the membrane resistance and capacitance; the membrane *time constant* is the resistance-capacitance product and a measure of the response time.

For a stimulus that exceeds threshold, an action potential develops and is propagated away from the stimulus site. The energy associated with an action potential comes from potential energy associated with the axon; the stimulus functions simply as a trigger mechanism. For this reason, increasing the stimulus intensity beyond threshold causes essentially no change in the response, hence the characterization as all or none.

The threshold for activation of a nerve or muscle is not a fixed value; it depends on the prior history of the cell. Thus, following the application of a brief subthreshold stimulus, there is a period of reduced threshold for activation. Under these circumstances, the threshold value can be investigated quantitatively by applying two successive stimuli, the first being a "conditioning" pulse and the second the "test" pulse. In this experiment, the variation in test-pulse magnitude required for excitation is observed to be a function of the amplitude and duration of the conditioning pulse. One would discover, for example, that as the conditioning-pulse duration increases, the threshold at first decreases slightly and then increases. In this instance, the decline in excitability is called *accommodation.* After the conditioning pulse ends, threshold increases beyond its normal value, a phenomenon known as *postcathodal depression.* As might be expected from its electrotonic property, the threshold modification caused by a conditioning current extends to regions on both sides of the stimulus site.

Excitation of a nerve fiber depends on both the amplitude and the duration of a square-wave stimulus. Activation becomes possible with a pulse of shorter duration if its amplitude is increased, and vice versa. A curve relating pulse amplitude to pulse duration barely producing an action potential is known as a *strength-duration* curve. As plotted in Fig. 1.14, it has a typically hyperbolic shape. If one assumes that the transmembrane voltage resulting from an application of a current pulse follows a resistance-capacitance ($RC$) charging curve, then parameters of interest can be readily calculated. Thus, if $E_t$ is the transmembrane threshold potential (assumed fixed) and $E_r$ is the resting potential, a depolarization of

$$\Delta E_d = E_t - E_r \tag{1.2}$$

is required for excitation. Letting the membrane resistance be $r_m$ and the capacitance $C_m$, so that the time constant $\tau$ equals $r_m C_m$, we have

$$\Delta E = I_s r_m (1 - e^{-t/\tau}) \tag{1.3}$$

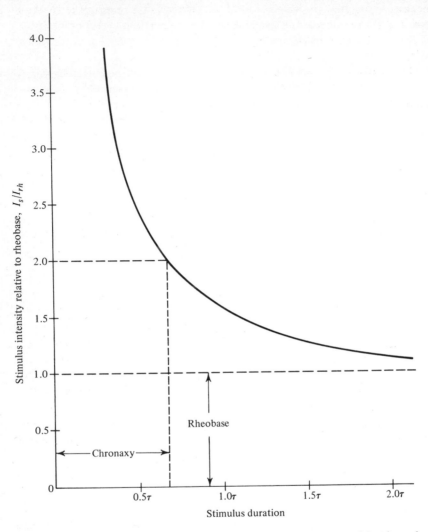

**Fig. 1.14** Strength-duration curve. The locus identifies a combination of stimulus strength and duration that is just sufficient for activation of an excitable cell.

where $I_s$ is the steady stimulus current, and $\Delta E$ is the increase in trans-membrane potential. Activation is achieved if the pulse duration $T$ permits $\Delta E = \Delta E_d$. That is, the following equation must be satisfied by a stimulus current $I_s$ of duration $T$:

$$I_s(T) = \frac{\Delta E_d}{r_m(1 - e^{-T/\tau})} \qquad (1.4)$$

The minimum stimulus current that is capable of achieving activation occurs for $T \to \infty$, and it is designated the *rheobasic current*, or *rheobase*, $I_{rh}$. From (1.4) we have

$$I_{rh} = \frac{\Delta E_d}{r_m} \tag{1.5}$$

so that

$$\frac{I_s}{I_{rh}} = (1 - e^{-T/\tau})^{-1} \tag{1.6}$$

and this is plotted in Fig. 1.14. The duration required for a current stimulus whose amplitude is twice rheobase is called *chronaxy*. From (1.6), chronaxy $T_c$ is given by

$$T_c = \tau(\ln 2) = 0.693\tau \tag{1.7}$$

Using the nerve chamber, one can follow the application of a suprathreshold stimulus by a second, delayed by a variable time. Such an experiment would reveal that for a brief period after activation it is impossible to elicit a response to the second stimulus, regardless of its strength. This interval is known as the *absolutely refractory* period. An interval follows this during which the threshold is finite, but very high. This is the *relatively refractory* period. During this time the threshold diminishes essentially exponentially to its resting value; this occurs while the membrane potential is resuming its normal value. Refractoriness gives us the reason local currents of a propagated action potential do not reexcite that nerve or muscle portion having just been activated. In addition, it imposes an upper limit on the axon's spike discharge frequency.

## 1.8 SUMMARY

In this chapter we have given a brief description of the structure and electrical functions of nerve and muscle. For example, a simple reflex arc involves the following bioelectric phenomena:

1. Transformation of a physical or chemical input to a generator potential by the receptor cell
2. Initiation of a spike discharge in the afferent nerve fiber
3. Transmission of the presynaptic potentials to the postsynaptic motoneuron by a chemical mechanism; generation of postsynaptic potentials
4. Initiation of a spike potential in the motor nerve fiber

5. Release of ACh at the neuromuscular junction, causing end-plate
   activation
6. Contraction of muscle, due to excitation-contraction coupling
   mechanism

This presentation has been necessarily brief and descriptive. The
remainder of the book is intended to be neither, at least with respect to
two components of the reflex arc—the electrical properties of axons
and muscle fibers.

## BIBLIOGRAPHY

Ackerman, E.: "Biophysical Science," Prentice-Hall, Inc., Englewood Cliffs, N.J.,
    1962.
Brachet, J., and A. E. Mirsky: "The Cell," Academic Press Inc., New York, 1960.
Brazier, M.: "The Electrical Activity of the Nervous System," The Macmillan Com-
    pany, New York, 1958.
Bures, J., M. Petran, and J. Zachar: "Electrophysiological Methods in Biological
    Research," Academic Press Inc., New York, 1967.
Davson, H.: "A Textbook of General Physiology," 3d ed., Little, Brown and Com-
    pany, Boston, 1964.
Hodgkin, A. L.: "The Conduction of the Nervous Impulse," Charles C Thomas,
    Publisher, Springfield, Ill., 1964.
Katz, B.: "Nerve, Muscle, and Synapse," McGraw-Hill Book Company, New York,
    1966.
Kennedy, D. (ed.): "The Living Cell," W. H. Freeman and Company, San Francisco,
    1965.
Ochs, S.: "Elements of Neurophysiology," John Wiley & Sons, Inc., New York, 1965.
Paul, W. M. et al. (eds.): "Muscle," Pergamon Press, New York, 1965.
Ruch, T. C., and H. D. Patton (eds.): "Physiology and Biophysics," W. B. Saunders
    Company, Philadelphia, 1965.
Stevens, C. F.: "Neurophysiology: A Primer," John Wiley & Sons, Inc., New York,
    1966.
Whitfield, I. C.: "Manual of Experimental Electrophysiology," Pergamon Press,
    New York, 1964.

# 2
# Electrochemistry and Electrodes

## 2.1 INTRODUCTION

Any investigation of bioelectric phenomena must be based on reliable quantitative data. In particular, it is necessary to be able to make accurate measurements of electric potential. As we shall see, the requisite electrode systems will normally involve metal-electrolyte interfaces, membrane phenomena, and electrolyte-electrolyte junctions. The measurement of electric potentials thus unavoidably involves consideration of a variety of electrochemical configurations. The design of systems for making potential measurements and the interpretation of the results play a vital role in electrophysiology. The purpose of this chapter is to develop the underlying electrochemistry and then apply it toward an understanding of the properties of electrodes in electrophysiological potential measurements. In addition to this immediate goal, the results of this chapter will find application of equal importance in the analysis of biological membrane phenomena to be undertaken in successive chapters.

## 2.2 THERMODYNAMICS OF CLOSED SYSTEMS

We begin with a summary of the basic principles of chemical thermodynamics.   Let us consider first a system which is capable of exchanging
heat and work with its surroundings but where transport of matter
across its boundaries is excluded.   Such a system is designated as
*closed*.   We assume that a property of the system is its *internal energy U*.
Corresponding to the addition to the system of an increment of heat
$dQ$ and work $dW$, the increase in energy $dU$ satisfies

$$dU = dQ + dW \tag{2.1}$$

which is a statement of the *first law of thermodynamics* (or the principle
of conservation of energy) and may be considered as the defining equation of $U$.   In (2.1) $dU$ is a total differential since a direct consequence
of the conservation law is that $U$ is a single-valued function of the thermodynamic variables (that is, $p$, $V$, $T$).   If the work done on the system
results in a decrease in its volume arising from a pressure $p$, we have
$dW = -p\,dV$, and

$$dU = dQ - p\,dV \tag{2.2}$$

Consideration of a change in state of a system from, say, $p_1 V_1$
to $p_2 V_2$ shows that this may occur in different ways (i.e., along different
paths, for example, $C_1$ and $C_2$ in Fig. 2.1).   Since the work done on the
system is the integral of $-p\,dV$, this value clearly depends on the path.
On the other hand, the change in energy is a function only of the end
points.   In view of (2.2) and the above remarks, $dQ$ must also be dependent on the path.   This means that while $dU$ is a true differential of
state variables, $dQ$ and $dW$ are not exact differentials.   More properly,
the latter quantities should be distinguished by notation such as $\delta Q$ and

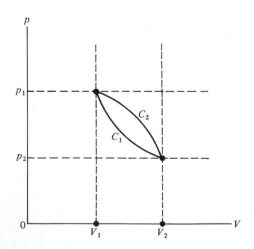

**Fig. 2.1**   $pV$ diagram for a change
of state.   Initial state $p_1 V_1$ and
final state $p_2 V_2$.

$\delta W$; however, there is normally no ambiguity and we shall not utilize this distinction.

For a system at temperature $T$, it turns out that the quantity $dQ/T$ is a function of the state of the system, i.e., an exact differential. This function is designated as the *entropy* and denoted by the symbol $S$.

$$dS = \left(\frac{\partial S}{\partial T}\right)_V dT + \left(\frac{\partial S}{\partial V}\right)_T dV = \left(\frac{\partial S}{\partial T}\right)_p dT + \left(\frac{\partial S}{\partial p}\right)_T dp$$

The relationship $dS = dQ/T$ assumes a *reversible* process; this is defined as one for which the initial states of both system and environment can be restored. In general, however, a real process is to some extent irreversible, and there will then be an *entropy production* $d_iS$, and $d_iS > 0$. That is, for a closed system at uniform $T$

$$d_iS = dS - \frac{dQ}{T} \geq 0 \tag{2.3}$$

and this is a statement of the *second law of thermodynamics*. For the equality in (2.3) and with the assumption that only $-p\,dV$ work is done on the system, then (2.1) takes on the form

$$dU = T\,dS - p\,dV \tag{2.4}$$

The importance of the second law of thermodynamics is that by means of (2.3) the direction in which a natural process will occur can be determined since it must satisfy $d_iS > 0$ $(dS > dQ/T)$. Reversibility can be thought of as the limiting case of a series of essentially infinitely prolonged states approximating equilibrium. For reversibility, and for equilibrium, $d_iS = 0$ $(dS = dQ/T)$.

**Example 2.1**  To illustrate the application of the second law to an irreversible process we consider two adjacent systems $a$ and $b$, each of constant composition, as illustrated in Fig. 2.2. During the time $\Delta t$, system $a$ and system $b$ each receive, respectively, the quantities of heat $d_eQ^a$ and $d_eQ^b$ from the environment. Over the same interval $\Delta t$, a quantity of heat $dQ^{ab}$ flows from $a$ to $b$. (A flow of heat from $b$ to $a$ is thus described by a negative value of $dQ^{ab}$.) We assume the temperature of $a$, $T^a$, to be greater than the temperature of $b$, $T^b$. The increase in entropy of $a$ and $b$ in the time $\Delta t$ is then

$$dS^a = \frac{dQ^a}{T^a} = \frac{d_eQ^a - dQ^{ab}}{T^a} \tag{2.5a}$$

$$dS^b = \frac{dQ^b}{T^b} = \frac{d_eQ^b + dQ^{ab}}{T^b} \tag{2.5b}$$

For the composite system

$$dS = dS^a + dS^b = \frac{d_eQ^a}{T^a} + \frac{d_eQ^b}{T^b} + dQ^{ab}\left(\frac{1}{T^b} - \frac{1}{T^a}\right) \tag{2.6}$$

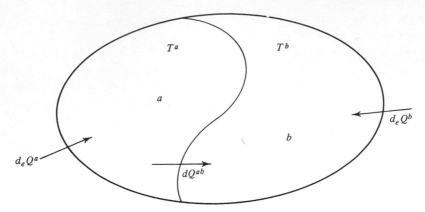

**Fig. 2.2**   Example of entropy production by heat transfer.

The change in entropy, $dS$, can be split into two components. The first, denoted $d_eS$, arises from interactions with the environment, while the second, $d_iS$, is due to internal changes. The latter must satisfy the requirement $d_iS \geq 0$, where the inequality holds for irreversible processes.

With respect to (2.6) we can identify $d_eS$ as

$$d_eS = \frac{d_eQ^a}{T^a} + \frac{d_eQ^b}{T^b}$$

The quantity $d_eS$ may be positive, negative, or zero (compensating entropy changes occur in the environment). The internal contribution to $dS$ in (2.6) is

$$d_iS = dQ^{ab}\left(\frac{1}{T^b} - \frac{1}{T^a}\right) \tag{2.7}$$

and this is not zero when $T^b \neq T^a$. This quantity is the entropy production due to the irreversible nature of the heat-flow process within the system, and the second law requires it to be positive or zero. Consequently, since we have assumed $T^a > T^b$, it is necessary that $dQ^{ab} > 0$. Thus the second law provides that heat flows from the region of higher to that of lower temperature, and is consistent with experience.

The above criteria for irreversible change and equilibria are not in the most desirable form (one specific shortcoming is the involvement of $dQ$, which is not a function of state). Several linear combinations of the parameters $U$, $S$, and $V$ have been introduced which are more useful. These combinations are called *thermodynamic potentials*. One important *thermodynamic potential* is the Gibbs free energy.

The Gibbs free energy $G$ is defined as

$$G = U - TS + pV \tag{2.8}$$

from which we also have

$$dG = dU - T\,dS - S\,dT + p\,dV + V\,dp \tag{2.9}$$

Note that $G$ is a function of state since each of the terms in the defining expressions is a state variable or function thereof. If (2.2) and (2.3) are substituted into (2.9), we obtain

$$dG = -S\, dT + V\, dp - T\, d_iS \qquad (2.10)$$

Since $d_iS \geq 0$, we can express the criteria of equilibrium and the direction of spontaneous change *under conditions of constant temperature and pressure* as follows:

For spontaneous change: $\qquad dG_{T,p} < 0 \qquad\qquad (2.11)$

For equilibrium: $\qquad\qquad dG_{T,p} = 0 \qquad\qquad (2.12)$

Although (2.12) is obtained on the assumption of constant temperature and pressure, it ensures equilibrium with respect to every possible change. One can see this by supposing that a spontaneous process could occur that produced changes in temperature and pressure. In this event, a subsequent process could occur to equalize these quantities with respect to the environment. The net result, however, corresponds to a process under constant temperature and pressure, hence subject to (2.12).

We shall be concerned with systems that involve conversion of energy into or from an electrical form. Consequently, if $dW_e$ represents the electrical work done *by* the system,

$$-dW = dW_e + p\, dV \qquad (2.13)$$

The previous equations, which assumed only $p\, dV$ type of work, need to be modified in the event $dW_e \neq 0$. In particular, under reversible conditions

$$dG = V\, dp - S\, dT - dW_e$$

so that for constant temperature and pressure

$$dG = -dW_e \qquad (2.14)$$

Equation (2.14) specifies that the *decrease* in Gibbs' free energy equals the reversible nonmechanical work done by a system at constant temperature and pressure. The decrease in $G$ is equal to what Gibbs called the "useful work." [Under irreversible conditions, the work done by the system is less than $(-dG)$.]

Another thermodynamic potential which has been defined, and which will be useful to us, is the *enthalpy $H$*. It is defined as

$$H = U + pV \qquad (2.15)$$

so that, using (2.2), we have

$$dH = dQ + V\,dp \tag{2.16}$$

The enthalpy is a measure of the heat added to a system under constant pressure conditions.

## 2.3 CHEMICAL EQUATIONS

If we consider a change in the $i$th substance in a system through the addition of $dn_i$ moles, then the internal energy will be changed by an amount proportional to $dn_i$. The constant of proportionality is designated the *chemical potential* $\mu_i$. For reversible conditions (2.4) then generalizes to

$$dU = T\,dS - p\,dV + \sum_i \mu_i\,dn_i \tag{2.17}$$

Such a change in composition might be brought about by the physical transport of material into or out of the system, i.e., an *open system*. In the case of a closed system, composition changes can occur as a result of a chemical reaction. In this case, $dn_i$ refers to either a reactant or a product. We shall be particularly interested in applications involving cell reactions in the study of electrode systems.

Equation (2.17) is known as the Gibbs equation. Although obtained here under the assumption of reversibility, it is also valid under irreversible conditions. *It constitutes the basic relationship for both equilibrium and nonequilibrium thermodynamics.* In the latter case it is assumed that the system can be subdivided in such a way that each part is large enough to define $T$, $p$, and $\mu_i$ without statistical fluctuations, yet small enough that these quantities do not vary within the subregion. Equation (2.17) is then said to apply "locally," that is, to each subregion. This condition amounts to the requirement that deviations from equilibrium not be "too great." (A discussion of this criterion can be found in de Groot and Mazur.[1]) From (2.17) it is seen that the internal energy is a function of the state variables $(S, V, n_1, n_2, \ldots, n_r)$ so that $U = U(S, V, n_1, n_2, \ldots, n_r)$.

The chemical potential $\mu_i$, which appears in (2.17), can be equated to the partial derivative of $U$ with respect to $n_i$, with $S$, $V$, and remaining components $n_j$ ($j \neq i$) held constant. That is,

$$\mu_i = \left(\frac{\partial U}{\partial n_i}\right)_{SVn_j} \tag{2.18}$$

[1] S. R. de Groot and P. Mazur, "Non-equilibrium Thermodynamics," North Holland Publishing Company, Amsterdam, 1962.

Substitution of (2.17) into (2.9) forms the basis of an alternative expression for chemical potential. We obtain

$$dG = -S\,dT + V\,dp + \sum_i \mu_i\,dn_i \tag{2.19}$$

so that

$$\mu_i = \left(\frac{\partial G}{\partial n_i}\right)_{Tpn_j} \tag{2.20}$$

Equation (2.20) provides more realistic constraints in the definition for $\mu_i$ than (2.18) since the addition of material to physical systems occurs rarely under isentropic conditions. The partial derivative of an extensive[1] quantity with respect to the number of moles of a constituent while the remaining constituents, temperature, and pressure are held constant is referred to as a *partial molar* quantity. The chemical potential is consequently a partial molar free energy.

The description of the quantity $\mu_i$ in (2.17) as a potential function may be seen to be quite appropriate on the basis of comparison with other such functions in the physical sciences. Thus, in mechanical or electrical systems, the potential at a point is the work required to carry a unit "quantity" (of appropriate substance) from a reference point to the particular point in question (field point). In electrostatics the "quantity" is the charge $q$ and the work is $q(\Phi - \Phi_0)$, where $\Phi$ is the potential at the field point and $\Phi_0$ is the reference potential. For gravitation, the "quantity" is the mass and the work equals $mg(h - h_0)$, where the product of the gravitational acceleration $g$ and difference in height $(h - h_0)$, that is, $g(h - h_0)$, is the potential difference. In the present instance (2.20) shows the "quantity" to be the mole, and the difference in potential times this quantity gives the (noncompressional)[2] work (constant temperature and pressure and reversible conditions assumed). Thus to transfer $dn_A$ moles of substance $A$ from a potential $\mu_0$ to a potential $\mu$ requires an amount of work $dW'$ given by $dW' = dn_A(\mu - \mu_0)$. In this illustration we are assuming, of course, that the potential is unaffected by the material transfer itself. This would always be true if the quantity transported were infinitesimal.

It will be useful for future applications to obtain an expression for the force per mole due to a potential gradient. Let a small quantity

---

[1] An *extensive* quantity is one that is a homogeneous function of degree one in the $n_i$'s. Examples include $G$, $U$, $S$, $V$. Quantities such as $T$ and $p$, which are homogeneous functions of zero degree in $n_i$, are *intensive*.

[2] In the study of systems such as solutions, solids, and living materials the work represented by $p\,dV$ is normally negligible compared with that represented by chemical reactions, electrical activity, contractile mechanisms, etc. Thus the use of the potential function $G$ is quite appropriate for our purposes.

of $\Delta n$ moles of a particular material be moved from a point at potential $\mu_1$ to an adjacent point at $\mu_2 = \mu_1 + \delta\mu$. By definition,

$$\mu_2 = \lim_{\Delta n \to 0} \left(\frac{\Delta G}{\Delta n}\right)_2 \qquad \mu_1 = \lim_{\Delta n \to 0} \left(\frac{\Delta G}{\Delta n}\right)_1$$

with constant temperature and pressure. Consequently,

$$\mu_2 - \mu_1 = \delta\mu = \frac{\Delta G_2 - \Delta G_1}{\Delta n} = \frac{\Delta G}{\Delta n}$$

If the separation between 2 and 1 is $\delta\mathbf{r}$, then we also have

$$\Delta G = -\Delta\mathbf{F} \cdot \delta\mathbf{r} = \delta\mu\,\Delta n$$

and $\Delta\mathbf{F}$ is the (noncompressional) force exerted on the quantity $\Delta n$ by the field. In the limit as $\Delta n \to 0$ we define

$$\mathbf{f} = \frac{\Delta\mathbf{F}}{\Delta n}$$

where $\mathbf{f}$ is the force per mole, and since $\delta\mu = -\mathbf{f} \cdot \delta\mathbf{r}$,

$$\mathbf{f} = -\nabla\mu \tag{2.21}$$

An illustration of the properties of the chemical potential is given in the following example. Consider a closed system consisting of component systems $a$ and $b$, which are open to each other. We assume both compartments at the same temperature $T$ and consider the conditions accompanying an assumed spontaneous transfer of $dn_i$ moles of substance $i$ from $a$ to $b$. Application of (2.17) to each compartment separately gives

$$dU^a = T\,dS^a - p^a\,dV^a - \mu_i{}^a\,dn_i \tag{2.22a}$$

$$dU^b = T\,dS^b - p^b\,dV^b + \mu_i{}^b\,dn_i \tag{2.22b}$$

Note that if $a$ and $b$ were at different pressures, a permeable membrane still could permit transfer of $dn_i$. The total work done on the system as a whole, $dW$, is given by

$$dW = -(p^a\,dV^a + p^b\,dV^b) \tag{2.23}$$

If Eqs. (2.22) are summed and (2.23) utilized, then

$$dU = T\,dS + dW - (\mu_i{}^a - \mu_i{}^b)\,dn_i \tag{2.24}$$

where $dU$ and $dS$ are the changes in the energy and entropy of the entire system. But from (2.1) $dQ = dU - dW$; thus if (2.3) is invoked, (2.24) becomes

$$T\,dS - dQ = T\,d_iS = (\mu_i{}^a - \mu_i{}^b)\,dn_i \geq 0 \tag{2.25}$$

When $\mu_i{}^a \neq \mu_i{}^b$, entropy production results. Since $d_iS \geq 0$, our assumption of flow of material from $a$ to $b$ requires that $\mu_i{}^a > \mu_i{}^b$. In other words, spontaneous flow takes place in a direction opposite to the potential gradient (i.e., from high to low potential).

If the above system were in equilibrium, the transfer of the small quantity $dn_i$ could be accomplished reversibly. In this case, the right-hand side of (2.25) must be zero, and $\mu_i{}^a = \mu_i{}^b$ is required. This tacitly assumes that movement of $dn_i$ from $a$ to $b$ is not prohibited by, say, an impermeable membrane. Subject to this restriction, a characteristic of systems in equilibrium is the constancy of chemical potentials of each substance throughout the system.

The nature of a physiological system is such that when potential measurements are made, one establishes, in effect, a galvanic cell. Accordingly, we shall need to consider the cell reactions that occur as a consequence of the measurement process. We proceed, therefore, to a study of chemical reactions with the intention of applying the results to electrode reactions. Consider the chemical reaction given by

$$aA + bB + \cdots = kK + lL + \cdots$$

where $A$, $B$, . . . are the reactants; $K$, $L$, . . . are the products; and $a$, $b$, . . . , $k$, $l$, . . . are the integral proportions in which they react (the stoichiometric coefficients). Since mass must be conserved, it is necessary that

$$-\frac{dn_A}{a} = -\frac{dn_B}{b} = \cdots = \frac{dn_K}{k} = \frac{dn_L}{l} = \cdots d\xi \qquad (2.26)$$

where $-dn_A$ is the number of moles of substance $A$ that reacts, etc., and $dn_K$ is the number of moles of substance $K$ that is produced, etc. The ratios in (2.26) equal $d\xi$, the *degree of advancement* of the reaction. In general, if we designate by $j$ a reactant or product and by $v_j$ its stoichiometric coefficient (negative for reactant and positive for a product), then (2.26) is simply summarized by

$$\frac{dn_j}{v_j} = d\xi \qquad (2.27)$$

For a closed system undergoing a chemical reaction, Eq. (2.17) transforms, by virtue of (2.27), into

$$dU = T \, dS - p \, dV + \left( \sum_i v_i\mu_i \right) d\xi \qquad (2.28)$$

In this form, the instantaneous state of the system is seen to be defined by three variables, namely, $S$, $V$, and $\xi$. This is a specific example of the relation $U = U(S,V,n_1,n_2,. . .,n_r)$, which asserts the internal energy to be completely defined by the state variables $(S, V, n_1, n_2, \ldots, n_r)$. If the chemical reaction were inhibited $(d\xi = 0)$, the system could nevertheless undergo changes in state in accordance with (2.4); hence it follows that $(\partial U/\partial S)_{V\xi} = T$ and $(\partial U/\partial V)_{S\xi} = -p$. This condition required by (2.4) is (correctly) satisfied by (2.28).

The change in internal energy can also be viewed as equal to the heat and work energy that crosses the boundary (i.e., a direct application of the first law); hence

$$dU = dQ - p\,dV = T\,dS - p\,dV - T\,d_iS$$

We can identify, from (2.28), the quantity

$$\left(\sum_i v_i\mu_i\right) d\xi = -T\,d_iS \tag{2.29}$$

As noted earlier, this relationship tacitly assumes the existence of homogeneous conditions.

Equation (2.29) is very important since it forms the basis for determining whether a reaction will take place and, if so, in what direction it will go. This is accomplished by reason of the constraint $d_iS \geq 0$, which is an expression of the second law. Accordingly, a reaction will proceed from left to right (that is, $d\xi > 0$) if

$$\sum_i v_i\mu_i < 0 \qquad \text{(spontaneous reaction left to right)} \tag{2.30}$$

Conversely, a reaction will proceed from right to left ($d\xi < 0$) if

$$\sum_i v_i\mu_i > 0 \qquad \text{(spontaneous reaction right to left)} \tag{2.31}$$

Finally, equilibrium is maintained if

$$\sum_i v_i\mu_i = 0 \qquad \text{(equilibrium)} \tag{2.32}$$

Substitution of (2.27) into (2.19) gives

$$dG = \left(\sum_i v_i\mu_i\right) d\xi - S\,dT + V\,dp \tag{2.33}$$

for a closed system in which a chemical reaction advances $d\xi$. At constant temperature and pressure the above conditions [i.e., (2.30) to (2.32)] correspond to those stated by (2.11) and (2.12).

A quantity designated as the *affinity* $A$ of the chemical reaction is often used in discussing reactions. It is defined by

$$A = -\left(\sum_i v_i\mu_i\right) \tag{2.34}$$

From (2.29) we also have

$$d_i S = \frac{A}{T} \, d\xi \tag{2.35}$$

## 2.4 CHEMICAL POTENTIALS

The chemical potential will play an important role in the analysis of electrode systems and also in the discussion of membranes. This section is devoted to a further consideration of chemical potentials.

For a closed system, if the pressure is held constant, then the addition of heat $dQ$ is accompanied by a temperature change $dT$, where $C_p$ (the specific heat at constant pressure) is the proportionality constant. For a change in pressure at constant temperature, the thermal coefficient relating $dQ$ and $dp$ is denoted $h_T$. The state of the system is completely specified by the thermodynamic variables $T$ and $p$. In particular, we have

$$dQ = C_p \, dT + h_T \, dp \tag{2.36}$$

where $dQ$ is not a total differential; this differential form is denoted the *Pfaffian*.

If (2.36) is substituted into (2.16), the resultant expression is

$$dH = C_p \, dT + (h_T + V) \, dp \tag{2.37}$$

Now $H$ is a function of the state variables, so that (2.37) may also be written as a total differential. We get

$$dH = \left(\frac{\partial H}{\partial T}\right)_p dT + \left(\frac{\partial H}{\partial p}\right)_T dp \tag{2.38}$$

Comparison of (2.38) and (2.37) shows that

$$C_p = \left(\frac{\partial H}{\partial T}\right)_p \qquad h_T = \left(\frac{\partial H}{\partial p}\right)_T - V \tag{2.39}$$

Now, if we substitute (2.39) into (2.36) and divide by $T$, then

$$\frac{dQ}{T} = dS = \frac{1}{T}\left(\frac{\partial H}{\partial T}\right)_p dT + \frac{1}{T}\left[\left(\frac{\partial H}{\partial p}\right)_T - V\right] dp \tag{2.40}$$

Since $dS$ is a total differential of the thermodynamic variables $T$ and $p$, we also require

$$dS = \left(\frac{\partial S}{\partial T}\right)_p dT + \left(\frac{\partial S}{\partial p}\right)_T dp \tag{2.41}$$

If (2.41) is compared with (2.40), and if, furthermore, we recall that the second derivatives satisfy the reciprocity conditions

$$\frac{\partial^2 S}{\partial T\,\partial p} = \frac{\partial^2 S}{\partial p\,\partial T}$$

we obtain

$$\frac{1}{T}\frac{\partial^2 H}{\partial T\,\partial p} = \frac{\partial}{\partial T}\left\{\frac{1}{T}\left[\left(\frac{\partial H}{\partial p}\right)_T - V\right]\right\}$$

Expanding the right-hand side of the above expression leads to

$$h_T = \left(\frac{\partial H}{\partial p}\right)_T - V = -T\left(\frac{\partial V}{\partial T}\right)_p \tag{2.42}$$

Consequently, if (2.42) is substituted into (2.37), we get

$$dH = C_p\,dT + \left[V - T\left(\frac{\partial V}{\partial T}\right)_p\right]dp \tag{2.43}$$

Furthermore, from (2.39), (2.40), and (2.42) we have

$$dS = \frac{C_p\,dT}{T} - \left(\frac{\partial V}{\partial T}\right)_p dp \tag{2.44}$$

For a *perfect gas* $(\partial V/\partial T)_p = nR/p = V/T$; hence

$$dH = C_p\,dT \qquad\qquad \text{perfect gas} \tag{2.45}$$

$$dS = \frac{C_p\,dT}{T} - \frac{nR\,dp}{p} \qquad \text{perfect gas} \tag{2.46}$$

We shall now utilize these expressions to obtain a relation for the Gibbs free energy at an arbitrary $T$ and $p$ in terms of its value at $T_0$ and $p_0$. This can be achieved for the perfect gas by utilizing (2.45) and (2.46) and the fact that

$$G = U - TS + pV = H - TS \tag{2.47}$$

If we integrate (2.45) from $p_0$, $T_0$ to $p$, $T$, we get

$$H(T,p) = H(T_0,p_0) + \int_{T_0}^{T} C_p\,dT \tag{2.48}$$

Integration of (2.46) from $p_0$, $T_0$ to $p$, $T$ yields

$$S(T,p) = S(T_0,p_0) + \int_{T_0}^{T} \frac{C_p}{T}\,dT - nR\ln\frac{p}{p_0} \tag{2.49}$$

We now apply (2.47) to the $i$th component of a mixture of perfect gases. If we take $(\partial/\partial n_i)_{Tpn_j}$ of both sides of (2.47), noting that $G$, $H$, and $S$

are extensive quantities, we get

$$\mu_i = \frac{\partial G}{\partial n_i}\bigg|_{Tpn_j} = \frac{\partial H_i}{\partial n_i}\bigg|_{Tpn_j} - T \frac{\partial S_i}{\partial n_i}\bigg|_{Tpn_j} \tag{2.50}$$

The right-hand side of (2.50) can be evaluated from (2.48) and (2.49). This leads to

$$\mu_i = h_i(T_0,p_0) + \int_{T_0}^{T} c_{pi}\, dT - T s_i(T_0,p_0) - T \int_{T_0}^{T} \frac{c_{pi}}{T}\, dT$$

$$+ RT \ln \frac{p_i}{p_0} \tag{2.51}$$

where $h_i$, $s_i$, and $c_{pi}$ are partial molar quantities ($c_{pi}$ is the heat capacity per mole) and $p_i$ is the partial pressure, of the $i$th substance. Now (2.51) can be written as

$$\mu_i = \mu_i^\circ(T) + RT \ln p_i \tag{2.52}$$

where $\mu_i^\circ(T)$ is a function of $T$ and depends, also, on the choice of reference state $(T_0,p_0)$. We can express this result in terms of the mole fraction $N_i$ of the $i$th substance since $p_i = N_i p$. Thus

$$\mu_i = \mu_i^\circ(T) + RT \ln p + RT \ln N_i$$

and

$$\mu_i = \mu_i^\circ(T,p) + RT \ln N_i \tag{2.53}$$

where

$$\mu_i^\circ(T,p) = \mu_i^\circ(T) + RT \ln p$$

For gas mixtures under conditions such that their behavior is not perfect, (2.52) must be modified. Normally this is done by introducing a function known as the *activity* $a_i$, defined by

$$\mu_i = \mu_i^\circ(T) + RT \ln a_i \tag{2.54}$$

An *ideal solution* is one in which all components satisfy Raoult's law

$$p_i = p_i^\circ N_i$$

where $p_i$ is the partial pressure of the $i$th component in the solution, $N_i$ is its mole fraction, and $p_i^\circ$ is the vapor pressure of the pure $i$th substance. Since the chemical potential of the solution equals that of its vapor phase, and assuming the latter to be a perfect gas, we have

$$\mu_i = \mu_i^\circ(T,p_i^\circ) + RT \ln N_i \tag{2.55}$$

For the more likely case that the solution is not ideal, we can use the above form by defining an activity $a_i$ and *activity coefficient* $\gamma_i$ which satisfy

$$p_i = p_i^\circ a_i = p_i^\circ N_i \gamma_i \tag{2.56}$$

in which case (2.55) is replaced by

$$\mu_i = \mu_i^\circ + RT \ln a_i = \mu_i^\circ + RT \ln N_i \gamma_i \tag{2.57}$$

By a suitable selection of $\mu_i^\circ$, the value of $\gamma_i$ tends to unity as the concentration becomes infinitely dilute. More precisely, $\mu_i^\circ$ is *defined* so that $\gamma_i \to 1$ as the mole fraction of *solvent* approaches unity (thus the same reference state is used for all components). The activity coefficient is seen to be a measure of the deviation of a real solute from ideal; this coefficient is often referred to as a *rational activity coefficient*.

It is usually more convenient to express the composition of a solution in terms of the concentration of its constituents represented as a molarity (or a molality). For dilute solutions, this concentration is proportional to the mole fraction. The chemical potential is written as

$$\mu_i = \mu_i^\circ + RT \ln C_i f_i \tag{2.58}$$

where $f_i$ is the activity coefficient appropriate to $C_i$ in moles per liter of solvent. As before, $\mu_i^\circ$ in (2.58) is a function of $T$ and the standard state; it differs, of course, from the value in (2.55) and (2.57). The activity $a_i = C_i f_i$ satisfies the requirement that $a_i/C_i \to 1$ as the mole fraction of solvent approaches unity. The expression for chemical potential in terms of $a_i$ is written as

$$\mu_i = \mu_i^\circ + RT \ln a_i \tag{2.59}$$

with no distinction in nomenclature from (2.57). The value of $a_i$ in (2.59) is related to that in (2.57) but in a way that depends on the density of the solvent (hence also on temperature). For this reason, concentration is usually expressed as a *molality* for which yet a different constant, $\mu_i^\circ$, appears in (2.59). Both measures will be utilized, and $\mu_i^\circ$ will be understood as the appropriate reference. For both cases the absence of a proportionality to the mole fraction for nondilute solutions causes the activity coefficient to differ from unity even for ideal solutions. This is in contrast to the rational activity coefficient, which would remain equal to 1. The activity coefficients of several electrolytes are given in Table 2.1.

Application of the form (2.59) for solid materials results in $\mu = \mu^\circ$; that is, the chemical potential is independent of physical changes. For a gas, (2.59) applies with $a_i = p_i$ as a good approximation. For a liquid,

**Table 2.1   Activity coefficients of electrolytes**

| Molality of solution | Activity coefficients (25°C) | | | | |
|---|---|---|---|---|---|
| | KCl | K₂SO₄ | CaCl₂ | NaCl | HCl |
| 0.0001 | 0.98 | 0.93 | .... | .... | .... |
| 0.0005 | 0.97 | 0.88 | .... | .... | 0.97 |
| 0.005 | 0.92 | 0.75 | 0.78 | 0.92 | 0.93 |
| 0.01 | 0.90 | 0.69 | 0.73 | 0.90 | 0.90 |
| 0.05 | 0.82 | 0.50 | 0.58 | 0.82 | 0.83 |
| 0.1 | 0.77 | 0.42 | 0.52 | 0.78 | 0.79 |
| 0.2 | 0.72 | .... | 0.49 | 0.73 | 0.76 |
| 0.3 | 0.69 | .... | .... | 0.71 | 0.76 |
| 0.5 | 0.65 | .... | 0.51 | 0.68 | 0.76 |

the activity $a_i$ relates to concentration by means of an activity coefficient, namely, $a_i = C_i f_i$.

## 2.5   DEBYE–HÜCKEL ANALYSIS FOR ACTIVITY COEFFICIENTS

In this section we shall consider the Debye-Hückel theory for the activity coefficient of spherical ions.[1]   Although the conditions under which this theory applies are somewhat restrictive, its development will be instructive because it helps illustrate the physical factors which affect the activity coefficient.   Furthermore, the preliminary developments could serve as a starting point for a more rigorous theory.

Using (2.58), we see that $RT \ln f_i$ is the difference between the partial molar free energy of the ideal (infinite dilution) material and its actual value.   This difference arises from the interaction of ions at the higher concentration, which results in a reduction in the electrostatic energy associated with the ionic fields, hence a reduction in its contribution to the chemical potential.   We shall proceed now to an evaluation of this quantity.   By equating the result to the term $RT \ln f_i$ we shall have obtained a desired relationship between the activity coefficient and the pertinent electrical and chemical parameters of the ionic solution.

The ionic model used in the derivation of the Debye-Hückel expressions is illustrated in Fig. 2.3.   The ion whose electrostatic free energy is to be determined is considered as being spherical with its net charge $q$

[1] C. Tanford, "Physical Chemistry of Macromolecules," John Wiley & Sons, Inc., New York, 1961; S. A. Rice and M. Nagasawa, "Polyelectrolyte Solutions," Academic Press Inc., New York, 1961.

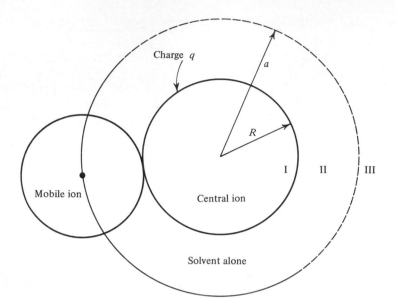

**Fig. 2.3**  Model of Debye-Hückel theory.  Region I is occupied by the central ion.  Since the mobile ion can approach no closer than its radius, region II is occupied by solvent only.  Region III contains surrounding volume charge density.

spread uniformly over a spherical surface of radius $R$.  $R$ corresponds, in a sense, to an ionic radius.  Mobile ions are considered as forming a volume charge distribution[1] exterior to the ion.  Because of the finite ionic diameter of the mobile ions, the "charged cloud" can approach no closer than the distance $a$ $(a > R)$ to the center of the ion under consideration, where $a$ is the sum of the central and mobile ion radii. Consequently, region II $(R < r < a)$ in Fig. 2.3 is occupied only by solvent.

The basic assumptions of the Debye-Hückel theory can be summarized as follows:

1. The ionic interaction between ions is assumed to be only coulombic.
2. The dielectric constant of the solution is taken as that of the solvent, the effect of the dissolved salts being neglected.
3. Ions are considered as having unpolarizable spherically symmetric charge distributions.  For simplicity, we consider the ionic charge $q$ as uniformly distributed (i.e., surface charge density $\sigma = q/4\pi R^2$).

[1] The continuous volume charge distribution is, in effect, a time average of the discrete charge entities.  An equivalent interpretation is that the volume charge density represents an instantaneous spatial average of the regions surrounding each central ion.

4. Near the central ion the concentration of the positive and negative mobile ions will differ, the former being greater near a negative central ion, the latter having a greater concentration near a positive central ion. We assume that the difference can be computed from the Boltzmann law. This asserts the distribution to be of the form $e^{-W/kT}$, where $W$ is the potential energy of the mobile ion and $k$ is Boltzmann's constant. According to assumption 1, the energy is electrostatic only.
5. As one recedes from the central ion, the density of each ion species approaches its bulk value. In this region microscopic electroneutrality is expected.

We let the desired electrostatic potential function of the central ion be designated $\Phi$. Then a mobile ion of species $i$ with a valence $z_i$ and with an electronic charge $q_e$ has a potential energy $W_i = z_i q_e \Phi$ in this field. Thus, if the bulk concentration of the $i$th ion species is denoted $c_i^0$ (ions per cubic meter), its concentration near the central ion under consideration will be

$$c_i = c_i^0 e^{-z_i q_e \Phi/kT} \tag{2.60}$$

Note that if the central ion is positively charged, $\Phi$ will be positive, and a negative ion density exceeds a positive ion density, as expected. For relatively distant points $\Phi \to 0$ and $c_i \to c_i^0$, as required.

The net charge density $\rho$ is given by

$$\rho = \sum_{i=1}^{r} c_i z_i q_e = q_e \sum_{i=1}^{r} z_i c_i^0 e^{-z_i q_e \Phi/kT} \tag{2.61}$$

Now Poisson's equation relates the potential function $\Phi$ to the charge density. That is, $\Phi$ must satisfy[1]

$$\nabla^2 \Phi = -\frac{\rho}{\epsilon} \tag{2.62}$$

where $\epsilon$ is the dielectric permittivity. From (2.61) and (2.62) we obtain

$$\nabla^2 \Phi = -\frac{q_e}{\epsilon} \sum_{i=1}^{r} z_i c_i^0 e^{-z_i q_e \Phi/kT} \tag{2.63}$$

which is the *Poisson-Boltzmann* equation.

As it stands, (2.63) is very difficult to solve mathematically. The Debye-Hückel solution is based on approximating the exponential in (2.63) by the leading term in the series expansion. Since $\sum_{i=1}^{r} z_i c_i^0 = 0$

---

[1] Mks units are used throughout this book.

because of electroneutrality, the linear term in the expansion is the dominant one, and we get

$$\nabla^2 \Phi = \kappa^2 \Phi \qquad (2.64)$$

where

$$\kappa^2 = \frac{q_e^2}{\epsilon k T} \sum_{i=1}^{r} c_i^0 z_i^2 \qquad (2.65)$$

subject to the approximation that $(z_i q_e \Phi / kT) \ll 1$. The linearizing approximation requires that the electrostatic energy be very much smaller than the thermal energy.

Because of the spherical symmetry $\nabla^2 \Phi$ can be written in terms of the radial variable alone, so that (2.64) becomes

$$\frac{1}{r} \frac{d^2(r\Phi)}{dr^2} = \kappa^2 \Phi \qquad (2.66)$$

or

$$\frac{d^2(r\Phi)}{dr^2} - \kappa^2(r\Phi) = 0 \qquad (2.67)$$

The solution of (2.67) is simply

$$r\Phi = A_1 e^{-\kappa r} + A_2 e^{\kappa r} \qquad (2.68)$$

and since $\Phi \to 0$ as $r \to \infty$, we require $A_2 = 0$; hence

$$\Phi = \frac{A_1 e^{-\kappa r}}{r} \qquad r \geq a \qquad (2.69)$$

Equation (2.69) applies only to region III (see Fig. 2.3); that is, $r \geq a$.

Application of Gauss' law at $r = a$ (remembering that the total charge on the central ion, $Zq_e$, has been assumed spread uniformly over the surface $r = R$) gives

$$\frac{\partial \Phi}{\partial r} = -\frac{Zq_e}{4\pi\epsilon a^2} \qquad \text{at } r = a \qquad (2.70)$$

Equation (2.70) can be used to evaluate $A_1$, that is,

$$\frac{-Zq_e}{4\pi\epsilon a^2} = -\frac{A_1 e^{-\kappa a}}{a^2}(1 + a\kappa) \qquad (2.71)$$

so that

$$\Phi = \frac{Zq_e}{4\pi\epsilon} \frac{e^{\kappa(a-r)}}{(1 + a\kappa)r} \qquad r \geq a \qquad (2.72)$$

In region II (see Fig. 2.3) Laplace's equation is satisfied, so that

$$\nabla^2 \Phi = \frac{1}{r} \frac{d^2(r\Phi)}{dr^2} = 0 \qquad (2.73)$$

and

$$\Phi = B_0 + \frac{B_1}{r} \qquad R \le r \le a \tag{2.74}$$

Now $\Phi$ and $\partial \Phi / \partial r$ must be continuous at $r = a$, and these two conditions applied to (2.72) and (2.74) suffice to determine $B_0$ and $B_1$. The result is that

$$\Phi = \frac{Zq_e}{4\pi\epsilon r}\left(1 - \frac{\kappa r}{1 + a\kappa}\right) \qquad R \le r \le a \tag{2.75}$$

The free energy associated with the central ionic charge of $Zq_e$ spread uniformly (or spherically symmetrically) over the surface $r = R$ is given by

$$W_e = \tfrac{1}{2}Zq_e\Phi\Big|_{r=R}$$

$$= \frac{Z^2 q_e{}^2}{8\pi\epsilon R}\left(1 - \frac{\kappa R}{1 + a\kappa}\right) \tag{2.76}$$

Now for infinitely dilute concentrations $c_i{}^0 \to 0$; hence $\kappa \to 0$. Accordingly, the change in free energy per ion due to a finite concentration is

$$\Delta W_e = -\frac{Z^2 q_e{}^2 \kappa}{8\pi\epsilon(1 + a\kappa)} \tag{2.77}$$

Consider, now, a solution containing only two ions, both of valence $z$. Then the bulk concentration $c^0$ of each must be the same. The central ion, in this case, is any of the solute ions. By definition $RT \ln f$ is the difference in the partial molar free energy between the actual and ideal state, the difference in free energy per particle being $(RT/L) \ln f = kT \ln f$.[1] But this quantity can be evaluated from (2.77) since that gives precisely the change in free energy per ion due to the ionic interactions in going from the state corresponding to the infinitely dilute to that for the actual solution. Equation (2.77) applied to the $z$-$z$ electrolyte gives

$$\Delta W_e = \frac{-z^2 q_e{}^2 \kappa}{8\pi\epsilon(1 + a\kappa)} \tag{2.78}$$

where

$$\kappa = \left(\frac{2q_e{}^2 c^0 z^2}{\epsilon k T}\right)^{\frac{1}{2}} \tag{2.79}$$

Accordingly,

$$kT \ln f = -\frac{z^2 q_e{}^2}{8\pi\epsilon} \frac{\kappa}{1 + \kappa a} \tag{2.80}$$

[1] $L$ is Avogadro's constant and $L = R/k$.

with $\kappa$ given by (2.79). If in place of $c^0$ the concentration is expressed in moles per liter, $C^0$, then

$$c^0 = 1,000LC^0 \qquad (2.81)$$

For multiply charged ions (2.80) is readily generalized. The activity coefficient for the $i$th ion, $f_i$, is found from

$$kT \ln f_i = \frac{-z_i^2 q_e^2}{8\pi\epsilon} \frac{\kappa}{1 + \kappa a_i} \qquad (2.82)$$

and $\kappa$ is given by (2.65).

The result in (2.80) and (2.82) is subject to the linearization approximation given in (2.64). For the $z$-$z$ electrolyte, this approximation is most poorly satisfied where the potential is a maximum, namely, for $r = a$. Using either (2.61) or (2.63), we must therefore examine the ratio

$$\frac{zq_e\Phi(r = a)}{kT}$$

to determine the conditions under which it would be small compared with unity. The severest test is for the condition $\kappa a \ll 1$, in which case, utilizing (2.72), we have

$$\frac{zq_e\Phi(r = a)}{kT} = \frac{z^2 q_e^2}{4\pi\epsilon a kT} \qquad (2.83)$$

and $zq_e\Phi/kT \ll 1$ everywhere, provided

$$a \gg \frac{z^2 q_e^2}{4\pi\epsilon kT} \approx 7A \; (\text{H}_2\text{O}, \; 25°\text{C}, \; z = 1) \qquad (2.84)$$

In general, $a < 7A$ and, consequently, deviations from the Debye-Hückel theory occur. What is important is the ratio of the electric field energy in the region of space over which $zq_e\Phi/kT \ll 1$ to the energy in all of space between ions. For ordinary univalent ions (2.80) appears satisfactory up to concentrations around 0.1 $M$. For higher valence, ionic strength nearer to 0.01 $M$ constitutes an upper limit. When (2.77) is applied to macroions of low charge density, such as proteins, the results are satisfactory if the solution contains univalent mobile ions up to an ionic strength of about 0.1 $M$. A comparison of activity coefficients measured experimentally by utilizing a concentration cell and computed from (2.80) is given in Table 2.2 for sodium and potassium chloride.[1]

[1] D. A. MacInnes, "The Principles of Electrochemistry," Dover Publications, Inc., New York, 1961.

**Table 2.2   Relation of activity coefficients to concentration**

| Concentration, moles/liter | Potassium chloride | | Sodium chloride | |
|---|---|---|---|---|
| | ƒ (observed) | ƒ (computed) | ƒ (observed) | ƒ (computed) |
| 0.005 | 0.9274 | 0.9276 | 0.9283 | 0.9281 |
| 0.01 | 0.9024 | 0.9024 | 0.9034 | 0.9036 |
| 0.03 | 0.8492 | 0.8490 | 0.8513 | 0.8515 |
| 0.05 | 0.8191 | 0.8183 | 0.8220 | 0.8220 |
| 0.08 | 0.7872 | 0.7874 | 0.7938 | 0.7923 |
| 0.10 | 0.7718 | 0.7720 | 0.7793 | 0.7776 |

## 2.6  THE GALVANIC CELL

The basic difficulty in making electrical measurements in physiological systems is that the desired potentials reside in an electrolytic medium, while the necessary electronic devices require metallic-wire connections. This means that there must be at least two metal-electrolyte interfaces, and, as we shall presently see, such junctions develop potentials of their own which add to the quantity measured.   Indeed, these junction emfs may be as large as or larger than the potential difference that one is trying to measure.   In this section, we shall investigate the factors that give rise to such electrode potentials so that they may be appropriately accounted for.

The phenomenon described above is exemplified by the galvanic cell.   Such a cell is illustrated in Fig. 2.4, and is seen to consist of a pair

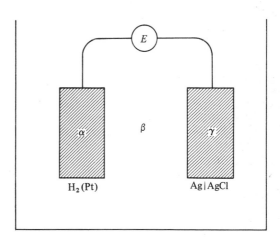

**Fig. 2.4**  Galvanic cell with hydrogen and silver–silver chloride electrodes.

of dissimilar electrodes $\alpha$ and $\gamma$ immersed in an electrolyte $\beta$. As an illustration, let $\alpha$ be a platinum electrode in contact with a stream of bubbling hydrogen, $\gamma$ be a silver electrode in contact with solid silver chloride, and $\beta$ be a solution of HCl. <u>If an external connection is made, a current will flow, and this requires that a reaction must occur at each electrode such that electrons are given up to the external circuit at one while being simultaneously absorbed at the other.</u> This fundamental characteristic is exemplified here by the electrode reactions

$$H_2 = 2H^+ + 2e^- \tag{2.85}$$

$$2AgCl + 2e^- = 2Ag + 2Cl^- \tag{2.86}$$

The reaction of the cell as a whole can be summarized as

$$H_2 + 2AgCl = 2HCl + 2Ag \tag{2.87}$$

The process is characterized by the hydrogen gas giving up electrons which are free to move around the external circuit to the silver–silver chloride electrode, whereupon the silver chloride is reduced to silver and a chloride ion. The ion combines with the hydrogen ions to increase the concentration of HCl. The cell reaction consequently involves hydrogen and silver chloride as reactants with silver and hydrochloric acid as products.

For the system under consideration (2.28) applies. The reaction in this case is, of course, the cell reaction. For a degree of advancement $d\xi$,

$$dU = T\,dS - p\,dV + \left(\sum_i v_i\mu_i\right)d\xi \tag{2.28}$$

This expression relates the change in internal energy directly to changes in the state variables, as has already been noted. The internal energy change can also be evaluated by an application of the first law, where, in this case, the expression must include a term for electrical work. The resultant expression is

$$dQ = dU + p\,dV + dW_e$$

where $dW_e$ is the electrical work done by the system. Combining the above equations gives a modification of (2.29) in the form of

$$-\left[\left(\sum_i v_i\mu_i\right)d\xi + dW_e\right] = T\,dS - dQ = T\,d_iS \tag{2.88}$$

Assuming reversible conditions ($d_iS = 0$), then we have

$$-dW_e{}^r = \left(\sum_i v_i\mu_i\right)d\xi \tag{2.89}$$

where the superscript on $dW_e$ denotes the assumption of reversibility.

If in the external circuit one places a battery so that its emf acts to oppose the cell reaction almost exactly, then a reversible process is established. In this case, we can utilize (2.89) to evaluate the cell emf (i.e., the open-circuit voltage). The standard sign convention associates the reaction $d\xi > 0$ with the flow of positive charge within the cell from left to right. A positive emf must then correspond to the potential of the right-hand electrode exceeding that of the left.[1] If we assume that when the degree of cell reaction has advanced (reversibly) by unity [that is, $d\xi = 1$ in (2.28)][2] $n$ moles of electrons pass through the circuit ($n$ equivalents), then

$$\Delta W_e = nEF \tag{2.90}$$

where $E$ is the emf of the cell and $F$ is Faraday's constant. Substitution of (2.90) into (2.89) gives

$$nEF = -\sum_i v_i \mu_i \tag{2.91}$$

for a reversible cell. For the specific problem introduced above, we note [see (2.86)] that $n = 2$; hence, from (2.87) and (2.91),

$$2EF = \mu_{H_2} + 2\mu_{AgCl} - 2\mu_{Ag} - 2\mu_{HCl} \tag{2.92}$$

Reversible conditions are provided above by placing an emf in the external circuit which almost exactly opposes that generated by the cell. This arrangement brings about equilibrium, which is, as noted earlier, a manifestation of reversibility. On the other hand, the resultant current flow and energy transfer are infinitesimal, a condition that is in conflict with the requirements of practical voltmeter circuits. Accordingly, some deviation from reversibility is inevitable. If (2.89) is substituted into (2.88), we get a limitation on the electrical work $dW_e$ that is performed in an irreversible process, namely,

$$dW_e{}^r - dW_e = T\,d_iS \geq 0 \tag{2.93}$$

Consequently, the measured emf under irreversible conditions will be smaller than that given by (2.91). The difference is known as *overpotential* and is a source of error in potential measurements.

The result given by (2.91) can be expressed directly in terms of the activities of constituents by utilizing (2.59). In the following formulation, we adopt the nomenclature that products of reactions be designated

[1] If the resulting emf were to come out negative, then flow of positive charge from left to right would require $d\xi < 0$.

[2] Since a virtual process is being considered, a differential degree of advancement is conceptually more appropriate. However, setting $d\xi = 1$ simplifies the algebra; the result is independent of $d\xi$.

by double primes while the reactants are labeled with single primes. The cell emf becomes

$$E = \frac{1}{nF} \left[ \sum_i \left( |v_i'|\mu_i' - v_i''\mu_i'' \right) \right]$$

$$= \frac{1}{nF} \left[ \sum_i \left( |v_i'|\mu_i^\circ - v_i''\mu_i^\circ \right) \right] - \frac{RT}{nF} \ln \frac{\prod_i (a_i'')^{v_i''}}{\prod_i (a_i')^{|v_i'|}} \tag{2.94}$$

We define the cell potential under standard conditions to be the *standard potential of the galvanic cell, $E_0$.* That is,

$$E_0 = \frac{1}{nF} \left( \sum |v_i'|\mu_i^\circ - v_i''\mu_i^\circ \right) \tag{2.95}$$

so that (2.94) takes the form

$$E = E_0 - \frac{RT}{nF} \ln \frac{\prod_i (a_i'')^{v_i''}}{\prod_i (a_i')^{|v_i'|}} \tag{2.96}$$

In discussing galvanic cells, the following nomenclature will be used. A semicolon (;) or vertical line (|) will be used to indicate a metal-electrolyte boundary. A liquid junction is shown by a colon (:). Two solutes in the same solution are separated by a comma, while the concentration is shown in parentheses following the substance. Other symbols will be defined as they are introduced.

**Example 2.2** We consider a cell having the following configuration:

$$H_2(g)p_{atm}, Pt|HCl|AgCl(s)|Ag$$

where the symbols $s$ and $g$ refer to solid and gaseous states and the hydrogen gas is at a pressure of $p$ atm. The vertical lines separate the different phases. The reaction taking place at the hydrogen electrode[1] is

$$\tfrac{1}{2}H_2 \rightarrow H^+ + e^-(Pt) \tag{2.97}$$

which corresponds to the flow of positive charge from left to right, from the hydrogen to the silver–silver chloride, within the cell. The electrons released flow in the opposite direction from the hydrogen to the silver electrode, through the external circuit. The electrons are absorbed at the silver–silver chloride

---

[1] The left-hand electrode consists of platinum over which hydrogen gas is bubbled. The reactions involve only the hydrogen, so that the electrode behavior is essentially a *hydrogen electrode.*

electrode, where the following reaction takes place:

$$Ag^+ + e^-(Ag) \rightarrow Ag \tag{2.98}$$

$$AgCl(s) \rightarrow Ag^+ + Cl^- \tag{2.99}$$

In addition, it is assumed that a copper (wire) connection is made to the platinum and silver electrodes (these interfaces are labeled Cu' and Cu'', respectively), so that the electron transfer takes the form

$$e^-(Pt) \rightarrow e^-(Cu') \tag{2.100}$$

$$e^-(Cu'') \rightarrow e^-(Ag) \tag{2.101}$$

Since $\mu_{e^-}(Cu'') = \mu_{e^-}(Cu')$,† the overall cell reaction is

$$\tfrac{1}{2}H_2 + AgCl \rightarrow H^+ + Cl^- + Ag \tag{2.102}$$

Thus, the operation of the cell is such as to convert hydrogen and silver chloride into hydrochloric acid and silver. Application of (2.96) to (2.97) yields for the emf of the cell, $E$, the value

$$E = E_0 - \frac{RT}{F} \ln \frac{(a_{H^+})(a_{Cl^-})}{(a_{H_2})^{\frac{1}{2}}} \tag{2.103}$$

where $a_{H_2} = p_{H_2}$, the gas pressure of hydrogen (assumed to follow the perfect-gas law).

The method of approach and the final result in (2.103) were based on a consideration of the galvanic cell as a whole. An alternative approach would be to sum the individual potential differences (known as single potentials) between adjacent phases in the cell. In such a formulation it is important to recognize that, as distinct from neutral substances, the total potential of an ion species, $i$, depends not only on the chemical potential but also on the electric potential of the phase. The electrochemical potential for an ion of valence $z_i$, defined as the sum of the aforementioned components, is denoted as $\bar{\mu}_i$ and expressed by

$$\bar{\mu}_i \equiv \mu_i + z_i F \phi$$

and $\phi$ is called the *Galvani* potential. The Galvani potential of a phase equals the electrostatic work required to bring a unit charge from infinity to within the phase. Since an ideal unit charge does not exist independently of the material carrier, the aforementioned work involves a chemical (configurational) part as well as an electrostatic one, and there is a difficulty in dividing the two unambiguously. Consequently, from an operational standpoint, the Galvani potential is not measurable. In its place one can consider the work required to bring a unit charge

† Even at very high electrostatic potentials, the electron concentration in copper is extremely low, so that the chemical potential is independent of such concentrations and the *chemical potential* is constant.

from infinity to a point which lies just external to the phase but beyond the effect of short-range surface forces (normally around $10^{-3}$ cm from the surface). This quantity, which is measurable (since only electrostatic work is involved), is called the *Volta* potential $\psi$. The discontinuity in potential across the surface interface, denoted $\chi$, is related to the above quantities since

$$\chi = \phi - \psi$$

Physically $\psi$ arises from excess charge, while $\chi$ is due to a surface double layer, the discontinuity in potential being proportional to the double-layer strength. If the electrochemical potentials in this example are summed around the entire circuit, the indeterminable potentials noted above drop out[1] and the final expression is identical to (2.96).


## 2.7  HALF–CELL POTENTIALS

The emf evaluated by (2.91) and exemplified in (2.103) is the emf of the complete cell. It proves convenient to think of this result as the sum of potentials associated with each electrode separately, the so-called *half-cell* potential. Such a separation is an artificial one, of course; physical significance can be ascribed in this case only to half-cell pairs. But an advantage in producing a list of, say, $M$ half-cell potentials is that $M(M-1)$ pairs may be derived therefrom.

In the implementation of this scheme, we note that the emf of a galvanic cell can be written as the sum of at least two Galvani potential differences, and that these individual contributions are not measurable. Thus, the resolution of the emf into components cannot be accomplished except by the arbitrary procedure of combining the given electrode with a standard (reference) electrode and assigning the emf of the resultant cell to the electrode under consideration (i.e., the reference thus is assigned a zero potential). This practical potential thus differs from the theoretical Galvani potential difference by the constant Galvani potential difference of the reference electrode. The standard reference has been chosen as the hydrogen electrode, and by convention, it is the electrode at the left in the cell.

As an illustration of this procedure, we divide the contribution to the hydrogen–silver chloride cell described in Example 2.2 into two parts, one due to the hydrogen electrode (and dependent on $a_{H^+}$ and $p_{H_2}$) and the other due to the silver–silver chloride electrode (and a function of

---

[1] D. J. Ives and G. J. Janz, "Reference Electrodes," pp. 3–9, Academic Press Inc., New York, 1961.

$a_{Cl^-}$). The result is that (2.103) is expressed as

$$E = \left[ E_0(Ag|AgCl) - \frac{RT}{F} \ln a_{Cl^-} \right]$$
$$- \left[ E_0(H_2) + \frac{RT}{F} \ln a_{H^+} - \frac{RT}{2F} \ln p_{H_2} \right] \quad (2.104a)$$

$$E = E(Ag|AgCl) - E(H_2) \quad (2.104b)$$

where the first bracketed term $E(Ag|AgCl)$ is the silver–silver chloride electrode potential,[1] and the second term $E(H_2)$ is the hydrogen electrode potential.

By convention $E_0(H_2)$, the standard potential of the hydrogen electrode, is chosen equal to zero at 1 atm pressure and $a_{H^+} = 1$, at all temperatures. Consequently, under nonstandard conditions, the hydrogen electrode potential is

$$E(H_2) = \frac{RT}{F} \ln \frac{a_{H^+}}{(p_{H_2})^{\frac{1}{2}}} \quad (2.105)$$

Since $E_0(H_2) = 0$, the separation of $E_0$ in (2.103) into $[E_0(Ag|AgCl) - E_0(H_2)]$ of (2.104) becomes defined, and $E_0(Ag|AgCl)$ is simply the standard potential of galvanic cell $H_2,Pt|HCl|AgCl(s)|Ag$.

To evaluate (2.105) requires, essentially, the chemical potential of the hydrogen ion. But it is not possible to make such a determination since one cannot vary the hydrogen-ion concentration alone, as is required by (2.20). It is only possible to determine the sum of chemical potentials of combinations of ions whose net charge is zero. For example, one can vary the concentration of HCl, hence of $(H^+ + Cl^-)$ taken as a sum. Or, equivalently, from (2.32) and the equilibrium $HCl \rightleftharpoons H^+ + Cl^-$, we have $\mu_{H^+} + \mu_{Cl^-} = \mu_{HCl}$, so that the sum of the hydrogen and chloride potential equals the measurable potential of HCl.

In general, if a mole of salt dissociates into $v_+$ moles of cation and $v_-$ moles of anion, then a measurable combination is

$$\mu_{salt} = v_+\mu_+ + v_-\mu_- \quad (2.106)$$

which is called the chemical potential of the salt. This can be written in terms of activities, giving

$$\mu_{salt} = (\mu°) + RT \ln (a_+{}^{v_+}a_-{}^{v_-}) \quad (2.107)$$

We define

$$a_\pm{}^{(v_++v_-)} = a_+{}^{v_+}a_-{}^{v_-} = a_{salt} \quad (2.108)$$

[1] The terminology *electrode potential* is equivalent to *half-cell potential*.

to be the *mean ionic activity*.  Similarly, we have

$$f_{\pm}^{(v_+ + v_-)} = f_+^{v_+} f_-^{v_-}$$

(2.109)

where $f_+$ is the *mean ion activity coefficient*.  Where ion activities such as $a_{H^+}$ in (2.105) arise as a result of splitting the total cell potential into its electrode components, the following arbitrary assignment is made:

$$a_+ = a_- = a_{\pm}$$

(2.110)

where $a_{\pm}$ satisfies (2.108).  This assignment is clearly consistent with (2.108), so that despite its arbitrariness, whole-cell potentials will always be correct.

The procedure for evaluating a half-cell potential requires, essentially, that one set up a hypothetical cell consisting of the given electrode and a standard hydrogen electrode and then determine the standard potential for the cell.  The procedure involves measurement of the cell potential experimentally under, in general, nonstandard conditions and then extrapolating to standard conditions.  We shall illustrate for two important classes of electrodes: electrodes of the first kind, which are metals in solutions of their salt; and electrodes of the second kind, which are the combination of a metal covered by a sparsely soluble salt.

### ELECTRODES OF THE FIRST KIND

We consider here the case of a metal $M$ in contact with a solution of its ions $M^{v+}$ of ion activity $a_{M^{v+}}$.  The half-cell configuration is as follows:

$$H_2(1 \text{ atm}), Pt|H^+(aq; a_{H^+} = 1)| \; |M^{v+}(aq)|M$$

(2.111)

where the double line assumes an ideal liquid junction[1] that introduces no emf.  From measurements of cell potential as a function of concentration of dilute solutions (making corrections to nonideal conditions based on Debye-Hückel theory), the standard electrode potential $E_0(M - M^{v+})$ can be determined.  This is usually stated for a molal concentration scale.  The electrode reaction at the metal is

$$M^{v+} + (v_+)e^- = M$$

(2.112)

so that the overall cell potential, which is also the electrode potential under nonstandard conditions, is

$$E(M|M^{v+}) = E_0(M|M^{v+}) + \frac{60}{v_+} \log a_{M^{v+}} \qquad mv$$

(2.113)

The standard electrode potential depends not only on the kind of metal but also on the valence of its ions in solution.

[1] The properties of a liquid junction will be discussed in the next section.

**Table 2.3   Standard potentials of electrodes of the first kind (at 25°C)**

| *Electrode reaction* | $E_0(M\,|\,M^{v+})$, *volts* |
|---|---|
| $Li^+ + e^- = Li$ | $-3.04$ |
| $K^+ + e^- = K$ | $-2.92$ |
| $Ca^{++} + 2e^- = Ca$ | $-2.86$ |
| $Na^+ + e^- = Na$ | $-2.71$ |
| $Mg^{++} + 2e^- = Mg$ | $-2.38$ |
| $Al^{3+} + 3e^- = Al$ | $-1.66$ |
| $Zn^{++} + 2e^- = Zn$ | $-0.76$ |
| $Cu^+ + e^- = Cu$ | $0.52$ |
| $Hg^{++} + 2e^- = Hg$ | $0.80$ |
| $Ag^+ + e^- = Ag$ | $0.80$ |
| $Au^+ + e^- = Au$ | $1.60$ |
| $H^+ + e^- = \frac{1}{2}H_2$ | $0.00$ |

The convention shown in (2.111) requires the hydrogen electrode to be on the left and the emf to be given, as already defined, as the right-hand electrode relative to the left (hydrogen) electrode. Table 2.3 lists some standard electrode potentials of metal electrodes. The sign given corresponds to the conventions already assumed, which we may summarize as follows: (1) The cell emf is obtained from the Gibbs free energy change within the cell by $\Delta G = -nFE$. (2) The emf represents the potential of the right-hand electrode with respect to the left. (3) The electrode potential of a given electrode equals the cell emf when the electrode is at the right and a standard hydrogen electrode is on the left. These rules form the basis of the *Stockholm convention*. Unfortunately, some of the older papers will be found with opposite signs.

## ELECTRODES OF THE SECOND KIND

For physiological measurements the condition demanded of electrodes of the first kind, that they be placed in electrolytes containing salts of their own, is unduly restrictive. Furthermore, the toxicity of metals to a variety of physiological preparations may exclude their use. Electrodes of the second kind consist of a metal $M$ that is in contact with a sparsely soluble salt $M_p A_q$ (where the anion $A$ is present in the solution) and have generally more satisfactory properties. An illustration is the silver–silver chloride electrode considered in Example 2.2, which is used in chloride solutions. This electrode finds much application in biological work. Another electrode of considerable importance is the calomel electrode, which consists of mercury covered by mercurous chloride.

For the silver–silver chloride electrode, the standard potential $E_0[\text{Ag}|\text{AgCl}(s)|\text{Cl}^-]$ at 25°C is 0.222 volt.  Consequently, the half-cell potential as defined in (2.104) will be

$$E(\text{Ag}|\text{AgCl}) = 0.222 - 0.060 \log a_{\text{Cl}^-} \qquad \text{volts} \qquad (2.114)$$

The hydrogen electrode potential as expressed by (2.104) and (2.105) is

$$E(\text{H}_2) = 0.060 \log \frac{a_{\text{H}^+}}{(p_{\text{H}})^{\frac{1}{2}}} \qquad \text{volts}$$

where $p_{\text{H}_2}$ is the hydrogen gas pressure in atmospheres.  By definition $\text{pH} = -\log(a_{\text{H}^+})$, so that

$$E(\text{H}_2) = -0.060[\text{pH} + \tfrac{1}{2}\log(p_{\text{H}_2})] \qquad (2.115)$$

The total cell potential, from (2.104), is then

$$E = E(\text{Ag}|\text{AgCl}) - E(\text{H}_2)$$
$$= 0.060\,\text{pH} + 0.030 \log(p_{\text{H}_2}) - 0.060 \log a_{\text{Cl}^-} + 0.222$$

The calomel electrode is represented as

$$:\text{KCl}(m)|\text{Hg}_2\text{Cl}_2|\text{Hg} \qquad (2.116)$$

where $(m)$ indicates the molal concentration, and $(:)$ the liquid junction to another half-cell.  The electrode reaction is

$$\text{Hg}_2\text{Cl}_2 + 2e^- \rightleftharpoons 2\text{Hg} + 2\text{Cl}^- \qquad (2.117)$$

so that the potential of the electrode $E(\text{Hg}|\text{Hg}_2\text{Cl}_2)$ is

$$E(\text{Hg}|\text{Hg}_2\text{Cl}_2) = E_0(\text{Hg}|\text{Hg}_2\text{Cl}_2) - \frac{RT}{F} \ln a_{\text{Cl}^-} \qquad (2.118)$$

**Table 2.4  Standard potentials of electrodes of the second kind**

| Electrode | Standard potential, volts (at 25°C) |
|---|---|
| $\text{Pb}(\text{Hg})|\text{PbSO}_4(s),\text{SO}_4^{--}$ | $-0.350$ |
| $\text{Ag}|\text{AgI}(s),\text{I}^-$ | $-0.152$ |
| $\text{Hg}|\text{Hg}_2\text{I}_2,\text{I}^-$ | $-0.040$ |
| $\text{Ag}|\text{AgBr}(s),\text{Br}^-$ | $0.071$ |
| $\text{Ag}|\text{AgCl}(s),\text{Cl}^-$ | $0.222$ |
| $\text{Hg}|\text{HgSO}_4(s),\text{SO}_4^{--}$ | $0.616$ |

where the reaction in (2.117) goes to the right when the hydrogen half-cell is on the left. For $T = 25°C$, $E_0 = 0.268$ volt. The standard potentials of other electrodes of the second kind are given in Table 2.4.

## 2.8  LIQUID–JUNCTION POTENTIALS

It is frequently necessary to connect an electrode to the system under study through an intermediate electrolytic solution. This is the case, for example, with the calomel electrode, where a pipette of KCl connects the KCl bathing solution to the preparation under study. As a consequence, electrolyte-electrolyte junctions arise, and these contribute additional emf components to the system. Accordingly, it is necessary to study the properties of liquid junctions so that they may be properly accounted for when potential measurements are interpreted.

When two electrolytes are brought into contact, there will be a region in which the composition will vary from that of one to that of the other of the two solutions. For reasons that are discussed in the next chapter, an electric field will be created in the transition region. We wish to determine the total emf across this junction under equilibrium conditions.

Let us assume the interface between regions I and II can be idealized as planar and of infinite transverse extent. We now divide the transition region into layers of infinitesimal thickness such that a jump variation in composition can be assumed across each segment. Figure 2.5 illustrates such an element of differential thickness which is defined by the bounding planes $A$ and $B$ that are parallel to the interface. On side $A$ we have the ions $1, 2, \ldots, i, \ldots, n$ at concentrations $C_1, C_2, \ldots, C_i, \ldots, C_n$; while at $B$ the ionic concentrations (which differ

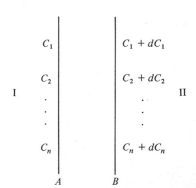

**Fig. 2.5** Lamina of a liquid junction with transverse variation in composition.

by an infinitesimal) are designated $C_1 + dC_1$, $C_2 + dC_2$, . . . , $C_i + dC_i$, . . . , $C_n + dC_n$. Let $z_1$, $z_2$, . . . , $z_i$, . . . , $z_n$ be the respective valences and $t_1$, $t_2$, . . . , $t_i$, . . . , $t_n$ the respective transference numbers.[1] Let us suppose that a Faraday of (positive) charge is caused to flow reversibly from $A$ to $B$. We designate by $dE_L$ the potential difference between $A$ and $B$. The flux, in moles, of the $i$th ion is given by $t_i/z_i$, the flow being from left to right for positive $z_i$ and in the reverse direction for negative $z_i$. The *increase* in potential from $A$ to $B$† must be countered by an emf in the external circuit (almost equal in strength) to ensure reversibility.

The total change in Gibbs free energy, $dG$, if we assume isothermal and isobaric conditions, is

$$dG = \sum_{i=1}^{n} \frac{t_i}{z_i} d\mu_i \tag{2.119}$$

where $d\mu_i$ is the difference in the chemical potential of the $i$th ion across the layer. We can confirm this in the following way. We make use of (2.19) and note that any gain by region $B$ of material crossing the element requires a loss of the same material by region $A$ (and vice versa). Consequently, for region $B$

$$[\Delta G]_B = \sum_i \frac{t_i}{z_i} [\mu_i]_B$$

while for region $A$

$$[\Delta G]_A = - \sum_i \frac{t_i}{z_i} [\mu_i]_A$$

The total increase in free energy of the entire system is the sum $[\Delta G]_A + [\Delta G]_B = \Delta G$. Consequently, we have

$$\Delta G = \sum_i ([\mu_i]_B - [\mu_i]_A) \frac{t_i}{z_i} = \sum_i \frac{t_i}{z_i} \Delta\mu_i \tag{2.120}$$

Equation (2.119) follows from the above result.

[1] The transference number $t_i$ is the relative quantity of the total current carried by the $i$th ion. Thus, $\sum_{i=1}^{n} t_i = 1$. One Faraday of charge requires the transport of 1 mole of a univalent substance or $1/z$ moles of a substance of valence $z$. For the assumed current flow, positive ions move from $A$ to $B$ while negative ions flow in the reverse direction.

† As before, positive $E_L$ corresponds to the potential of the right-hand side exceeding that of the left-hand side, just as for $E$.

Since we assume the process to be isothermal and isobaric, $dG$ also equals the nonmechanical (in this case electrical) work done on the system. This is precisely the same reasoning as that which led to (2.91), except that in this case, a physical rather than a chemical process is responsible for the changes in composition. Thus, from (2.119) and (2.120) we get

$$dG = -dW_e = -F\, dE_L = \sum_i \frac{t_i}{z_i}\, d\mu_i \tag{2.121}$$

under constant temperature and pressure and reversibility.[1] If (2.59) is substituted into (2.121), the result can be put into the following form:

$$-dE_L = \frac{dG}{F} = \frac{RT}{F} \sum_i \frac{t_i}{z_i}\, d(\ln a_i) \tag{2.122}$$

The total liquid-junction potential is obtained by integration,

$$-E_L = \frac{RT}{F} \int_{\mathrm{I}}^{\mathrm{II}} \sum_i \frac{t_i}{z_i}\, d(\ln a_i) \tag{2.123}$$

where I is the solution on the left and II that on the right. Subject to the assumption of stratified conditions this result is quite general.

If it is assumed that the variation in concentration of the $i$th ion is linear from the value at the left to that at the right, and if the activities are set equal to the concentrations, and further if the mobilities[2] are assumed constant throughout the junction, then (2.123) can be integrated. (This is referred to as Henderson's integration.) To sketch in the details, if we let $x$ represent the fraction of mixing, then the concentration of the $i$th substance as a function of $x$ is

$$C_i = C_i' + (C_i'' - C_i')x \tag{2.124}$$

and $0 \le x \le 1$. In (2.124) $x = 0$ corresponds to the left-hand edge (at concentration $C_i'$) and $x = 1$ the right-hand edge (concentration $C_i''$) of the liquid junction. The transference number of the $i$th ion at a

---

[1] An alternative point of view is to equate the increase in Gibbs free energy to the increase in *all* noncompressional energy (e.g., chemical and electrical). Thus,

$$dG = F\, dE_L + \sum_i \frac{t_i}{z_i}\, d\mu_i$$

and since equilibrium is postulated, $dG = 0$; hence (2.121) results. In this formulation $F\, dE_L$ is considered as a potential energy term arising within the system, while in the argument leading to (2.121) it is thought of as an energy leaving the system, hence equal to the decrease in internal energy under reversible (equilibrium) conditions.

[2] The mobility of the $i$th ion, $u_i$, is defined to be the velocity per unit field.

given $x$ is given by the ratio of the contribution of the $i$th ion to the total current, divided by the total current.

An infinitesimal deviation of the electric field from that which results in equilibrium constitutes the driving force for the (infinitesimal) flow of current. If this field is designated $\Delta E(x)$, then the electric current carried by the $i$th ion is $C_i u_i \, \Delta E(x)$, while the total current is $\sum_i C_i u_i \, \Delta E(x)$, where $C_i$ is in equivalents.[1] Note that the electric field, while in general a function of $x$, acts equally on all ion species. Since positive ions move to the right and negative ions to the left, the total net current is simply the absolute sum of the components as given. Accordingly,

$$t_i = \frac{C_i u_i}{\sum\limits_i C_i u_i} = \frac{C_i u_i}{\sum\limits_i C_i' u_i + x \sum\limits_i (C_i'' - C_i') u_i} \tag{2.125}$$

Substituting (2.124) and (2.125) into (2.123) with the assumption that $a_i = C_i$ and that $u_i$ are constant (independent of $x$), gives

$$-E_L = \frac{RT}{F} \int_0^1 \sum_i \frac{(u_i/z_i)(C_i'' - C_i') \, dx}{\sum\limits_i C_i' u_i + x \sum\limits_i u_i (C_i'' - C_i')} \tag{2.126}$$

and after performing the integration, we have the result

$$E_L = \frac{RT}{F} \frac{\sum\limits_i (u_i/z_i)(C_i'' - C_i')}{\sum\limits_i u_i (C_i'' - C_i')} \ln \frac{\sum\limits_i C_i' u_i}{\sum\limits_i C_i'' u_i} \tag{2.127}$$

Because this result was obtained along rather formal lines, the physical basis for the liquid-junction potential is somewhat obscure. In the next chapter we shall show that the potential arises as a balance of both electrostatic and diffusion forces. The latter is quantitatively related to the gradient of the chemical potential.

The assumption that $a_i = C_i$ in leading to (2.126) is quite severe, and one can expect results to be satisfactory only for dilute solutions. For the case that all ions are univalent (2.127) simplifies. We define the following parameters ($C$ now refers to molar concentrations):

$$U_1 = C_1^{+\prime} u_1^{+\prime} + C_2^{+\prime} u_2^{+\prime} + \cdots \tag{2.128a}$$

$$W_1 = C_1^{-\prime} u_1^{-\prime} + C_2^{-\prime} u_2^{-\prime} + \cdots \tag{2.128b}$$

$$U_2 = C_1^{+\prime\prime} u_1^{+\prime\prime} + C_2^{+\prime\prime} u_2^{+\prime\prime} + \cdots \tag{2.128c}$$

$$W_2 = C_1^{-\prime\prime} u_1^{-\prime\prime} + C_2^{-\prime\prime} u_2^{-\prime\prime} + \cdots \tag{2.128d}$$

---

[1] The concentration, symbolized here by $C$, has the dimensions of equivalents for Eqs. (2.125) to (2.127) only.

In terms of (2.128), (2.127) becomes

$$E_L = \frac{RT}{F} \frac{(U_1 - W_1) - (U_2 - W_2)}{(U_1 + W_1) - (U_2 + W_2)} \ln \frac{U_1 + W_1}{U_2 + W_2} \tag{2.129}$$

In the next chapter we shall consider a more accurate membrane model which does not depend on an assumed linear mixing. This is the *constrained diffusion* membrane and was originally formulated by Planck. This result for univalent ions is expressed by the equation

$$\frac{\xi U_2 - U_1}{W_2 - \xi W_1} = \frac{\ln(C_2/C_1) - \ln \xi}{\ln(C_2/C_1) + \ln \xi} \cdot \frac{\xi C_2 - C_1}{C_2 - \xi C_1} \tag{2.130}$$

where $C_1$ and $C_2$ are the total concentration of solutions in I and II, respectively; $U_1$, $U_2$, $W_1$, $W_2$ are as defined by (2.128); and $\xi$ is related to the desired emf by

$$E_L = \frac{RT}{F} \ln \xi \tag{2.131}$$

Numerical evaluation of (2.130) and (2.131) is more difficult than using (2.129). In spite of the fact that the Planck solution depends on fewer assumptions, the two methods give very similar results, so that the Henderson model is often used. As an illustration of the application of (2.129) we consider the liquid junction

$$HCl(C):KCl(C)$$

in which two solutions at the same concentration have a common ion. From (2.129) [in this case (2.127) is actually simpler to use] we get

$$E_L = \frac{RT}{F} \ln \frac{u_{H^+} + u_{Cl^-}}{u_{K^+} + u_{Cl^-}} \tag{2.132}$$

The junction

$$HCl(C_1):HCl(C_2)$$

yields a potential

$$E_L = \frac{u_{H^+} - u_{Cl^-}}{u_{H^+} + u_{Cl^-}} \frac{RT}{F} \ln \frac{C_1}{C_2} \tag{2.133}$$

upon application of (2.130) and (2.131).

**Example 2.3** **(Salt Bridge)** It is sometimes desirable to eliminate a liquid-junction potential in a cell. This can be accomplished through the use of a salt bridge. Suppose, for example, the junction potential we wish to eliminate is caused by

$$HCl(C_1):NaCl(C_2)$$

We interpose a salt bridge consisting of saturated KCl between the two components, resulting in

$$HCl(C_1):KCl(sat):NaCl(C_2) \qquad\qquad (2.134)$$

Since saturated KCl is approximately 4.2 $N$ (at 25°C), then the potential of the left-hand junction, by (2.129), is

$$E_L = \frac{RT}{F} \frac{C_1(u_{H^+} - u_{Cl^-}) - 4.2(u_{K^+} - u_{Cl^-})}{C_1(u_{H^+} + u_{Cl^-}) - 4.2(u_{K^+} + u_{Cl^-})} \ln \frac{C_1(u_{H^+} + u_{Cl^-})}{4.2(u_{K^+} + u_{Cl^-})} \qquad (2.135)$$

If $C_1 \ll 4.2$, the coefficient of the logarithm is approximately

$$\frac{4.2(u_{K^+} - u_{Cl^-})}{4.2(u_{K^+} + u_{Cl^-})} \qquad\qquad (2.136)$$

and since $u_{K^+} \approx u_{Cl^-}$, this quantity is very small. A similar result is obtained from a consideration of the other liquid junction in (2.134). The result is not quite as satisfactory as indicated above because of the many approximations involved. In particular, the equivalence of potassium and chloride mobility is not a particularly good approximation. As a consequence, an uncertain residual potential of 1 to 2 mv must be expected in the use of the salt bridge. A further illustration of the approximate nature of the above result is given in the following example.[1]

We consider the cell

$$Pt, H_2(1 \text{ atm}); HCl(a_{HCl}):KCl(3.5\ N):NaCl(10^{-4}\ N); AgCl; Ag$$

If the liquid junctions can be neglected, then the emf of the cell is the difference in electrode potentials as evaluated in (2.104) and (2.115). Assuming a dilute solution, so that the chloride activity may be set equal to its concentration, we have

$$E = E_0(Ag|AgCl) - \frac{RT}{F} \ln C_{Cl^-} - \frac{RT}{F} \ln a_{H^+} \qquad (2.137)$$

$$E = 0.060 \text{ pH} + \left[ E_0(Ag|AgCl) - \frac{RT}{F} \ln C_{Cl^-} \right] \qquad (2.138)$$

Now since $E$ can be measured while $E_0(Ag|AgCl)$ and $C_{Cl^-}$ are presumed known, we can calculate $a_{H^+}$. Thus, the assumption of zero emf for a salt bridge implies that one could determine single ion activity. From a thermodynamic point of view, only mean activities have any significance, and this will not be exactly equal to the single activity. That is, in this example

$$a_{H^+} \neq (a_{HCl})^{\frac{1}{2}} = (a_{\pm})_{HCl} \qquad\qquad (2.139)$$

Despite the above remarks the cell described by (2.138) serves to determine from a practical standpoint the pH of the HCl solution. One may, in fact, view (2.138) as the definition of pH from an operational standpoint.

## 2.9   MEMBRANE POTENTIALS

We have, thus far, described some of the electrochemical considerations in the measurement of electric potentials in biological systems. To

---

[1] K. Spiegler and M. Wyllie, Electrical Potential Differences, in "Physical Techniques in Biological Research," vol. II, Academic Press Inc., New York, 1956.

accomplish this goal, we have found it necessary to understand the behavior of an electrode system involving liquid junctions and metal-electrolyte interaction.   Such effects are summarized by

$$E = E_{0a} - \frac{RT}{F} \sum_i v_{ia} \ln a_{ia} - \frac{RT}{F} \int_c^a \sum_i \frac{t_i}{z_i} d(\ln a_i)$$

$$- \frac{RT}{F} \sum_i v_{ic} \ln a_{ic} - E_{0c} \qquad (2.140)$$

where $E_{0a}$ is the standard electrode potential at the anode, $E_{0c}$ that at the cathode, while $v_{ia}$ and $v_{ic}$ are the number of moles of component $i$ participating in the reaction at the anode and cathode per equivalent. The sign of $v_i$ is positive for a product and negative for a reactant, assuming positive current flow from left to right; this convention explains the negative sign in the fourth term on the right-hand side of (2.140).

The remainder of this section is devoted to some preliminary remarks on membrane phenomena.   This will be of interest as a further development of electrochemical principles.   It will also serve the specific purpose of providing an understanding of the glass electrode.   A more detailed study of membranes will be undertaken in Chap. 3.

**Example 2.4**   Let us consider the cell consisting of two solutions of HCl separated by a diaphragm, where the potential difference is measured between the solutions by hydrogen electrodes.   The reaction on the left (compartment 1) is

$\frac{1}{2}H_2 \to H^+ + e^-$

while that on the right (compartment 2) is

$H^+ + e^- \to \frac{1}{2}H_2$

We assume that the HCl on the right has an activity $a''$ associated with a concentration $C''$, while that on the left has an activity $a'$ and concentration $C'$.   When a Faraday of charge has been transported from left to right, 1 mole of hydrogen ions is produced at the electrode in compartment 1.   Of these, a quantity $t^+$ moles of $H^+$ will be transported to compartment 2, and at the same time $t^-$ moles of $Cl^-$ will be transported to the left.   Since the total transport is 1 Faraday, we require

$$1 = t^- + t^+ \qquad (2.141)$$

The net result is the transfer of $t^-$ moles of HCl from compartment 2 to 1. Accordingly,

$$-\Delta G = EF = -t^-(\mu_1 - \mu_2) = t^- RT \ln \frac{a''_{HCl}}{a'_{HCl}} \qquad (2.142)$$

Note that the assumed flow of positive charge from left to right within the cell, hence that the right-hand terminal is positive with respect to the left (that is, $E > 0$), requires $a'' > a'$, hence $C'' > C'$.   The spontaneous cell reaction results in a decrease of $C''$ and an increase in $C'$ and is consistent with the second law of thermodynamics.

If we utilize (2.141), this result can also be written

$$EF = t^-RT \ln \frac{a''}{a'} = RT \ln \frac{a''_{H^+}}{a'_{H^+}} - \left( t^+RT \ln \frac{a''_{H^+}}{a'_{H^+}} - t^-RT \ln \frac{a''_{Cl^-}}{a'_{Cl^-}} \right) \qquad (2.143)$$

The second form above corresponds to (2.140), where the emf of the cell separates into two terms, the first being the electrode potential and the second the liquid-junction potential. This separation is convenient conceptually, but, as noted earlier, individual ionic activities such as appear on the right-hand side of (2.143) have no thermodynamic meaning.

The results stated in (2.140) must be modified when the membrane is not freely permeable to all constituents since in this case, interfacial potentials are set up between the membrane and solution. We can see how these potentials arise if we consider the behavior of a membrane whose anions are essentially fixed in position but whose cations can move freely (cation exchanger).[1] If such a membrane were placed in a dilute electrolyte, cations from the exchanger would tend to diffuse out in view of their greater concentration, while anions would tend to diffuse in. (Despite the greater anion concentration within the membrane, since they are fixed, they do not participate in diffusion.) This process is sharply arrested by an accumulation of a layer of positive ions on the solution side and negative ions on the membrane side of the interface. This constitutes an electric double layer, and it sets up a strong electrostatic field opposing further migration. We shall consider this process in greater detail in the next chapter, but the physical basis for this boundary potential, known as a *Donnan* potential, should be apparent.

For an ion exchange membrane, such as described above, there will also be a membrane potential difference. This may be formulated by the approach taken in Sec. 2.8 for liquid junctions. In the next chapter we shall consider such membranes in greater detail so that the influence of the fixed charges may be elucidated. For the present, we shall apply the Henderson treatment of liquid junctions, although the assumptions concerning the variation in composition are not strictly valid. The net membrane potential can, therefore, be expressed as

$$E_m = (E_{Don})_1 + (E_{Don})_2 + E_J \qquad (2.144)$$

where $(E_{Don})_1$ and $(E_{Don})_2$ are the Donnan potentials at interfaces 1 and 2, respectively, and $E_J$ is the diffusion potential within the membrane.

A multi-ionic potential is one that arises when a selective membrane separates two solutions that contain more than one salt species. The

[1] In the next chapter we shall show that such a membrane is selectively permeable to cations.

configuration of such a cell is (using a cation exchanger as an illustration)

$$\text{Electrode 1} \left| \begin{array}{c} \text{Solution contains cations} \\ a_1', a_2', \ldots, a_i', \ldots, a_n' \end{array} \right| \begin{array}{c} \text{Ideal cation} \\ \text{exchanger membrane} \end{array} \right|$$

$$\left. \begin{array}{c} \text{Solution contains cations} \\ a_1'', a_2'', \ldots, a_i'', \ldots, a_n'' \end{array} \right| \text{Electrode 2}$$

The junction potential is given by (Henderson's conditions)

$$E_J = \frac{RT}{F} \frac{\Sigma(\bar{u}_i/z_i)(\bar{C}_i'' - \bar{C}_i')}{\Sigma \bar{u}_i(\bar{C}_i'' - \bar{C}_i')} \ln \frac{\Sigma \bar{C}_i' \bar{u}_i}{\Sigma \bar{C}_i'' \bar{u}_i} \tag{2.145}$$

where $\bar{u}_i$ is the mobility of the $i$th species *within the membrane* and $\bar{C}_i$ is the concentration *within the membrane*.

An expression for the Donnan potential at the interface between the left-hand solution and the membrane is developed in Chap. 3; the result is that

$$(E_{\text{Don}})_1 = \frac{RT}{zF} \ln \frac{a_i'}{\bar{a}_i'} \tag{2.146}$$

where the bar refers to membrane quantities, while its absence denotes the solution. In a similar way

$$(E_{\text{Don}})_2 = -\frac{RT}{zF} \ln \frac{a_i''}{\bar{a}_i''} \tag{2.147}$$

is the Donnan potential contributed by the interface on the right-hand side. Equations (2.146) and (2.147) evaluate the equilibrium potential which must be satisfied by each ionic species $i$.

For the case where the left-hand compartment has a single cation $A$ with activity $a_A'$ while the right-hand side contains a mixture of $n$ cations of equal valence $z$ and activity $a_i''$, then $E_m$ computed from (2.145) to (2.147) comes out

$$E_m = -\frac{RT}{zF} \ln \frac{\sum_n a_i''(\bar{U}_i''/\bar{U}_A')}{a_A'} \tag{2.148}$$

In this equation $\bar{U}$ is defined as the ratio of the mobility to the activity coefficient ($\bar{U} = \bar{u}/f$) within the membrane, the double prime referring to the right-hand edge and the single prime to the left-hand edge. Note that $a_i''$ and $a_A'$ are the activity in the bulk solution.

An important application of (2.148) is to a membrane of glass composition. A property of the glass membrane is that it behaves as if it were permeable only to hydrogen ions. In an application, the membrane separates a standard solution, such as an acid of hydrogen-ion activity $(a_0)_{H^+}$, from the solution whose pH is desired. If the unknown solution contains $n$ univalent cation species in addition to the hydrogen-ion of activity $a_{H^+}$, then from (2.148) we have

$$E = -\frac{RT}{F} \ln \frac{a''_{H^+} + \Sigma a''_i (\bar{U}''_i / \bar{U}'_{H^+})}{(a_0)'_{H^+}} \tag{2.149}$$

Since $\bar{U}''_i / \bar{U}'_{H^+}$ is very small for most ions for glass membranes, this reduces to

$$E = -\frac{RT}{F} \ln \frac{a''_{H^+}}{(a_0)'_{H^+}} = (E_0)_g + 2.30 \frac{RT}{F} (\text{pH})$$

$$= (E_0)_g + 60(\text{pH}) \qquad \text{mv} \tag{2.150}$$

where $(E_0)_g$ is a constant and independent of the pH. Errors may arise in the above formulation if the activity of sodium ions in the unknown solution is very large compared with the hydrogen ion. In this case one should use

$$E = -\frac{RT}{F} \ln \frac{a''_{H^+} + a''_{Na^+} (\bar{U}''_{Na^+} / \bar{U}'_{H^+})}{(a_0)'_{H^+}} \tag{2.151}$$

## 2.10  DESCRIPTION OF ELECTRODES

In this section we discuss the physical form of a number of electrode types which are useful in electrophysiology. These electrodes embody the principles developed in the previous sections.

### THE GROSS METAL ELECTRODE

This electrode is metal and is formed into a plate, wire, mesh, disk, etc. It is often placed directly in contact with the preparation under study, as, for example, surface electrodes or implanted electrodes. Electrodes of the second kind are the preferred material since they are nonpolarizable.

Polarization describes the interference with normal cell reaction due to electrolysis of the medium and buildup of products at the electrodes (such as gas bubbles). These effects occur whenever sizable and prolonged currents flow through the circuit. The larger and longer this condition, the more likely it is that polarization effects will be encountered. The effect is to produce an opposing potential (a *polarization potential*).

Electrodes of the second kind experience fewer polarization effects, hence are preferable. The most generally used electrode is the silver–silver chloride type. It is relatively nontoxic, and the silver chloride is reasonably insoluble. Another commonly used electrode is platinum coated with platinum chloride. Actually, platinum without a coating is satisfactory where very small currents are involved.

For heavy currents it is preferable to utilize a metal electrode in contact with a soluble salt of the metal. For this purpose, zinc–zinc sulfate has been commonly used. The disadvantage of this electrode is that it is highly toxic to biological materials, and some means of isolation must be used.

For many applications, particularly where currents of very small magnitude or short duration are involved, plain metal electrodes are used. Platinum is not readily polarized and is used quite frequently. Silver is also used fairly extensively; it will develop a silver chloride surface if permitted to contact chloride fluids, in which case it behaves as an Ag|AgCl electrode.

For recording from deeply placed locations, stainless steel and tungsten are utilized because of their strength. When implanted electrodes must be left in place for long periods, the tissue reaction may be an important factor in the choice of material. Chemically and mechanically inert materials are preferable; commonly used materials are platinum, platinum-iridium alloy, and to a lesser extent stainless steel, tungsten, and even silver.

## THE GROSS BRIDGE ELECTRODE

This electrode consists of a salt bridge which connects the point of measurement to a half-cell such as calomel. The salt bridge is usually a saturated solution of KCl. For greater ease of handling, a KCl–saturated agar gel is used. Upon inserting the bridge into the solution, the liquid-junction potential at the solution-bridge interface may change as much as 2 mv within a half hour or so. Consequently, it is usually necessary to wait for a time until conditions become somewhat stabilized. In addition to a calomel electrode, a zinc–zinc sulfate electrode or a silver–silver chloride electrode may be used. The latter, however, is soluble in KCl and should, therefore, be avoided under stringent conditions.

## THE WICK–TYPE ELECTRODE

This is of either type described above except that the actual connection to the solution is by means of a wick. The latter is usually a strand of cotton fibers, and contains a physiological saline solution.

## THE GLASS MICROELECTRODE

Electrodes for investigation of intracellular potentials require tip diameters of the order of 1 micron or less if they are to be small compared with cell size. Electrodes with tip diameters in the order of microns are known as microelectrodes. Used most frequently is the glass micropipette.

The tip of the electrode is formed by heating and then drawing out glass capillary tubing with an initial diameter of 1 to 2 mm. Tip diameters of 1 micron and smaller may be readily achieved. Such electrodes are capable of entering cells of 5 microns or greater without causing serious damage; the wound apparently seals around the glass. The tip is filled with a 3 $M$ KCl solution which forms a salt bridge to a reference half-cell. The latter is usually a silver–silver chloride or calomel cell.

Because of its high concentration KCl will diffuse from the micro-electrode tip into the medium in which it is placed. Ordinarily, the rate is not sufficient to poison the nerve or muscle fiber. A small tip is desirable to minimize this flow as well as for minimum cell damage. On the other hand, the resistance of the electrode, ordinarily in the order of $10^7$ to $10^8$ ohms, goes up rapidly with decreasing tip diameter. Because of the adverse effects of high resistance, a taper of 1/10 is usually acceptable, although smaller values cause somewhat less membrane damage.

In operation the indifferent electrode is constructed from the same materials as is the microelectrode. For example, both may utilize silver–silver chloride (or both utilize calomel) half-cells. The purpose of this is to be able to balance out the electrode potentials in the circuit.

### THE METAL MICROELECTRODE

The glass microelectrodes have one major and obvious disadvantage, namely, that they are very fragile. It is possible to use metal-tipped electrodes in their place, although in other respects such electrodes are inferior.

One type of such electrode is made from stainless-steel wire which can be prepared with tips down to 10 microns or less. Such electrodes, however, are rather brittle and have a high impedance. The point of the steel wire is reduced in size by electrolysis in hydrochloric acid. By periodically raising and lowering the wire into the electrolyte a desired taper can be produced.

A more popular metal microelectrode is made of tungsten. This is much stiffer than steel and has good mechanical properties even at 1 micron diameter. It is etched by electrolysis in a cell containing saturated aqueous potassium nitrite. Resistances of the order of 25 to 200 M$\Omega$ can be expected for this type of electrode.

Insulation of the metal electrode is usually achieved by dipping it into varnish. The tungsten electrode may be coated with vinyl lacquer or coated with glass by a fire-polishing technique. Only the very tip of the electrode is left exposed. This in turn is plated with a metal with more satisfactory properties. For example, Baldwin et al.[1] suggest first plating with gold followed by a second plating of platinum black from a chloroplatinic acid solution.

## THE GLASS ELECTRODE

This electrode is based on the properties of glass, which acts as if it were a membrane permeable only to hydrogen ions. It is usually constructed in the form of a bulb with the interior filled with a standard solution of hydrochloric acid. A connection to the interior of the bulb is made with a silver–silver chloride wire which is brought out through a seal in the neck of the bulb. The size of the electrode depends on its application; glass electrodes for intracellular pH measurement are available with conical-shaped tips with diameters of 50 microns or less.

## 2.11 SOURCES OF ERROR IN ELECTRODE MEASUREMENTS

In this chapter we have been concerned with the measurement of electric potentials that arise in biological processes. In order to do this the generation of additional emfs by the electrodes themselves is unavoidable, as we have seen. It has been necessary, therefore, to characterize the emfs arising out of the measurement system itself so that they could be eliminated, reduced, or taken into account.

**Example 2.5** An illustration of seeming ambiguity in potential measurement is given by Davies and Ogston.[2] They consider a two-compartment system with a permselective membrane separation. Solution 1 consists of 0.012 $M$ NaCl made up to 0.12 $M$ with NaNO$_3$ and at pH 5.60. Solution 2 consists of bicarbonate saline (0.095 $M$ chloride) at pH 7.40. Voltages of solution 2 compared with solution 1 depend on the electrode system as follows:

1. With a pair of calomel electrodes connected through KCl salt bridges $E = 22$ mv.
2. With a pair of Ag|AgCl electrodes $E = -30$ mv.
3. With a pair of glass electrodes $E = -84$ mv.

On the basis of the analytic techniques of the previous sections, we shall show that all readings are consistent. But clearly it is necessary to provide an appropriate interpretation of these readings.

Thus, we may assume that the standard potential of the calomel electrodes would balance out and that the liquid-junction potentials of the salt bridge could be neglected. Accordingly, the transmembrane potential $E_m$ of solution 2 with respect to 1 may be taken as essentially 22 mv. Thus

$$E_m = 22 \text{ mv} \qquad (2.152)$$

[1] H. A. Baldwin, S. Frenk, and J. Y. Lettvin, Glass Coated Tungsten Microelectrodes, *Science*, **148**:1462–1464 (June 11, 1965).

[2] R. E. Davies and A. G. Ogston, *Biochem. J.*, **46**:324 (1950).

Using the pair of Ag|AgCl electrodes, we may assume that the standard potentials would cancel out. However, a net electrode potential arises from the difference in chloride concentration, and this adds to the membrane potential which is desired. Applying (2.140) or (2.114), we have

$$E = E_m - 60 \log \frac{(C_{Cl^-})_2}{(C_{Cl^-})_1}$$

$$= 22 - 60 \log \frac{0.095}{0.012} \qquad \text{mv} \qquad\qquad (2.153)$$

where we use the concentration, rather than activities, of the chloride. The result of the above computation is

$$E = 22 - 52 = -30 \text{ mv}$$

which corresponds to the measured value. Thus, based on the measurement of $-30$ mv and the information on chloride concentration, the membrane potential (that is, 22 mv) could be obtained provided the correction for chloride sensitivity of the Ag|AgCl system were taken into account.

The pair of glass electrodes responds to the pH in addition to the membrane emf. Accordingly, upon application of (2.150) (but note that the electrode inserted in solution 2 will have the standard solution on the *right*), one expects

$$E = E_m - 60(\text{pH}_2 - \text{pH}_1)$$

$$= 22 - 108 = -86 \qquad\qquad \text{mv} \qquad\qquad (2.154)$$

This result is consistent with the measurement and again serves to emphasize the dependence of the measured result on the particular electrode system utilized. Conversely, it is only through analysis based on the electrochemical properties of the measurement system that correct interpretations of potential measurements can be achieved.

In deriving (2.89) for the emf of a cell, the assumption of reversibility was made. Since subsequent formulas for electrode emf depend on this condition, the interpretation of potential measurements is subject to error if the system is not in equilibrium. However, it is impossible to measure potentials without some flow of current and hence some degree of irreversibility. If we let $E_i$ be the emf under conditions for which the current is nonzero, then the difference between $E_i$ and the reversible emf $E_{rev}$ is denoted the *overpotential* $\eta$. Thus,

$$\eta = E_i - E_{rev} \qquad\qquad (2.155)$$

If the system is not in thermodynamic equilibrium at zero current, then

$$\eta = E_i - E(i = 0) \qquad\qquad (2.156)$$

is called the *polarization*. However, the terms *polarization* and *overpotential* are so frequently interchanged that no further distinction will be made between the two.

It is clearly a matter of the greatest importance in making interpretations of potential measurements either that conditions ensure negligible overpotential or that appropriate corrections be made.   Three kinds of overpotential will be discussed here.   We summarize them below.[1]

1. As a result of passage of current, changes in concentration (activity) of reactants may take place in the vicinity of the electrodes.   The result is, in effect, the creation of a concentration cell; an alteration of the equilibrium electrode potential results therefrom.   This effect will be referred to as *concentration overpotential* $\eta_c$.
2. For low current density, concentration effects are relatively small. Under these conditions, the transition reaction may be the rate-limiting step.   The overpotential associated with the transition of charge. carriers across the interface is known as *transition overpotential* $\eta_t$.
3. Finally, a *resistance polarization* $\eta_\Omega$ arises because of an ohmic drop at the electrodes.   It may be caused by a variety of phenomena, such as presence of surface films and absorption of inhibitors.

An analysis of the concentration overpotential was given by Nernst[2] under the following assumptions:

1. Fixed layer of potential determining ions at each electrode of thickness $\delta$, where the concentration differs from the surrounding bulk solution.   (This condition arises because the ions resulting from the electrode reaction are assumed to diffuse relatively slowly so that they accumulate to form the boundary layer under discussion.)
2. Concentration of the aforementioned ions in the diffusion layer varies linearly with distance from the electrode.
3. Electrochemical equilibrium at each electrode is assumed.   This implies that the electrode reaction proceeds more rapidly than the transport reaction, a condition implied by (1).

The procedure can be outlined by means of an example.   We consider a pair of silver electrodes in a silver nitrate electrolyte which contains a large concentration of $KNO_3$.   An external emf drives a cell reaction (electrolysis) where silver is dissolved at the anode and is deposited at the cathode.   The $NO_3^-$ ions, being in great supply, carry most of the current in the electrolyte and, furthermore, change in con-

[1] The following material on overpotential is based largely on the treatment in G. Kortum, "Treatise on Electrochemistry," 2d ed., Elsevier Publishing Company, New York, 1965.
[2] N. Z. Nernst, *Physik. Chem.*, **47**:52 (1904).

centration to a negligible extent. Accordingly, the reaction results in an increase in silver-ion concentration in the vicinity of the anode and a decrease near the cathode. The cell emf $E_g$, under resultant steady-state conditions, can be found by assuming the momentary existence of reversible conditions,[1] whereupon (2.140) may be applied:

$$E_g = \frac{RT}{F} \ln \frac{(a_{Ag^+})_{\text{anode}}}{(a_{Ag^+})_{\text{cathode}}} \tag{2.157}$$

Only the electrode reaction contributes to the emf; the absence of a liquid-junction potential is explained by the $NO_3^-$ carrying most of the current and having, to first approximation, a uniform concentration.

If, say, the cathode is considered and if $(a_{Ag^+})_0$ denotes the silver-ion activity in the bulk solution and $(a_{Ag^+})$ the activity at the cathode, then the concentration overpotential is

$$\eta_c = \frac{RT}{z_{Ag}F} \ln \frac{a_{Ag^+}}{(a_{Ag^+})_0} \tag{2.158}$$

(Although $z_{Ag} = 1$, the factor is included here to facilitate generalization to multivalent electrode materials.) Since the liquid-junction potential is very small, the movement of silver ions in the boundary layer is by diffusion. The flow can be computed by using Fick's law[2] and the assumed linear concentration gradient. This yields

$$j_x = -D \frac{dC_{Ag^+}}{dx} = D \frac{C_{0,Ag^+} - C_{Ag^+}}{\delta} \tag{2.159}$$

where $j_x$ is the ion flow rate per unit area, $D$ is Fick's constant, and $x$ is the distance *to* the electrode. Solving for $C_{Ag^+}$ from (2.158) on the assumption that $f_{0,Ag^+} = f_{Ag^+} = 1$† and noting that the current density $J = z_{Ag}Fj_x$ yields

$$J = \frac{z_{Ag}FD}{\delta} (C_{0,Ag^+} - C_{Ag^+}) = \frac{z_{Ag}FDC_{0,Ag^+}}{\delta} \left(1 - \exp \frac{z_{Ag}F\eta_c}{RT}\right) \tag{2.160}$$

At high currents $C_{Ag^+} \to 0$, so that $J$ approaches a limiting value in (2.160) since $C_{Ag^+}$ becomes negligible compared with $C_{0,Ag^+}$. In this case

$$J_{\lim} = \frac{z_{Ag}FDC_{0,Ag^+}}{\delta} = KC_{0,Ag^+} \tag{2.161}$$

---

[1] In this case, at the anode we have $Ag^+ + e^- \to Ag$ and at the cathode $Ag \to Ag^+ + e^-$, which is just the reverse from what occurs during electrolysis.

[2] This is the subject of detailed study in Chap. 3.

† This assumption is a good one in view of the high $NO_3^-$ concentration.

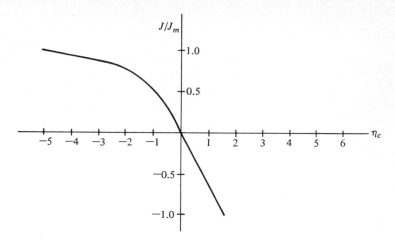

**Fig. 2.6** Concentration overpotential as a function of relative current density based on the Nernst analysis.

From (2.160) and (2.161) we get

$$C_{\text{Ag}^+} = C_{0,\text{Ag}^+}\left(1 - \frac{J}{J_{\lim}}\right) \tag{2.162}$$

and

$$\eta_c = \frac{RT}{z_{\text{Ag}}F} \ln \frac{KC_{0,\text{Ag}^+} - J}{KC_{0,\text{Ag}^+}} \tag{2.163}$$

The relation expressed by (2.163) is plotted in Fig. 2.6. Note that for small values of $J$ (2.163) is approximated by

$$\eta_c \approx -\frac{RT}{z_{\text{Ag}}F} \frac{J}{J_{\lim}} \tag{2.164}$$

showing a linear dependence of overpotential on current density for small $(J/J_{\lim})$. Since $J_{\lim}$ is proportional to the silver-ion concentration, the concentration overpotential is an effect which mainly arises when the latter is small. The overpotential can be reduced by increasing $J_{\lim}$, and this in turn can be accomplished by decreasing $\delta$ by stirring or shaking.

Equations (2.162) to (2.164) apply in general; one simply replaces the silver ion in the example by the actual potential limiting ion in the problem for which the concentration overpotential is being evaluated. Since the diffusion coefficients of metal ions do not differ greatly, the result expressed by (2.163) may be considered as giving the correct magnitude for any simple metal electrode.

At sufficiently low current densities, the rate-determining process is no longer the transport reaction noted above, and other effects become relatively prominent. The most important is the potential limiting reaction associated with the charge transfer process across the electrode-solution interface. This is referred to as the transition reaction, and the corresponding potential required to drive it is the *transition overpotential*. In considering this effect, we shall assume that concentrations of substances participating in the electrode reaction remain constant and independent of current density (i.e., no concentration overpotential). Furthermore, the counterelectrode is assumed nonpolarizable, and an excess of supporting electrolyte is assumed.

We consider the electrode reaction of the form

$$O + ze^- \leftrightarrows R \tag{2.165}$$

where the substances $O$ and $R$ are assumed soluble and where the forward process (reduction) is *cathodic* while the reverse (oxidation) is *anodic*. Since the transition of charge carriers across the electrode interface takes place in both directions, the resultant current can, in principle, be decomposed into constituent anodic $(J_a)$ and cathodic $(J_c)$ partial current densities. Each current component characterizes the rate of the respective partial reaction. The total current density $J$ is given by the sum of the components, that is,

$$J = J_a - |J_c| \tag{2.166}$$

At equilibrium $J_a = |J_c| = J_0$, where $J_0$ is the exchange current density (i.e., the anodic component when the total current is zero).

To first order, the reaction rates will be proportional to the concentration of the reactants. Then if the activation energies of anodic and cathodic reactions were $\Delta G_a$ and $\Delta G_c$, respectively, utilizing the Arrhenius equation, we have

$$-j_c = k_c C_0 \exp\left(-\frac{\Delta G_c}{RT}\right) \tag{2.167}$$

$$-j_a = k_a C_R \exp\left(-\frac{\Delta G_a}{RT}\right) \tag{2.168}$$

for the cathodic and anodic fluxes. These expressions are valid for pure chemical processes and must be modified as a consequence of the existence of the potential difference $E_i$. Its effect is to change the activation energy, inhibiting it in one direction and favoring it in the other direction. If we assume that $\Delta G_a$ is diminished by the fraction $\alpha$ of the energy barrier $zFE_i$, then $\Delta G_c$ must be increased by $(1 - \alpha)zFE_i$. The way in which these factors combine to form the net activation energies

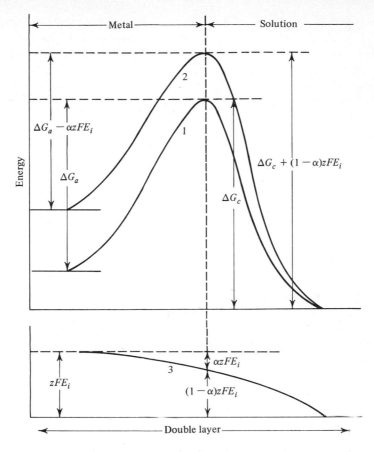

**Fig. 2.7**  Effect of Galvani potential differences on activation energies
of electrode reactions.   Curve 1 is the chemical free energy.   Curve 2
includes the contribution from electric energy.   Curve 3 is the electric
potential alone.   All curves are functions of the distance from the
electrode.   (*From G. Kortum, "Treatise on Electrochemistry," Elsevier
Publishing Company, New York, 1965.*)

is illustrated in Fig. 2.7.   The parameter $\alpha$ is known as the *transition
factor*.

The resultant partial current densities are then

$$J_a = zFk_a C_R \exp\left(-\frac{\Delta G_a - \alpha zFE_i}{RT}\right) \tag{2.169}$$

and

$$J_c = -zFk_c C_0 \exp\left[-\frac{\Delta G_c + (1-\alpha)zFE_i}{RT}\right] \tag{2.170}$$

Under equilibrium conditions $E_i = E_{rev}$, and (2.169) provides an expression relating the exchange current density to $E_{rev}$, namely,

$$J_0 = zFk_aC_R \exp\left(-\frac{\Delta G_a - \alpha zFE_{rev}}{RT}\right)$$

$$= zFk_cC_0 \exp\left[-\frac{\Delta G_c + (1-\alpha)zFE_{rev}}{RT}\right] \qquad (2.171)$$

Since the transition overpotential $\eta_t = E_i - E_{rev}$, we can combine (2.169) to (2.171) to obtain

$$J = J_a - |J_c|$$

$$= J_0 \left\{\exp\left(\frac{\alpha zF}{RT}\eta_t\right) - \exp\left[-\frac{(1-\alpha)zF}{RT}\eta_t\right]\right\} \qquad (2.172)$$

In this form, the equilibrium state and the corresponding exchange current density $J_0$ are used as reference points.

Equation (2.172) is correct provided that the concentrations $C_0$ and $C_R$ are independent of $J$ or $E$ (a pure transition overpotential). A plot of both (2.171) and (2.172) is given in Fig. 2.8, based on the assumption that $\alpha = 0.5$ and $C_R = C_0$. For small values of current density, the exponentials in (2.172) can be replaced by the first term of their

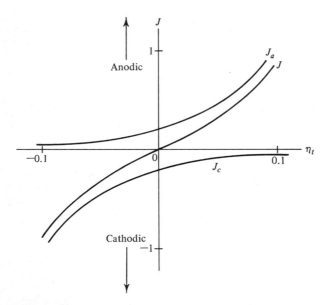

**Fig. 2.8** Relationship of transition overpotential $\eta_t$ to steady-state current density. (*From G. Kortum, "Treatise on Electrochemistry," Elsevier Publishing Company, New York, 1965.*)

Taylor series expansion to yield the linear relation

$$J = \frac{J_0 z F \eta_t}{RT} \tag{2.173}$$

Thus, the overpotential is proportional to the current density for small currents. For example, platinum will develop a polarization potential of 2 mv for $J = 0.1 \ \mu a/cm^2$. Such a current density would arise at a 10-micron electrode which had a total current flow of approximately $0.1 \times 10^{-12}$ amp (0.1 pico-amp). It is also clear from (2.173) that the overpotential is proportional to $J/J_0$ and will be lower for higher exchange currents. Consequently, it is desirable that $J_0$ be as large as possible; it may be unacceptably small in very dilute solutions.

The remaining mechanism to be discussed which leads to over-potential is described as *resistance polarization* $\eta_\Omega$. It arises basically from the resistance of anodic films or adsorption layers to the transfer of ions or electrons. This effect is established very rapidly and plays an important role in the behavior of the electrode under time-varying conditions. Its magnitude may take on values that are comparable to that of transition or concentration overpotential. Its effect is to impede the anodic dissolution of the metal, and the term *passivity* is applied to the case where this rate is reduced to an infinitesimal.

The films appear to be formed by interaction of the electrode metal with anions in the electrolyte. Their properties vary a great deal; they show ionic, electronic, and in some cases almost no conductance. Illustrative of these three conditions are the potential-vs.-time curves in Fig. 2.9 for constant-current conditions. As the film covers more and more of the surface area of the electrode, the inhibition of current flow increases. For the nonconducting film, with $\tau$ the time for complete coverage, an approximate expression for $E_i$ can be derived,[1] which gives

$$E_i = E_{\text{rev}} + \frac{k}{\tau - t} \tag{2.174}$$

that is, $E_i(t)$ is hyperbolic. In the case of ionic conducting films, the behavior following complete coverage (when the potential has the value $E_{ia}$) is characterized by a growing film thickness, and the linear relationship[2]

$$E_i = E_{ia} + k't \tag{2.175}$$

is approximately true. Electronic conduction takes place in oxide coatings and has been of particular importance in the investigations of the anodic passivation of iron.

[1] Kortum, *op. cit.*, pp. 513–527.

[2] *Ibid.*

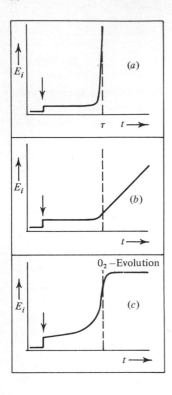

**Fig. 2.9** Response of anodic layers to a constant-current step. (*a*) Nonconducting layer; (*b*) ion-conducting layer; (*c*) electron-conducting layer. (*From G. Kortum, "Treatise on Electrochemistry," Elsevier Publishing Company, New York, 1965.*)

One can characterize the net electrical effect of a poorly conducting film as that of a leaky capacitance, or parallel combination of resistance and capacitance. For a film of thickness $d$ and cross-sectional area $A$, the resistance $R$ would be

$$R = \frac{\sigma d}{A} \tag{2.176}$$

and the capacitance $C$ would be

$$C = \frac{\epsilon A}{d} \tag{2.177}$$

where $\sigma$ is the specific resistivity of the film and $\epsilon$ its dielectric constant.[1] The time constant for this phenomenon is given by

$$\tau = RC = \epsilon \sigma \tag{2.178}$$

Typical values are in the order of $10^{-5}$ sec. One can characterize concentration and transition polarization-time curves as resulting from an $RC$ network with much larger values of time constant (larger capacitance

[1] R. Plonsey and R. Collin, "Principles and Applications of Electromagnetic Fields," p. 181, McGraw-Hill Book Company, 1961.

values).    Thus, in Fig. 2.10, the polarization-time curve of an electrode due to a step current ($i = 5$ ma) shows the effect of resistance polarization as occurring in the initial phase ($t \sim 10^{-5}$ sec), while other types of polarization ($\eta_p$) are responsible for the ensuing behavior.

A nonpolarizable electrode is one for which there is unhindered ionic exchange between electrode and solution.    Such an ideal electrode does not exist in practice.    If, for example, a so-called nonpolarizable electrode is placed in a very dilute solution, then the exchange current can be so small that other processes dependent on the presence of impurities become controlling.    In this case, the electrode will behave in an unstable mode.    Furthermore, for moderate to high current densities, a significant concentration overpotential can develop.    In general, as is clear from (2.164)·and (2.173), some overpotential is an inevitable concomitant of a finite electrode current.    Thus, while it is desirable to use electrodes whose coefficient relating overpotential to current density is small, a generally effective move to reduce polarization difficulties is to reduce the current density in the circuit through the use of preamplifiers with very high input impedance.

A variety of amplifiers have been designed which have very high input resistances and also means for compensating the input capacitance (usually by positive feedback).    Such compensation is particularly important when recording from microelectrodes; as a consequence of their high resistance, an otherwise poor frequency response is likely to result.    Normally, the preamplifier operates as a cathode follower, or a

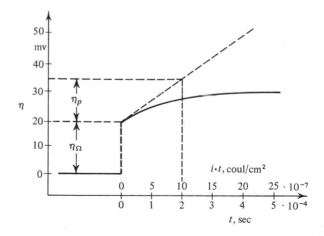

**Fig. 2.10**  Electrode response to a current step ($i = 5$ ma/cm$^2$); a polarization-time curve. (*From G. Kortum, "Treatise on Electrochemistry," Elsevier Publishing Company, New York, 1965.*)

variation thereof, with a gain in the order of unity. The effect of input grid current may be significant with high-impedance microelectrodes, but this can be counteracted through the use of special electrometer tubes or compensating (feedback) circuits.[1]

Under ac conditions[2] the electrode current and polarization potential will be linearly related if the current density is not too high. The effect of the electrode polarization can thus be accounted for by a suitable impedance. Platinum electrodes are commonly used because of their low impedance, and a porous coating of platinum black is often applied since it will reduce the impedance by up to four orders of magnitude. For a platinum-black electrode, the linear condition is maintained up to a current-density value of approximately 1 ma/cm at 1 kHz; lower "limit values" apply at lower frequencies.

The impedance can be represented by a series resistance $R_p$ and capacitance $C_p$. While independent of current density, as noted above, these quantities do depend on frequency. Within the linear range superposition can be used provided the polarization impedance appropriate to the frequency is used with each frequency component. Both $C_p$ and $R_p$ decrease with increasing frequency according, roughly, to a power law. However, the exponent itself may be a slowly varying function of frequency. Thus, according to Schwan,[3] for platinum-black electrodes of surface area in excess of 1 mm$^2$, $C_p$ decreases as $f^{-0.3}$ at 10 Hz, changing to $f^{-0.5}$ at 100 kHz. On the other hand, $R_p$ varies as $f^{-0.5}$ over most of this range. For electrodes with area less than 1 mm$^2$, the behavior may be quite different, and a 50 percent variation in exponent can occur.

Where electrodes are used for stimulating, high current densities may be unavoidable. An electrode of 1 mm diameter and insulated to the tip, carrying 0.5 ma of current, will have an electrode current density of approximately 50 ma/cm$^2$. In considering the performance of the electrode system under these conditions, the temporal behavior is equally important. Advantage can be taken of the time for buildup of overpotential, and, where possible, polarity may be switched before such increase becomes too significant. Analysis of the electrode behavior is, however, complicated by the nonlinearity of the representative $RC$ circuit; some approaches to this problem are described by Weinmann and Mahler.[4]

[1] R. L. Schoenfeld, Biological Amplifiers, in W. L. Nastuk (ed.), "Physical Techniques in Biological Research," vol. V, Academic Press Inc., New York, 1964.

[2] H. Schwan, Determination of Biological Impedances, in W. L. Nastuk (ed.), "Physical Techniques in Biological Research" vol. VI, Academic Press Inc., New York, 1963.

[3] *Ibid.*

[4] J. Weinmann and J. Mahler, An Analysis of Electrical Properties of Metal Electrodes, *Proc. Fifth Intern. Conf. Med. Electronics, Univ. of Liege*, pp. 487–508, 1963.

The purpose of this section has been to point out possible sources of error in making potential measurements and to indicate how such errors might be evaluated quantitatively and also reduced. This material, of necessity, has been treated relatively briefly. Details on preparation of electrodes, stability, techniques for measurements of electrode properties, etc., can be found in references cited.[1]

## BIBLIOGRAPHY

de Groot, S. R., and P. Mazur: "Non-equilibrium Thermodynamics," North Holland Publishing Company, Amsterdam, 1962.

Helfferich, F. G.: "Ion Exchange," McGraw-Hill Book Company, New York, 1962.

Ives, D. J., and G. J. Janz: "Reference Electrodes," Academic Press Inc., New York, 1961.

Katchalsky, A., and P. F. Curran: "Nonequilibrium Thermodynamics in Biophysics," Harvard University Press, Cambridge, Mass., 1965.

Kleinzeller, A., and A. Kotyk: "Membrane Transport and Metabolism," Academic Press Inc., New York, 1960.

Kortum, G.: "Treatise on Electrochemistry," Elsevier Publishing Company, New York, 1965.

Lyklema, J.: The Measurement and Interpretation of Electrical Potentials from a Physico-chemical Point of View, *Proc. Fifth Intern. Conf. Med. Electronics, Univ. of Liege*, pp. 458–479, 1963.

MacInnes, D. A.: "The Principles of Electrochemistry," Dover Publications, Inc., New York, 1961.

Pitzer, K., and L. Brewer: "Thermodynamics," 2d ed., McGraw-Hill Book Company, New York, 1961.

Reid, C.: "Principles of Chemical Thermodynamics," Reinhold Publishing Corporation, New York, 1960.

Shedlovsky, T.: "Electrochemistry in Biology and Medicine," John Wiley & Sons, Inc., New York, 1955.

Spiegler, K., and M. Wyllie: Electrical Potential Differences, in "Physical Techniques in Biological Research," vol. IIA, 2d ed., Academic Press Inc., New York, 1968.

Van Rysselberghe, P.: "Thermodynamics of Irreversible Processes," Blaisdell Publishing Co., New York, 1963.

---

[1] K. Frank and M. C. Becker, Microelectrodes for Recording and Stimulation, in "Physical Techniques in Biological Research," vol. V, pp. 23–88, Academic Press Inc., New York, 1964. J. M. Delgado, Electrodes for Extracellular Recording and Stimulation, in "Physical Techniques in Biological Research," vol. V, pp. 89–143, Academic Press Inc., New York, 1964. H. P. Schwan, Determination of Biological Impedances, in "Physical Techniques in Biological Research," vol. VI, Academic Press Inc., New York, 1963.

# 3
# Subthreshold Membrane Phenomena

## 3.1  INTRODUCTION

If a microelectrode is inserted into an individual nerve or muscle fiber of a living organism, a dc potential will be found to exist with respect to an external reference electrode.  For most biological materials the magnitude will lie between 50 and 100 mv with the inside negative with respect to the outside.  The immediate basis for the resting potential is the difference in ionic composition of the internal and external electrolytes.  We leave until the next chapter a discussion of the role of the membrane in producing the observed concentrations.  Interest in this chapter is centered mainly on the passive membrane phenomena which, under the given conditions, result in the observed resting potential.

While interest in this chapter is concerned mainly with membrane behavior that may be described as electrically passive and linear, we should keep in mind that we shall also have to account for its nonlinear behavior when it is activated.  This most marvelous phenomenon under which the membrane properties undergo very great changes in response to an adequate stimulus accounts for the ability of neurons to transmit signals through the central nervous system.

## 3.2 IONIC COMPOSITION OF CELLS

As noted above, the explanation of the resting potential is to be found in the difference in composition between the extracellular and intracellular fluid. Table 3.1 gives representative values for mammalian muscle. We note a high concentration of potassium ion inside and a high concentration of sodium and chloride ion outside the cell. Table 3.2 compares the ion concentration ratios in a number of species, and it is interesting to note that the same qualitative relationships exist. A further comparison between various species is given in Table 3.3, which lists the observed resting potential. Table 3.4 lists comparable values for plant cells.

The cell membrane acts to separate the inside and outside regions, as a result of which a system with two compartments is created. Such a

**Table 3.1  Steady-state ion concentrations ($\mu$M/cm$^3$) in mammalian muscle cells***

| Extracellular fluid | | | | Intracellular fluid | | | |
|---|---|---|---|---|---|---|---|
| Cations | | Anions | | Cations | | Anions | |
| Na$^+$ | 145 | Cl$^-$ | 120 | Na$^+$ | 12 | Cl$^-$ | 4 |
| K$^+$ | 4 | HCO$_3^-$ | 27 | K$^+$ | 155 | HCO$_3^-$ | 8 |
| Others | 5 | Others | 7 | | | A$^-$ | 155 |
| (Ca$^{++}$,Mg$^{++}$, . . .) | | (SO$_4^{--}$, . . .) | | | | | |
| | pH | 7.43 | | | pH | 6.9 | |

* J. W. Woodbury, Biophysics of the Cell Membrane, in T. C. Ruch and H. D. Patton (eds.), "Physiology and Biophysics," W. B. Saunders Company, Philadelphia, 1965.

**Table 3.2  Ionic concentration ratios in various species**

| Species and tissue | Potassium ratio [K]$_i$/[K]$_o$ | Sodium ratio [Na]$_o$/[Na]$_i$ | Chloride ratio [Cl]$_o$/[Cl]$_i$ |
|---|---|---|---|
| *Loligo* axon | 40 | 9.2 | 12.5 |
| *Sepia* axon | 36 | 10.7 | |
| *Carcinus* axon | 38 | | |
| Frog nerve | 60 | 3.0 | 3.9 |
| Frog sartorius muscle | 56 | 13.0 | 33.0 |
| Rat cardiac muscle | 52 | 11.5 | |
| Cat myocardium | 31.5 | 24.5 | |
| Eel electroplaque | 30 | 18 | |

**Table 3.3   Comparison of measured resting potentials and calculated Nernst potentials, mv**

| Preparation | Potassium Nernst potential | Chloride Nernst potential | Observed resting potential |
|---|---|---|---|
| *Loligo* axon | 96 | 66 | 61 |
| *Sepia* axon | 93 | . . . | 62 |
| *Carcinus* axon | 95 | . . . | 82 |
| Frog myelinated axon | 106 | 35 | 71 |
| Frog sartorius muscle: | | | |
| In Ringer's | 101 | 91 | 94 |
| In natural circulation | 98 | 104 | 97 |
| Cat papillary muscle | 90 | . . . | 88 |
| Eel electroplaque | 89 | . . . | 78 |

**Table 3.4   Ionic concentration ratios (plant cells)***

| Plant cells | $[Na]_i/[Na]_o$ | $[Na]_o/[Na]_i$ | $[K]_i/[K]_o$ | $[Cl]_i/[Cl]_o$ |
|---|---|---|---|---|
| *Valonia macrophysa* | 0.18 | 5.5 | 41.6 | 1.03 |
| *V. ventricosa* | 0.07 | 14.1 | 48.0 | 1.05 |
| *Halicystis Osterhoutii* | 1.12 | 0.9 | 0.53 | 1.04 |
| *Nitella clavata* | 46.1 | 0.02 | 1,065 | 100.5 |
| *Chara ceratophylla* | 2.4 | 0.42 | 63 | 3.1 |

* W. J. V. Osterhout, *Botan. Rev.*, **2**:283 (1936).

system can give rise to a potential difference between the two compartments for a variety of reasons. We shall discuss here the following, and most likely, possible mechanisms.

1. A static diffusion potential arises when certain charged particles cannot pass through the interface between the two regions.
2. A diffusion potential is set up by ions passing through the boundary (ion flux); this potential can be maintained only by expenditure of energy to replenish the ions that flow.
3. If current is passed across the membrane wall because of an external source of emf, then an ohmic potential drop will occur.

In order to proceed with a mathematical analysis of several membrane models we shall need to develop suitable formulas for the evaluation of ion flow due to each of the forces that are likely to occur. One of the most important of these is that due to diffusion, and a quantitative description can be obtained either from thermodynamic considerations or by application of kinetic principles. Although we could immediately utilize the results of the previous chapter, we develop first the kinetic viewpoint since it will provide physical concepts that are not apparent from thermodynamics. A second force of importance is that due to the electric field, and this will be discussed in a subsequent section.

## 3.3 KINETIC APPROACH TO PASSIVE TRANSPORT (DIFFUSION)

A wide variety of diffusion processes exist in physics and chemistry. All such processes have in common the flow of a substance due to thermal energy when its density is nonuniform. Examples include diffusion of gases in regions where unequal concentrations exist, flow of heat in a region of nonuniform temperature distribution, and flow of current in a semiconductor where an unequal distribution of charge carriers exists. We shall be specifically interested in diffusion of ions in solutions where there are nonuniform concentrations.

If we consider a gas under equilibrium conditions, each particle moves about experiencing a sequence of collisions with other particles in the container. While a specific particle may make a net displacement from a given initial position, the time-average flow across a given plane will be zero. If a quantity of a different gas is added and confined initially to a relatively small region within the container, then, in time, the added gas will become uniformly distributed. The process by which this is brought about is called diffusion. We consider, now, a semi-quantitative description of this process.

For simplicity we assume that the concentration of the added gas is uniform in the $yz$ plane but varies smoothly in the transverse, $x$, direction; the problem is therefore one-dimensional. Let us consider the plane $x = X_0$. There will be a flow of particles across this plane from the "left" in the positive $x$ direction and a flow in the opposite direction from the gas at the "right." One can think of an infinitely thin slab at any arbitrary $x$, where, as a consequence of collisions within this slab, a flow to the right and to the left results whose time average is equal. Since the thermal energy is uniform, the resulting one-dimensional model assumes, in effect, that following each collision a given particle will have a fixed velocity of magnitude $v_x$ and an equiprobable positive or negative $x$ direction. Consequently, the net flow across a unit of area in the plane $x = X_0$ over an interval $\Delta t$ is

$$j_x(X_0) = \frac{C_l v_x \, \Delta t - C_r v_x \, \Delta t}{2 \Delta t} \tag{3.1}$$

where $C_l$ and $C_r$ represent characteristic concentrations at the left and right of $X_0$, respectively.[1]  The factor of 2 in (3.1) takes into account that only half the particles are traveling toward $X_0$.  Dimensional analysis suggests that $C_l$ be evaluated at $X_0 - l$ while $C_r$ be taken at $X_0 + l$, where $l$ is the mean free path.   Since $l$ is a very small quantity, we can express the difference in concentration $C_l - C_r$ by a linear term alone, that is,

$$C_r - C_l = 2l \frac{dC(x)}{dx} \bigg|_{x = X_0} \tag{3.2}$$

where $C(x)$ expresses the concentration as a function of $x$.   Consequently,

$$j_x = -l v_x \frac{dC(x)}{dx} \tag{3.3}$$

A somewhat more rigorous formulation and derivation of (3.3) is as follows.   If we let $X_0 = 0$, then

$$j_x = \frac{\int_0^{-\infty} C(x) p_l(x) v_x \, \Delta t \, dx - \int_0^{\infty} C(x) p_r(x) v_x \, \Delta t \, dx}{2\Delta t} \tag{3.4}$$

where $p_r(x)$ is the probability that a particle crossing $x = 0$ from the right started (had its most recent collision) between $x$ and $x + dx$ and $p_l(x)$ is a similar probability density function for the region $x < 0$.   If we expand $C(x)$ by a Maclaurin series and the quadratic and higher terms are neglected, then

$$2j_x = v_x C(0) \int_0^{-\infty} p_l(x) \, dx + v_x \frac{dC}{dx} \int_0^{-\infty} x p_l(x) \, dx$$
$$- v_x C(0) \int_0^{\infty} p_r(x) \, dx - v_x \frac{dC}{dx} \int_0^{\infty} x p_r(x) \, dx \tag{3.5}$$

By definition

$$\int_0^{-\infty} p_l(x) \, dx = \int_0^{\infty} p_r(x) \, dx = 1$$

while

$$\int_0^{\infty} x p_r(x) \, dx = - \int_0^{-\infty} x p_l(x) = l \tag{3.6}$$

where $l$ is the mean free path.[2]  This leads immediately to the result

---

[1] The dimension of $j_x$ is in particles per (length)$^2$ per unit time and corresponds to that of $C$ [particles per (length)$^3$].   The flow magnitude depends on the product of concentration and velocity, of course.

[2] This is not immediately obvious.   If the distance between successive collisions of a *given particle* is $y$ and its probability density function is $p_0(y)$, then the mean free

given by (3.3). The contribution to the above integrals comes mainly from the interval at and near the origin, since $p(x) \propto \exp(-|x|/l)$; consequently, little error results if the infinite limits are replaced by, say, $5l$. Over this restricted range the variation of $C(x)$ can be expected to be small and the linear approximation for $C(x)$, as well as the constancy of $l$, is justified.

For the general three-dimensional case it turns out that a factor of $\frac{1}{3}$ is required in (3.3) so that

$$j_x = -\frac{lv}{3}\frac{dC}{dx} \tag{3.7}$$

where $v$ is the average velocity (arbitrary direction).[1] Similar equations can be written for the $y$ and $z$ components, so that we have, in general,

$$\mathbf{j} = -\frac{lv}{3}\nabla C \tag{3.8}$$

The experimentally determined coefficient in (3.8) is called the *diffusion coefficient D*, that is,

$$\mathbf{j} = -D\nabla C \tag{3.9}$$

and for a gas we have

$$D = \tfrac{1}{3}lv \tag{3.10}$$

---

path, by definition, is

$$\int_0^\infty y p_0(y)\, dy = l$$

If at a given *instant of time* the collision interval for *each particle* whose path will cross $X_0$ is the random variable $z$, then the probability density function $p_1(z)$, taken across the ensemble, equals $kzp_0(z)$ ($k$ is a constant which must equal $1/l$). The factor $z$ appears because the longer interval is proportionately more likely to occur at a given instant. For a Poisson process it turns out that

$$\int_0^\infty z p_1(z)\, dz = 2l$$

From symmetry considerations

$$\int_0^\infty x p_r(x)\, dx = -\int_0^{-\infty} x p_l(x)\, dx = 1 \int_0^\infty x p_1(x)\, dx = l$$

[1] The factor of $\frac{1}{3}$ for three dimensions can be thought of as arising from the equal probability of a particle's having an $x$, $y$, or $z$ velocity. A rigorous formulation based on particles with uniform velocity $v$ confirms this result. The numerical coefficient will differ from that given if, in addition, a velocity distribution is considered. For particles of similar mass and with a maxwellian distribution the numerical factor is 0.35 instead of 0.33.

Equation (3.9) is known as Fick's law, and $D$ is sometimes referred to as Fick's constant. For particles under highly concentrated conditions interaction effects become significant and the chemical activities, rather than the concentration gradient, are the driving force for diffusion. This conclusion was noted in the thermodynamic treatment in the previous chapter.

Under an external force the average velocity acquired by a particular species is a measure of its mobility. A close connection exists between mobility and diffusion coefficient. Since the mean thermal velocity normally exceeds greatly the *drift velocity* due to an applied force, then following each collision the particle direction is random and the average drift velocity is zero. If we assume a fixed force $f$ to be acting (say electrostatic or gravitational) and that the mass of each particle is $m$, then since the initial velocity is zero, the mean drift velocity of an ensemble of particles equals the acceleration $f/m$ times an average of each particle's elapsed time since its most recent collision. Now the latter is the same as the averaged interval until the following collision, and this in turn is the same as the mean time $\tau$ between collisions for a single particle (see footnote 2, page 82). Thus we have

$$v = \frac{f\tau}{m} \tag{3.11}$$

We shall define mobility here as[1]

$$u' = \frac{v}{f} \tag{3.12}$$

Since $l = v\tau$ and $\tau = mu'$, then if we consider particles in a gas, where (3.10) applies, we have

$$D = \tfrac{1}{3}mu'v^2 \tag{3.13}$$

But $mv^2/2$ depends only on the temperature and, according to kinetic theory, equals $\tfrac{3}{2}kT$. Consequently,

$$D = u'kT \tag{3.14}$$

Although the result in (3.14) is based on the validity of (3.10), which was obtained for a gas, it is actually true in general (for example, it applies to ions in solution, particles in suspension, etc.). To show this we imagine that the particles of interest (e.g., ions in solution) have a concentration gradient in the $x$ direction only. We consider the application of a fixed force $f_x$ in the $x$ direction, so that the drift current

---

[1] This definition differs by a factor from that adopted in Sec. 2.8 for ions. To distinguish the two, a prime is employed in (3.12). At the end of this section the two forms will be related.

due to $f_x$ just balances the diffusion current. Designating the diffusion current $j_{\text{diff}}$ and the drift current $j_{\text{dr}}$, we require

$$j_{\text{dr}} = C(x)u'f_x \qquad\qquad \mu'f_x = U_x \ (3.12) \qquad (3.15)$$

where $C(x)$ is the concentration. [Dimensionally $C(x)$ is in particles per unit volume, in which case $j_{\text{dr}}$ and $j_{\text{diff}}$ are expressed as the number of particles crossing a unit area per second.] The diffusion current is given by (3.9) as

$$j_{\text{diff}} = -D\frac{dC}{dx} \qquad (3.16)$$

Consequently, because $j_{\text{diff}} = -j_{\text{dr}}$, we get

$$\frac{dC}{dx} = \frac{u'f_x}{D}C \qquad (3.17)$$

Since the conditions described by (3.17) are equilibrium conditions, the concentration distribution defined by $C(x)$ must satisfy the laws of statistical mechanics. That is,

$$C(x) = C_0 e^{-U/kT} \qquad (3.18)$$

where $U$ is the potential energy. In this case since $f_x$ is in the $x$ direction, $U = -f_x x + \text{const}$ and

$$\frac{dU}{dx} = -f_x \qquad (3.19)$$

If (3.18) is substituted into (3.17) and (3.19) is used, one obtains

$$f_x\frac{C_0 e^{-U/kT}}{kT} = \frac{u'f_x}{D}C_0 e^{-U/kT} \qquad (3.20)$$

from which we verify that

$$D = u'kT \qquad (3.14)$$

This expression is known as the *Einstein relation*. Since $R = kL$, where $R$ is the gas constant and $L$ is Avogadro's number, and $F = eL$, where $F$ is the Faraday and $e$ the electronic charge, (3.14) can also be expressed as

$$D = \frac{u'eRT}{F} \qquad (3.21)$$

We shall be interested exclusively in movement of ions, and for this class of problem the definition of mobility is the *velocity per unit field*.

Adopting the notation of an unprimed $u$ for this case, then for a field $E$ and corresponding velocity $v$ we have

$$u = \frac{v}{E} \;=\; \textit{electric field mobility} \tag{3.22}$$

Consequently, for the $i$th ion species with valence $z_i$ we have

$$\frac{u_i}{|z_i|e} = \frac{v}{|z_i|eE} = \frac{v}{f} \;=\; \mu_i'$$

so that

$$u_i' = \frac{u_i}{|z_i|e} \;=\; \textit{force mobility} \tag{3.23}$$

Note that by definition the mobility is a positive quantity.  The Einstein relation (3.21) can also be written

$$D_i = \frac{u_i RT}{|z_i|F} \tag{3.24}$$

for the $i$th ion species with valence $z_i$, in view of (3.23).

### 3.4  NERNST–PLANCK EQUATIONS

For biological membranes the major forces which cause ion movement are those due to diffusion and to an electric field.  In this section we develop appropriate formulas for the ion flux due to the combination of both factors.

With respect to the ion flow resulting from diffusion we have, from (3.9),

$$(\mathbf{j}_i)_{\text{diff}} = -D_i \nabla C_i \tag{3.25}$$

as the flux of the $i$th ion species.  In this and following work we shall express the concentration in moles per unit volume, whereupon $j_i$ is in moles per cross section per second.  The diffusion coefficient for the $i$th species has been designated $D_i$.

Now if an electric potential field $\Phi$ is present, then, since ion velocity equals the field times the mobility, the ion flux due to the electric forces is

$$(\mathbf{j}_i)_{\text{el}} = -u_i \frac{z_i}{|z_i|} C_i \nabla \Phi \tag{3.26}$$

The factor $z_i/|z_i|$ in (3.26) takes care of the sign of the ion flux, which is in the direction of the negative gradient for positive ions and in the opposite direction for negative ions. The total flux when both diffusion and electric forces are present is then

$$\mathbf{j}_i = -D_i \left( \boldsymbol{\nabla} C_i + \frac{z_i C_i F \boldsymbol{\nabla}\Phi}{RT} \right) \qquad (3.27)$$

when (3.24) is used to replace $u_i$. The units of $j_i$ are moles per unit cross section per unit time. Equation (3.27) and variations of it are referred to as the *Nernst-Planck equations.*

We can convert (3.27) into an expression for electric current density by recognizing that each mole carries a charge $Fz_i$; thus

$$\mathbf{J}_i = -D_i F z_i \left( \boldsymbol{\nabla} C_i + \frac{z_i C_i F \boldsymbol{\nabla}\Phi}{RT} \right) \qquad (3.28)$$

and $\mathbf{J}_i$ has the dimensions of amperes per unit area. Equation (3.28) can be rewritten to show the explicit dependence on mobility as follows:

$$\mathbf{J}_i = -\left( u_i RT \frac{z_i}{|z_i|} \boldsymbol{\nabla} C_i + u_i |z_i| C_i F \boldsymbol{\nabla}\Phi \right) \qquad (3.29)$$

*BECAUSE DIFFUSION IS DRIVEN BY THE CONCENTRATION GRADIENT, FLOW IS INDEPENDENT OF CHARGE*

Note that the sign of the diffusion term depends on whether the ion is positive or negative, but the electric current is in the direction of the negative potential gradient for either case.

In the thermodynamic treatment, diffusion is characterized by the movement of material from regions of high to low chemical potential. Let us assume for the moment that the chemical potential is a function of $x$ only. Then if a change of potential $\Delta\mu$ occurs over an interval $\Delta x$, an effective force of $-\Delta\mu/\Delta x$ acts on 1 mole of the substance under consideration, as noted in (2.21). The velocity acquired by a particle equals the force per particle [that is, $f_x = -(1/L)(d\mu/dx)$] times the mobility $u'$ as defined in (3.12). Thus,

$$v_x = -\frac{u'}{L}\frac{d\mu}{dx} \qquad (3.30)$$

If we substitute (3.23) and generalize to the case where the spatial variation of $\mu$ is arbitrary, then for the $i$th ion species

$$\mathbf{v}_i = -\frac{u_i}{|z_i|eL} \boldsymbol{\nabla}\mu \qquad (3.31)$$

The above result may be compared with (3.9) by substituting the expression for $\mu_i$ given in (2.58). This leads to

$$\mathbf{v}_i = -\frac{u_i RT}{|z_i|F} \left( \frac{1}{C_i} \boldsymbol{\nabla} C_i + \frac{1}{f_i} \boldsymbol{\nabla} f_i \right)$$

Finally,

$$\mathbf{j}_i = \mathbf{v}_i C_i = -\frac{u_i R T}{|z_i| F} \left[ 1 + \frac{C_i (\boldsymbol{\nabla} f_i)}{f_i (\boldsymbol{\nabla} C_i)} \right] \boldsymbol{\nabla} C_i$$

$$\text{or} \quad \mathbf{j}_i = -\frac{u_i R T}{|z_i| F} \left[ 1 + \frac{\boldsymbol{\nabla} (\ln f_i)}{\boldsymbol{\nabla} (\ln C_i)} \right] \boldsymbol{\nabla} C_i \tag{3.32}$$

A comparison of (3.32) with (3.9) shows that the more accurate coefficient $D_i$ is

$$D_i = \frac{u_i R T}{|z_i| F} \left[ 1 + \frac{\boldsymbol{\nabla} (\ln f_i)}{\boldsymbol{\nabla} (\ln C_i)} \right] \tag{3.33}$$

The result in (3.33) shows that the diffusion constant depends on the concentration since the activity coefficient has such a dependence. Equation (3.33) reduces to the ideal value given by (3.24) for low concentrations (when $f \to 1$). In the work to follow we shall assume that $f_i = 1$ and hence that (3.24) and (3.27) are valid. This means that results are strictly correct only in the case of dilute solutions.

In the analysis of membrane models we shall neglect the possible movement of water across the membrane. Such movement could arise from hydrostatic or osmotic pressure (or both). Osmotic effects occur at membranes that freely transmit water but not solute particles. Such membranes are *semipermeable*, and this property derives from a pore size that is adequate to pass solvent but is too small for the solute. If the mole fraction of water differs on the two sides of the membrane, then diffusion of water from the region of high concentration to low will take place. The "driving pressure" is called the *osmotic* pressure, and it can be evaluated by applying a hydrostatic pressure in the reverse direction so that equilibrium is produced.

The osmotic pressure of a given solution can be evaluated by considering it to be on one side of a semipermeable membrane with pure water on the other side. A hydrostatic pressure $P$ will be required on the solution side for equilibrium, and by definition this equals the osmotic pressure $\pi$. Considering Raoult's law and assuming a low concentration, one can show[1] that

$$\pi = \frac{n R T}{V}$$

where $n$ is the number of moles of solute in $V$ volumes of solvent. This approximate relationship is due to van't Hoff. For dilute solutions on both sides of a membrane the osmotic pressure depends on the difference

[1] E. J. Harris, "Transport and Accumulation in Biological Systems," p. 19, Butterworth Scientific Publications, London, 1960.

in particle density $\Delta C$ given by

$$\pi = \Delta C R T \tag{3.34}$$

If the interstitial region is chosen as a reference, then the intracellular medium is considered isotonic when $\Delta C$ in (3.34) is zero, and *hypertonic* or *hypotonic* according to whether $\Delta C$ is positive or negative, respectively. In neglecting water flow in the following work, isotonicity and the absence of hydrostatic forces are tacitly assumed.

## 3.5 ELECTRONEUTRALITY CONDITION

An important constraint on ions in a solution is the condition of *electroneutrality*. This requires that over a large volume the positive ionic charge per unit volume not deviate appreciably from the negative ionic density. If this were not the case, then the electrostatic forces would create a potential energy per particle so much greater than the mean thermal energy that the charged particles would move very strongly to reduce such potential differences, i.e., to restore electrical neutrality. As an illustration, if *all* negative ions within a sphere of radius $R$ were pushed into a surrounding shell of like volume, one could calculate the energy in the electrostatic field so created. This energy could only come from thermal energy, and the limiting case (maximum-size shell) is for complete conversion. In this (unlikely) event $R_m = 19(\kappa T/C)^{\frac{1}{2}}$ cm, where $T$ is absolute temperature, $\kappa$ is the relative permittivity, and $C$ is the ion concentration in particles per cubic centimeter.[1] The quantity $R_m$ is the *Debye shielding distance*, its magnitude for physiological systems is on the order of several angstroms.

The above considerations are satisfactory for an essentially uniform, infinite region. When phase boundaries and fixed-charge membranes are involved, additional considerations arise. We shall pursue this question further (and in a quantitative way) by utilizing a one-dimensional model[2] as illustrated in Fig. 3.1. As shown, the region $x \leq 0$ represents a membrane containing an immobile uniform fixed charge density $N$ (ions per unit volume). The region $x > 0$ contains the electrolyte solution chosen to contain only univalent ions and which is assumed to freely permeate the membrane region as well. The entire medium is assumed to have a uniform dielectric permittivity $\epsilon$. It is furthermore assumed that, despite $q_e\Phi \gg kT$, *Poisson's equation* holds.[3]

[1] J. G. Linhart, "Plasma Physics," p. 63, North Holland Publishing Company, Amsterdam, 1960.

[2] A. Mauro, Space Charge Regions in Fixed Charge Membranes and the Associated Property of Capacitance, *Biophys. J.*, **2**:179 (1962).

[3] *Ibid.* Electronic charge is symbolized here as $q_e$.

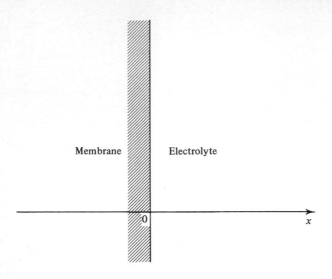

**Fig. 3.1**  Semi-infinite fixed-charge membrane.

Thus, since only variations with respect to $x$ can occur,

$$\frac{d^2\Phi}{dx^2} = -\frac{\rho}{\epsilon} \tag{3.35}$$

where $\Phi$ is the scalar potential and $\rho$ is the charge density. Utilizing (3.18) and letting the total density of negative and positive charge be denoted $n$ and $p$, respectively, yields

$$n = n_0 e^{q_e \Phi / kT} \tag{3.36a}$$

$$p = p_0 e^{-q_e \Phi / kT} \tag{3.36b}$$

The quantities $n_0$ and $p_0$ are the densities at $\Phi \to 0$ (this may be chosen, arbitrarily, as occurring for $x \to \infty$) and must be equal. The net charge density $\rho$ is given by

$$\rho = (p - n + N)q_e$$
$$= q_e n_0 (e^{-q_e \Phi / kT} - e^{q_e \Phi / kT}) + q_e N \tag{3.37}$$

Then substituting (3.37) into (3.35) yields

$$\frac{d^2\Phi}{dx^2} = \frac{2q_e n_0}{\epsilon} \left[ \sinh\left(\frac{q_e \Phi}{kT}\right) - \frac{N}{2n_0} \right] \tag{3.38}$$

This equation can be normalized by defining

$$\xi = \frac{x}{R_m} \qquad \text{where } R_m = \sqrt{\frac{kT\epsilon}{2n_0 q_e{}^2}} \tag{3.39}$$

and

$$y = \frac{q_e \Phi}{kT} \tag{3.40}$$

in which case (3.38) reduces to

$$\frac{d^2 y}{d\xi^2} = \sinh y - \frac{N}{2n_0} \tag{3.41}$$

Equation (3.41) is a one-dimensional form of the Poisson-Boltzmann equation that includes a term for the presence of fixed charges. The normalization factor $R_m$ is the Debye shielding distance as previously defined for the three-dimensional space.

For $x \geq 0$ (3.41) can be integrated. We first write it as

$$2\frac{dy}{d\xi}\left[\frac{d}{d\xi}\left(\frac{dy}{d\xi}\right)\right] = 2\frac{dy}{d\xi}\sinh y \tag{3.42}$$

and integration with respect to $\xi$ gives

$$\left(\frac{dy}{d\xi}\right)^2 = 2\cosh y + A \tag{3.43}$$

where $A$ is a constant. Since $y$ and $dy/d\xi \to 0$ as $\xi \to \infty$, $A = -2$. Hence

$$\frac{dy}{d\xi} = \sqrt{2(\cosh y - 1)} = -2\sinh\frac{y}{2} \tag{3.44}$$

In (3.44) the negative square root is used since the gradient should be negative for the assumed positive fixed-charge distribution. Integration of (3.44) now gives

$$\xi = -\ln\frac{\tanh (y/4)}{\tanh (y_0/4)} \tag{3.45}$$

where $y_0$ is the value of $y$ at $x = 0$. Equation (3.45) can be rearranged to give

$$y = 4\tanh^{-1}\left(e^{-\xi}\tanh\frac{y_0}{4}\right) \tag{3.46}$$

For large $\xi$ we can approximate (3.46) by

$$y \approx 4\tanh\left(\frac{y_0}{4}\right)e^{-\xi} \tag{3.47}$$

Since $\rho = -2q_e n_0 \sinh y$, while $y \to y_0$ as $\xi \to 0$, we see that

$$\rho \leq -(2q_e n_0 \sinh y_0)\frac{y}{y_0}$$

and utilizing (3.47), we have

$$\rho \leq \frac{-8q_e n_0 \sinh y_0}{y_0} \tanh\left(\frac{y_0}{4}\right) e^{-\xi} \tag{3.48}$$

Thus, from (3.48) we see that $R_m$ is a measure of the extent of the space charge. For regions which are large compared with $R_m$, therefore, electroneutrality can be assumed.

Analysis of (3.38) for $x \leq 0$ is more difficult and must be accomplished numerically. For $N/n_0 \gg 1$ an approximate solution can be found,[1] namely,

$$y \approx y_{-\infty} - \exp\left(\xi \sqrt{\frac{N}{2n_0}}\right) \qquad (\xi \leq 0) \tag{3.49}$$

where $y_{-\infty}$ is the maximum value of $y$ which occurs for $x \to -\infty$. By using (3.49) an expression for $\rho$ can be found from (3.37) and (3.41), which is

$$\rho \approx N q_e \exp \frac{x}{R_m \sqrt{2n_0/N}} \tag{3.50}$$

In this case the characteristic length is $R_m \sqrt{2n_0/N}$, so that, since $n_0/N \ll 1$, the space-charge extent is less than in the solution phase.

For ionic concentrations of the order given in Table 3.1, which are typical of physiological conditions, and with a relative dielectric constant for water ($\kappa = 80$), $R_m \approx 8$ Å. For membranes of biological thickness of the order 100 Å and for space-charge effects under 8 Å in extent, electroneutrality is a good approximation. We shall utilize this assumption in the following work, recognizing the tentative nature of the approximation.[2]

## 3.6  THE DONNAN MEMBRANE

The goal of this chapter is to seek to explain the basis for the resting potential of a cell. As a means to this end we consider at this point the properties of an idealized membrane with selective permeability to different ion species. When such a membrane is used to separate two electrolytes, a steady-state resting potential can result.

As a simple illustration, consider the two-compartment system shown in Fig. 3.2. Initially compartment I contains NaCl, while II has NaCl in a different concentration plus a sodium proteinate, NaP.

---

[1] Mauro, *ibid.*

[2] In addition to Mauro, see D. Agin, Electroneutrality and Electrodiffusion in the Squid Axon, *Proc. Natl. Acad. Sci.*, **57**:1232 (1967).

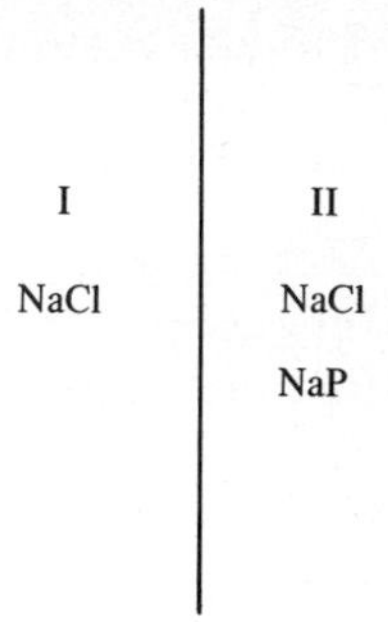

|   I   |   II  |
|-------|-------|
| NaCl  | NaCl  |
|       | NaP   |

**Fig. 3.2**  Two-compartment system.

We assume that the membrane is freely permeable to sodium and chloride ions but is impermeable to the very large protein ion.

If the initial concentration of sodium ions is the same in each compartment, then the chloride concentration in compartment I exceeds that in II. As a consequence Cl⁻ will diffuse from I to II. But this ion movement creates an electric field (directed from I to II) and results in the flow of Na⁺ from I to II with a concomitant effect of reducing the field. A net movement of NaCl results, along with a slow buildup of electric field. Ultimately, a steady-state condition is reached when no further flow occurs, and this is characterized by a balance of diffusion and electric forces acting on the sodium and chloride ions. The transmembrane potential that is reached under these conditions is called a *Donnan* potential.

The above steady-state condition is a true equilibrium, and it characterizes what is meant by the Donnan potential (or the Gibbs-Donnan system). Under this condition each ion species for which the membrane is permeable must satisfy $J_i = 0$. By utilizing (3.29) this means that

$$\nabla\Phi = -\frac{RT\nabla C_i}{z_i F C_i} = -\frac{RT}{z_i F}\nabla(\ln C_i) \tag{3.51}$$

In the more accurate expression, using (3.31), we have

$$\nabla\Phi = -\frac{RT}{z_i F}\nabla(\ln a_i)$$

We can integrate (3.51) to obtain

$$E_{\text{Don}} = \Phi_{\text{II}} - \Phi_{\text{I}} = -\frac{RT}{z_i F}\ln\frac{(C_i)_{\text{II}}}{(C_i)_{\text{I}}} \tag{3.52}$$

For the specific example above, application of (3.52) results in

$$E_{\text{Don}} = \frac{RT}{F}\ln\frac{(C_{\text{Na}})_{\text{I}}}{(C_{\text{Na}})_{\text{II}}} = \frac{RT}{F}\ln\frac{(C_{\text{Cl}})_{\text{II}}}{(C_{\text{Cl}})_{\text{I}}} \tag{3.53}$$

The following relationship must consequently be satisfied:

$$(C_{Na})_I (C_{Cl})_I = (C_{Na})_{II} (C_{Cl})_{II} \tag{3.54}$$

More generally, if all ions are, say, univalent, then (3.52) requires that for each $i$

$$\frac{(C_i^+)_{II}}{(C_i^+)_I} = \frac{(C_i^-)_I}{(C_i^-)_{II}} = r \tag{3.55}$$

where $r$ is a constant.

The above results can also be derived directly from the Boltzmann distribution. Using (3.36), we have for $x \to \infty$ that

$$p_1 = p_0 e^{-q_e E_{Don}/kT} \tag{3.56}$$

$$n_1 = n_0 e^{q_e E_{Don}/kT} \tag{3.57}$$

where $E_{Don} = \Phi(\infty) - \Phi(-\infty) = \Phi_1 - 0$. The quantities $p_1$ and $n_1$ denote positive and negative ion density at a large (actually infinite) distance to the *right* of the interface of the two compartments, while $p_0$ and $n_0$ are the values on the *left*. (Note that the positive fixed-charge membrane discussed in Sec. 3.5 is analogous to the immobile protein ion being investigated in the current problem.) Equations (3.56) and (3.57) yield the same value for $E_{Don}$ as (3.52). In addition, the discussion in Sec. 3.5 makes it clear that $E_{Don}$ is the correct potential difference between points that are beyond the transition zone at the phase interface. Nominally this requires a separation of, at least, several Debye lengths.

In the example under consideration the relationship of the Donnan potential to the concentration of the P ion can be calculated by imposing the condition of electroneutrality. Applied on sides I and II, it gives

$$(C_{Na})_I = (C_{Cl})_I \qquad (C_{Na})_{II} = (C_{Cl})_{II} + n(P^{-n})_{II} \tag{3.58}$$

where the proteinate is shown as a multivalent ion. Then, using (3.54) and (3.58), we get

$$(C_{Na})_{II} = \frac{n P^{-n} + \{ n(P^{-n})^2 + 4[(C_{Na})^2]_I \}^{\frac{1}{2}}}{2} \tag{3.59}$$

Consequently, (3.53) may be written

$$E_{Don} = -\frac{RT}{F} \ln \frac{n P^{-n} + \{ n(P^{-n})^2 + 4[(C_{Na})^2]_I \}^{\frac{1}{2}}}{2(C_{Na})_I} \tag{3.60}$$

We note that the potential is zero in the absence of P, but otherwise a potential is developed which causes side II to be negative with respect to side I.

If the transmembrane potential of a cell is explained by a Donnan equilibrium, then all (unconstrained) permeable ions must satisfy an equation of the form given by (3.52). Thus, for potassium

$$E_{\text{Don}} = \frac{RT}{F} \ln \frac{(C_{K^+})_o}{(C_{K^+})_i} \tag{3.61}$$

where subscript $o$ is the outside of the cell and $i$ is the inside of the cell and, consequently, $E_{\text{Don}}$ is the potential of the inside with respect to the outside. Putting in the constants $R$, $T$, and $F$ and changing the logarithm so that its base is 10 gives

$$E_{\text{Don}} = 60 \log \frac{(C_{K^+})_o}{(C_{K^+})_i} \qquad \text{mv} \tag{3.62}$$

and this result is shown in Table 3.3. Similarly, Table 3.3 gives the result for Cl⁻. We note fairly good agreement, which confirms the hypothesis of electrochemical equilibrium for both these ions; the electrochemical potential for both species is approximately the same on both sides of the membrane.

If $E_{\text{Don}}$ is computed for the sodium ion concentration, the result does not agree at all with the actual observed resting potential. Of course, it may be that the membrane is impermeable to the sodium ion, which would then play the same role as the protein ion of the previous example. However, while potassium and chloride pass through the membrane much more readily, a sodium flux can nevertheless be detected. Consequently, the assumption of a Donnan system is, by itself, inadequate to explain the biological observations.

In the above, we have thought of a two-compartment system separated by a physical membrane. It is also possible to apply the same arguments to the interface between differing phases, as, for example, the model described in Sec. 3.5. For more complex systems it is possible to have a Donnan equilibrium at phase boundaries even if the system itself is not in static equilibrium, provided that it is not the rate-determining step. We shall have occasion to use this consideration when we consider the "fixed-charge" membrane model in Sec. 3.9, and the concentration cell to be discussed in the next section.

## 3.7 FUNDAMENTAL RELATIONSHIPS (CONCENTRATION–CELL EXAMPLE)

The Donnan system, discussed in the previous section, is unduly restrictive in its assumption of static equilibrium conditions. In preference to this we shall simply require steady-state behavior. Specifically, we

shall assume the bulk solutions to be sufficiently extensive that during
the time it takes an ion to cross the membrane the concentration of all
components is essentially unchanged. Then within and at the boundaries of the membrane a steady-state condition will be reached, in which
case the concentration is time-independent and the continuity equation
becomes

$$\nabla \cdot \mathbf{j}_i = - \frac{\partial C_i}{\partial t} = 0 \tag{3.63}$$

If no return path is provided, then under steady-state conditions the
total current crossing the membrane must be equal to zero. Were this
not the case, charge would pile up, hence violating the conditions of
electroneutrality. This requirement may be formulated as

$$J = F \sum_i z_i j_i = 0 \tag{3.64}$$

Electroneutrality further requires that

$$\sum_i z_i C_i + \omega X = 0 \tag{3.65}$$

where $X$ is the concentration of fixed ion groups within the membrane in
equivalents per unit volume, and $\omega$ is $+1$ for positive and $-1$ for negative ions.

In the next sections we shall investigate the properties of a general
planar membrane separating multi-ionic solutions by integrating (3.28),
subject to the constraints of (3.63) to (3.65). Since the results are
somewhat complex, we consider first the simple example of the *concentration cell*. This is formed by a membrane which separates two solutions of the same electrolyte but of different concentration. For generality we assume the membrane to contain a fixed ionic charge
characteristic of *ion exchangers*. The ion exchanger may be thought of
as a sponge of fixed charges with ions of opposite sign freely floating
in the pores. The permeable ions are known as *counterions*. Much less
important, usually, is a smaller quantity of ions of the same sign as the
fixed charges which are known as *co-ions*. In the system to be studied,
the counterion is designated $A$ and the co-ion $Y$. The geometry is
shown in Fig. 3.3.

An expression for the membrane potential can be obtained from
(3.28) subject to (3.63) to (3.65). However, in this example it is somewhat simpler to use (2.123), for we immediately obtain

$$E_{\text{diff}} = - \frac{RT}{F} \int_{\text{II}}^{\text{III}} \left[ \frac{t_A}{z_A} d(\ln a_A) + \frac{t_Y}{z_Y} d(\ln a_Y) \right] \tag{3.66}$$

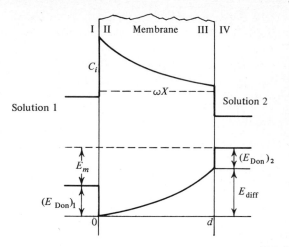

**Fig. 3.3**  Potential and concentration profiles in a concentration cell.  Membrane lies in the region $0 \leq x \leq d$.

where $E_{\text{diff}}$ is the (diffusion) potential across the membrane.  All quantities in (3.66) are defined *within* the membrane; the left-hand edge is designated II and the right-hand edge III (see Fig. 3.3).  We can simplify (3.66) by first noting that

$$t_A + t_Y = 1 \qquad \qquad \text{(3.67)}$$

*from butlown § pg 54*

$$\sum_{L=1}^{N} t_i = 1$$

which leads to

$$E_{\text{diff}} = -\frac{RT}{z_A F}\left\{ \ln \frac{(a_A)_{\text{III}}}{(a_A)_{\text{II}}} \right.$$

$$\left. + \int_{\text{II}}^{\text{III}} \frac{t_Y}{z_Y}\left[z_A d(\ln a_Y) - z_Y d(\ln a_A)\right] \right\} \qquad \text{(3.68)}$$

If we assume that 1 mole of salt dissociates into $v_A$ moles of $A$ and $v_Y$ moles of $Y$, then to preserve electroneutrality

$$z_A v_A + z_Y v_Y = 0 \qquad \qquad \text{(3.69)}$$

Then with this result we can express

$$z_A d(\ln a_Y) - z_Y d(\ln a_A) = \frac{z_A}{v_Y}\left[v_Y d(\ln a_Y) + v_A d(\ln a_A)\right]$$

$$= \frac{z_A}{v_Y}\left\{d[\ln\,(a_Y{}^{v_Y} a_A{}^{v_A})]\right\}$$

$$= z_A \left(1 + \frac{v_A}{v_Y}\right)[d(\ln a_\pm)] \qquad \text{(3.70)}$$

where the last result incorporates the mean ionic activity as defined in
(2.108).  Substitution of (3.69) and (3.70) into (3.68) gives

$$E_{\text{diff}} = -\frac{RT}{z_A F} \ln \left[ \frac{(a_A)_{\text{III}}}{(a_A)_{\text{II}}} + (z_A - z_Y) \int_{\text{II}}^{\text{III}} \frac{t_Y}{z_Y} d(\ln a_{\pm}) \right] \qquad (3.71)$$

We assume, now, that interfaces I-II and III-IV are in thermodynamic equilibrium; that is, no resistance to flow arises at either interface.
As a consequence the diffusion and electric forces are balanced, and a
Donnan phase-boundary potential is established at each interface in
accordance with (3.52).  The membrane potential is the sum of the
Donnan potentials and the diffusion potential as given in (3.71).  The
resultant membrane potential $E_m$ is

$$E_m = E_{\text{diff}} + (E_{\text{Don}})_1 + (E_{\text{Don}})_2 \qquad (3.72)$$

Since, from (3.52),

$$(E_{\text{Don}})_1 = \Phi_{\text{II}} - \Phi_{\text{I}} = -\frac{RT}{z_A F} \ln \frac{(a_A)_{\text{II}}}{(a_A)_{\text{I}}}$$

$$(E_{\text{Don}})_2 = \Phi_{\text{IV}} - \Phi_{\text{III}} = -\frac{RT}{z_A F} \ln \frac{(a_A)_{\text{IV}}}{(a_A)_{\text{III}}}$$

$$E_m = -\frac{RT}{z_A F} \ln \frac{(a_A)_{\text{IV}}}{(a_A)_{\text{I}}} + (z_A - z_Y) \int_{\text{II}}^{\text{III}} \frac{t_Y}{z_Y} d(\ln a_{\pm}) \qquad (3.73)$$

With ideal permselective membranes the co-ion flux vanishes, that is,
$t_Y = 0$.  In this case the second term in (3.73) goes out, and

$$E_m = -\frac{RT}{z_A F} \ln \frac{(a_A)_{\text{IV}}}{(a_A)_{\text{I}}} \qquad (3.74)$$

The above expression is the *Nernst equation*.

A physical interpretation of the concentration-cell potential is
readily given and is based on the membrane's having a greater permeability to the counterion than to the co-ion.  This means that the diffusion
of the counterion from the region of high to low concentration is not
completely offset by a similar diffusion of the co-ion.  As a result,
counterion charges accumulate at the interface with the low-concentration region until an electric field "slows down" the counterion and
"speeds up" the co-ion so that the net current is zero and electroneutrality
is maintained.  Thus for a cation exchanger (counterions are cations)
and with the assumption that region 2 is less concentrated than region 1,
the membrane potential $\Phi_2 - \Phi_1 = E_m$ is positive.  This result is also
clear from (3.74).

If the results in this section had been obtained by a direct integration of (3.28) subject to (3.63) to (3.65), then in addition to deter-

mining $E_m$, it would also be possible to calculate the potential and concentration profiles as shown in Fig. 3.3. This procedure will be discussed in the next section.

## 3.8  THE PLANCK MEMBRANE

The Planck membrane model is a planar, neutral membrane of infinite transverse extent which separates two solutions which are assumed to be multi-ionic. We list below the specific assumptions which are to be made in the investigation of the properties of this type of membrane.

1. The membrane is of constant thickness, is homogeneous, and is infinite in extent in the transverse plane. This last assumption validates a one-dimensional treatment of the problem.
2. The solutions on both sides of the membrane are well mixed, hence uniform. Furthermore, their volume is sufficiently great that changes in concentration can be neglected during the transit time through the membrane of any permeable ion.
3. Ion mobilities and activity coefficients are constant over the range considered and are the same within the membrane as in the free solution.
4. No net flow of solvent through the membrane takes place (i.e., osmotic and hydrostatic pressures are assumed to be negligible).
5. Ion flow due to convection is negligible, so that (3.28) correctly evaluates electric current due to each ion species.
6. For simplicity all ions are assumed to be univalent.

A diagram of the system is given in Fig. 3.4. The membrane could, physically, consist of a porous inanimate material such as sintered glass

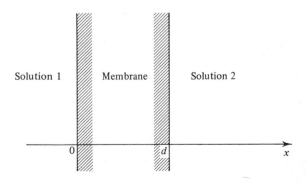

**Fig. 3.4**  The Planck membrane model.

or cellophane. The coordinate system is as shown, with the left-hand edge at $x = 0$ and the right-hand edge at $x = d$.

We are now ready to proceed with a mathematical analysis to determine the properties of this membrane. The first objective will be to determine the concentration profile. This will be followed by the details for the determination of the potential profile and the total membrane potential. The mathematics are somewhat laborious, but they are included because they are illustrative of a broad class of membrane problems.

To begin with, we note that in view of the one-dimensionality (3.63) can be written

$$\frac{\partial j_i}{\partial x} = 0 \tag{3.75}$$

since $\partial/\partial y = \partial/\partial z = 0$. The ion flux $j_i$ is in the $x$ direction, of course. Integration of (3.75) shows that $j_i$ is a constant (i.e., it does not depend on $x$). For convenience we specify the constant separately for anions and cations as follows:

$$J_i^+ = -A_i u_i^+ \qquad J_i^- = B_i u_i^- \tag{3.76}$$

where $J_i^+$ is the current density in the positive $x$ direction due to the $i$th cation flux, while $J_i^-$ is the same but for the $i$th anion flux. If (3.76) is substituted into (3.29) and the assumed condition of univalence recalled, then one obtains

$$A_i = RT \frac{dC_i^+}{dx} + FC_i^+ \frac{d\Phi}{dx} \tag{3.77a}$$

$$B_i = RT \frac{dC_i^-}{dx} - FC_i^- \frac{d\Phi}{dx} \tag{3.77b}$$

We now define $A$ and $B$ as

$$A = \sum_{i=1}^{k} A_i \qquad B = \sum_{i=1}^{m} B_i \tag{3.78}$$

where $k$ is the number of cation, and $m$ the number of anion, species. As a consequence of electroneutrality, the total number of cation moles equals that of anions, so that

$$C = \sum_{i=1}^{k} C_i^+ = \sum_{i=1}^{m} C_i^- \tag{3.79}$$

If we sum (3.77) over the index $i$, we obtain

$$A = RT \frac{dC}{dx} + FC \frac{d\Phi}{dx} \qquad B = RT \frac{dC}{dx} - FC \frac{d\Phi}{dx} \tag{3.80}$$

Adding the two equations in (3.80) gives

$$2RT \frac{dC}{dx} = A + B$$

which, upon integration, yields

$$C(x) = \frac{(A + B)x}{2RT} + \text{const}$$

Thus, the *total* concentration must vary linearly through the membrane.[1] If $C = C_0$ at $x = 0$ and $C = C_d$ at $x = d$, then

$$C(x) = \frac{C_d - C_0}{d} x + C_0 \tag{3.81}$$

By taking the difference of the equations in (3.80) we get

$$\frac{d\Phi}{dx} = \frac{A - B}{2FC} = \frac{(A - B)d}{2F[(C_d - C_0)x + dC_0]} \tag{3.82}$$

and we see that the potential profile will be logarithmic. The potential difference across the entire membrane is obtained by integrating (3.82) with respect to $x$, from zero to $d$, whereupon

$$E_m = \Phi(d) - \Phi(0) = \frac{(A - B)d}{2F(C_d - C_0)} \ln \frac{C_d}{C_0} \tag{3.83}$$

This result is not very useful since it depends on the constants $A$ and $B$, which are not yet related to the quantities that are known.

Since we have assumed steady-state conditions and a closed path for current is not provided, the net current must be zero. That is,

$$\sum_{i=1}^{k} J_i^+ + \sum_{i-1}^{m} J_i^- = 0 \tag{3.84}$$

Consequently, from (3.29) we have

$$\sum u_i^+ RT \frac{dC_i^+}{dx} + \sum u_i^+ FC_i^+ \frac{d\Phi}{dx} - \sum u_i^- RT \frac{dC_i^-}{dx}$$
$$+ \sum u_i^- FC_i^- \frac{d\Phi}{dx} = 0 \quad (3.85)$$

If we define

$$\Sigma u_i^+ C_i^+ = U \quad \text{and} \quad \Sigma u_i^- C_i^- = W \tag{3.86}$$

[1] This result is consistent with, but not so restrictive as, the Henderson assumption of linear mixtures, where the concentration of *each* ion species is assumed to vary linearly. [See (2.124).]

then (3.85) simplifies to

$$\frac{d\Phi}{dx} = -\frac{RT}{F}\frac{d(U - W)/dx}{U + W} \tag{3.87}$$

Substitution of (3.76) into (3.84) shows that

$$\Sigma A_i u_i^+ = \Sigma B_i u_i^- = K \tag{3.88}$$

where $K$ is a constant (since all terms in the summation are fixed). If (3.77a) is multiplied by $u_i^+$ while (3.77b) is multiplied by $u_i^-$, and if the resulting expressions are summed over the index $i$ then, by utilizing the definitions in (3.86) and (3.88), the following are obtained:

$$K = RT\frac{dU}{dx} + FU\frac{d\Phi}{dx} \tag{3.89a}$$

$$K = RT\frac{dW}{dx} - FW\frac{d\Phi}{dx} \tag{3.89b}$$

The value of $d\Phi/dx$ is now substituted from (3.82), resulting in the formulation of the following differential equations:

$$\frac{dU}{dx} + \frac{U(A - B)d}{2RT[(C_d - C_0)x + C_0 d]} = \frac{K}{RT} \tag{3.90a}$$

$$\frac{dW}{dx} - \frac{W(A - B)d}{2RT[(C_d - C_0)x + C_0 d]} = \frac{K}{RT} \tag{3.90b}$$

The above equations correspond to the standard form

$$\frac{dy}{dx} + f(x)y = g(x) \tag{3.91}$$

which can be converted into a perfect integral by multiplying by the integrating factor $e^{\int f dx}$. The resulting solution is in the form

$$y = e^{-\int f dx}[\alpha + \int e^{\int f dx}g(x)\,dx] \tag{3.92}$$

where $\alpha$ is an arbitrary constant.

The integrating factor for (3.90a) is

$$\begin{aligned}
e^{\int f dx} &= \exp\left[\frac{(A - B)d}{2RT}\int\frac{dx}{(C_d - C_0)x + C_0 d}\right]\\
&= \exp\left\{\frac{(A - B)d}{2RT(C_d - C_0)}\ln[(C_d - C_0)x + C_0 d]\right\}\\
&= [(C_d - C_0)x + C_0 d]^{(A-B)d/[2RT(C_d-C_0)]} \tag{3.93}
\end{aligned}$$

With (3.92) as a basis and using (3.93), we get

$$U = \frac{K}{RT}[(C_d - C_0)x + C_0 d]^{-(A-B)d/[2RT(C_d-C_0)]}$$

$$\frac{2RT(C_d - C_0)}{(A-B)d + 2RT(C_d - C_0)} \frac{[(C_d - C_0)x + C_0 d]^{1+(A-B)d/[2RT(C_d-C_0)]}}{C_d - C_0}$$
$$+ \alpha[(C_d - C_0)x + C_0 d]^{-(A-B)d/[2RT(C_d-C_0)]} \quad (3.94)$$

$$U = \frac{2K[(C_d - C_0)x + C_0 d]}{2RT(C_d - C_0) + (A-B)d}$$
$$+ \alpha[(C_d - C_0)x + C_0 d]^{-(A-B)d/[2RT(C_d-C_0)]} \quad (3.95)$$

The solution of (3.90$b$) is similar except that $(A - B)$ is replaced by $-(A - B)$. The result is

$$W = \frac{2K[(C_d - C_0)x + C_0 d]}{2RT(C_d - C_0) - (A-B)d}$$
$$+ \beta[(C_d - C_0)x + C_0 d]^{(A-B)d/[2RT(C_d-C_0)]} \quad (3.96)$$

where $\beta$ is an arbitrary constant.

If we insert the boundary conditions that $U = U_1$ when $x = 0$ and $U = U_2$ when $x = d$ (and, similarly, $W = W_1$ at $x = 0$ and $W = W_2$ at $x = d$), we obtain

$$U_2 = \frac{2KC_d d}{2RT(C_d - C_0) + (A-B)d} + \alpha(C_d d)^{-(A-B)d/[2RT(C_d-C_0)]} \quad (3.97)$$

$$U_1 = \frac{2KC_0 d}{2RT(C_d - C_0) + (A-B)d} + \alpha(C_0 d)^{-(A-B)d/[2RT(C_d-C_0)]} \quad (3.98)$$

Defining

$$\xi = \left(\frac{C_d}{C_0}\right)^{(A-B)d/[2RT'(C_d-C_0)]} = \left(\frac{C_0}{C_d}\right)^{(A-B)d/[2RT(C_d-C_0)]} \quad (3.99)$$

then

$$\xi U_2 = \frac{2KC_d \xi d}{2RT(C_d - C_0) + (A-B)d}$$
$$+ \alpha(C_0 d)^{-(A-B)d/[2RT(C_d-C_0)]} \quad (3.100)$$

and

$$\xi U_2 - U_1 = \frac{2Kd(\xi C_d - C_0)}{2(C_d - C_0)RT + (A-B)d} \quad (3.101)$$

In a similar way, we show that

$$W_2 - \xi W_1 = \frac{2Kd(C_d - \xi C_0)}{2(C_d - C_0)RT - (A-B)d} \quad (3.102)$$

If (3.101) is divided by (3.102), the result is

$$\frac{\xi U_2 - U_1}{W_2 - \xi W_1} = \frac{2(C_d - C_0)RT - (A-B)d}{2(C_d - C_0)RT + (A-B)d} \frac{\xi C_d - C_0}{C_d - \xi C_0} \quad (3.103)$$

Taking the logarithm of both sides of (3.99) results in

$$\ln \xi = \frac{(A - B)d}{2(C_d - C_0)RT} \ln \frac{C_d}{C_0}$$

from which (3.103) can be put into the form

$$\frac{\xi U_2 - U_1}{W_2 - \xi W_1} = \frac{\ln (C_d/C_0) - \ln \xi}{\ln (C_d/C_0) + \ln \xi} \frac{\xi C_d - C_0}{C_d - \xi C_0} \tag{3.104}$$

Equation (3.104) is the desired result and is a transcendental equation for $\xi$ in terms of the known quantities of the problem. From the solution of (3.104) for $\xi$ we can compute the membrane potential. We note that by virtue of (3.83) a simple relation exists linking $E_m$ to $\xi$, namely,

$$E_m = \frac{RT}{F} \ln \xi \tag{3.105}$$

The above results are those that were utilized in Sec. 2.8 in the solution of two illustrative examples.

If (3.83) is substituted into an expression for $\Phi(x)$† and the condition $(C_d - C_0)/C_0 = \delta \ll 1$ assumed, then we get

$$\Phi(x) = \Phi(0) + \frac{E_m}{\ln (1 + \delta)} \ln \left(1 + \frac{\delta x}{d}\right)$$

The logarithmic expressions can be approximated by the first term of their Taylor expansions, so that

$$\Phi(x) = \Phi(0) + \frac{E_m x}{d} \tag{3.106}$$

Thus, a linear potential variation occurs for $C_d \approx C_0$. If this field is substituted into the one-dimensional Poisson equation (3.35), we get

$$\frac{\rho}{\epsilon} = - \frac{\partial^2 \Phi(x)}{\partial x^2} = 0$$

so that the resultant field is consistent with the assumed electroneutrality of (3.79).

If $(C_d - C_0)/C_0$ is not small, then the logarithmic potential variation results, in which case a net charge density is required. This appears inconsistent with the assumption of electroneutrality. However, if $\rho$ is computed from (3.35) and (3.82), it will be found to be very small compared with $FC_0$, so that (3.79) remains essentially satisfied. The assumption of electroneutrality, rather than the more rigorous Poisson equation, is conventional in membrane studies and appears satisfactory in most cases.

† Integration of (3.82) gives $\Phi(x) - \Phi(0) = \dfrac{(A - B)d}{2F(C_d - C_0)} \ln \left[\dfrac{(C_d - C_0)x + dC_0}{dC_0}\right]$.

The double layer which forms at the membrane-solution interface can be considered by assuming the existence of a Donnan equilibrium. Since biological membranes are in the order of a Debye length in thickness, it is not certain that this approach will always be satisfactory. That is, the boundary effects, so to say, extend throughout the membrane. In this case, space-charge effects must be considered in evaluating the membrane properties.[1]

Application of a step change in current to a membrane produces, normally, two relaxation processes. The first, with a short time constant, arises from a dielectric *charging* process during which time the membrane resistance remains stationary. The second process occurs because the newly established electric field causes a *redistribution* of ionic concentration within the membrane, in order to meet the new equilibrium requirements. Efforts to quantitatively explain the charging and redistribution processes observed in the squid axon, utilizing the electrodiffusion model developed in this chapter, have met with only limited success. Other predictions of the single-ion electrodiffusion model to steady-state voltage-current relationships do not correspond to experiments with the squid axon. A summary of this work can be found in a paper by Cole,[2] who concludes that without significant modification the electrodiffusion model is not adequate to explain observations on the squid-axon membrane.

## 3.9 FIXED-CHARGE MEMBRANE

The physiological membrane contains complex compounds such as proteins and lipoids. These have intrinsic electrolytic properties which include the presence of *fixed ions*.[3] Consequently, the assumption of an electrically neutral membrane, which was made in the Planck model, requires modification. In this section, the properties of membranes containing a uniform distribution of fixed charges will be discussed. Such membranes are known as ionic membranes and have ion-exchanger properties.

In addition to the modification of the *membrane process*, the presence of fixed membrane charges will cause the ionic composition to change abruptly at the solution-membrane interface. As before, it will be assumed that this interface offers essentially no resistance to ion flow; the rate-limiting step is assumed to be the membrane diffusion process.

[1] D. Agin, Electroneutrality and Electrodiffusion in the Squid Axon, *Proc. Natl. Acad. Sci.*, **57**:1232 (1967).

[2] K. S. Cole, Electrodiffusion Models for the Membrane of Squid Axon, *Physiol. Rev.*, **45**:340 (1965).

[3] H. Davson, "A Textbook of General Physiology," chap. 8, 3d ed., Little, Brown and Company, Boston, 1964.

On this basis, the *boundary process* will be described by a Donnan equilibrium.

The above specifications may be formalized as follows, it being understood that for this membrane model, assumptions listed at the beginning of Sec. 3.8, unless modified, are otherwise as originally stated:

1. The membrane matrix is composed of fixed ions of a constant charge density.
2. The membrane interstices are permeable to all ions in the external solutions and are homogeneous throughout the membrane. All ions are assumed univalent. (See Schlögl[1] for the case of multivalent ions.)
3. The membrane surfaces are in a state of permanent, *instantaneously* established Donnan equilibrium. This assumption implies that the membrane process itself completely controls the rate of ion flux.

The above membrane model is often referred to as the *fixed-charge* model and was developed by Teorell, Meyer, and Sievers;[2] it is illustrated in Fig. 3.5.

[1] R. Schlögl, Elektrodiffusion in freier Lösung und geladenen Membranen, *Z. Physik. Chem. (Frankfurt)*, **1**:305 (1954).
[2] T. Teorell, Transport Processes and Electrical Phenomena in Ionic Membranes, *Progr. Biophys. Biophys. Chem.*, **3**:305 (1953).

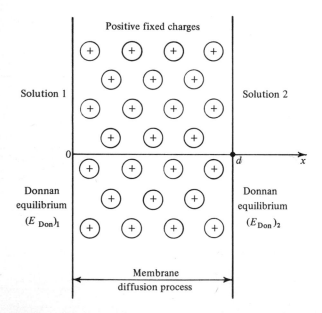

**Fig. 3.5** Fixed-charge membrane model with positive fixed ions.

The general procedure for the analysis of this membrane is essentially the same as for the Planck membrane. Specifically, (3.28) is integrated subject to (3.63) to (3.65), but, in this case, we assume that $X \neq 0$ in (3.65). The details follow closely those given for the Planck membrane. The final result is that the potential difference *within* the membrane can be expressed as

$$E_{\mathrm{diff}} = \frac{RT}{F} \ln \xi \tag{3.107}$$

where $\xi$ is a solution of

$$\frac{\xi U_2 - U_1}{W_2 - \xi W_1} = \frac{\ln K - \ln \xi}{\ln K + \ln \xi} \frac{\xi \bar{C}_2{}^+ - \bar{C}_1{}^+}{\bar{C}_2{}^- - \xi \bar{C}_1{}^-} \tag{3.108}$$

and

$$U_1 = \sum_i u_i{}^+ \bar{C}_i{}^+(x = 0) \qquad U_2 = \sum u_i{}^+ \bar{C}_i{}^+(x = d)$$

$$W_1 = \sum_i u_i{}^- \bar{C}_i{}^-(x = 0) \qquad W_2 = \sum u_i{}^- \bar{C}_i{}^-(x = d) \tag{3.109}$$

The notation corresponds to that defined earlier except for the use of a bar (for example, $\bar{C}_i$) to denote ion concentrations *within* the membrane); the absence of the bar refers to concentrations in the bulk solution. In addition,

$$\bar{C}_2{}^+ = \Sigma \bar{C}_i{}^+(x = d) \qquad \bar{C}_1{}^+ = \Sigma \bar{C}_i{}^+(x = 0)$$

$$\bar{C}_2{}^- = \Sigma \bar{C}_i{}^-(x = d) \qquad \bar{C}_1{}^- = \Sigma \bar{C}_i{}^-(x = 0) \tag{3.110}$$

The parameter $K$ is defined implicitly by the equation

$$K = \frac{\bar{C}_2{}^- - 0.5[\ln (K/\xi)/\ln K]\omega X}{\bar{C}_1{}^- - 0.5[\ln (K/\xi)/\ln K]\omega X}$$

$$= \frac{\bar{C}_2{}^+ + 0.5[\ln (K\xi)/\ln K]\omega X}{\bar{C}_1{}^+ + 0.5[\ln (K\xi)/\ln K]\omega X} \tag{3.111}$$

where $X$ is the fixed ion concentration in equivalents per unit volume. For $X = 0$, it can be verified that (3.111) and (3.108) lead to (3.104), as expected.

In solving for the membrane potential it is necessary to determine $\xi$ from the transcendental equation (3.108). This is done by trial and error. One first guesses at a value for the ratio $\ln \xi/\ln K$. When this value is substituted into (3.111), it is possible to determine $K$, hence $\ln \xi$ and $\xi$. These values are then substituted into (3.108), and if they are correct, an identity is obtained. Otherwise a new trial for $\ln \xi/\ln K$ is made.

In order to relate the ionic concentrations at the inner and outer edge of the membrane to that in the bulk solutions, the Donnan theory is utilized, as has been noted. If we designate by $C_i$ the concentration of the $i$th ion in the external solution, then the Donnan equilibrium ratio [see (3.55)] at the left-hand membrane-solution interface (designated with subscript 1) is

$$r_1 = \frac{(\bar{C}_1{}^+)_1}{(C_1{}^+)_1} = \frac{(\bar{C}_2{}^+)_1}{(C_2{}^+)_1} = \cdots \frac{(\bar{C}_k{}^+)_1}{(C_k{}^+)_1} = \frac{\Sigma(\bar{C}_i{}^+)_1}{\Sigma(C_i{}^+)_1} = \frac{\bar{C}_1{}^+}{C_1{}^+} \qquad (3.112)$$

Similarly, the right-hand membrane-solution interface (subscript 2) gives

$$r_2 = \frac{(\bar{C}_1{}^+)_2}{(C_1{}^+)_2} = \frac{(\bar{C}_2{}^+)_2}{(C_2{}^+)_2} = \cdots \frac{(\bar{C}_k{}^+)_2}{(C_k{}^+)_2} = \frac{\Sigma(\bar{C}_i{}^+)_2}{\Sigma(C_i{}^+)_2} = \frac{\bar{C}_2{}^+}{C_2{}^+} \qquad (3.113)$$

In the above equations, we have defined

$$\bar{C}_1{}^+ = \Sigma(\bar{C}_i{}^+)_1 \qquad C_1{}^+ = \Sigma(C_i{}^+)_1 \qquad \bar{C}_2{}^+ = \Sigma(\bar{C}_i{}^+)_2 \qquad C_2{}^+ = \Sigma(C_i{}^+)_2$$

From (3.55), we also have

$$r_1 = \frac{(C_1{}^-)_1}{(\bar{C}_1{}^-)_1} = \frac{(C_2{}^-)_1}{(\bar{C}_2{}^-)_1} = \cdots \frac{(C_k{}^-)_1}{(\bar{C}_k{}^-)_1} = \frac{\Sigma(C_i{}^-)_1}{\Sigma(\bar{C}_i{}^-)_1} = \frac{C_1{}^-}{\bar{C}_1{}^-} \qquad (3.114)$$

$$r_2 = \frac{(C_1{}^-)_2}{(\bar{C}_1{}^-)_2} = \frac{(C_2{}^-)_2}{(\bar{C}_2{}^-)_2} = \cdots \frac{(C_k{}^-)_2}{(\bar{C}_k{}^-)_2} = \frac{\Sigma(C_i{}^-)_2}{\Sigma(\bar{C}_i{}^-)_2} = \frac{C_2{}^-}{\bar{C}_2{}^-} \qquad (3.115)$$

where

$$\bar{C}_1{}^- = \Sigma(\bar{C}_i{}^-)_1 \qquad C_1{}^- = \Sigma(C_i{}^-)_1 \qquad \bar{C}_2{}^- = \Sigma(\bar{C}_i{}^-)_2 \qquad C_2{}^- = \Sigma(C_i{}^-)_2$$

For electroneutrality in the external solution

$$\begin{aligned} C_1{}^+ = C_1{}^- = C_1 \\ C_2{}^+ = C_2{}^- = C_2 \end{aligned} \qquad (3.116)$$

Using (3.65) and (3.112) to (3.115), we can solve for $r$ as

$$r_{1,2} = \sqrt{1 + \left(\frac{\omega X}{2C_{1,2}}\right)^2} - \frac{\omega X}{2C_{1,2}} \qquad (3.117)$$

This result is also a generalization of (3.60). The total membrane potential difference is then given by the sum of the contributions from the two Donnan boundary potentials $(E_{\text{Don}})_1$ and $(E_{\text{Don}})_2$ plus that due to the membrane diffusion-potential difference as expressed by (3.107). That is, the total potential difference is given by

$$\Phi_2 - \Phi_1 = (E_{\text{Don}})_1 + (E_{\text{Don}})_2 + E_{\text{diff}} = E_m \qquad (3.118)$$

Note that positive potential difference is defined as the potential "on the right" minus that "on the left." In evaluating $(E_{\text{Don}})_1$ from (3.52) we get $-(RT/F)\ln r_1$, but in evaluating $(E_{\text{Don}})_2$, the solution-membrane ratio (for positive ions) should be substituted in (3.52) or $-(RT/F)\ln(1/r_2)$. The sum of the Donnan potentials is then

$$(E_{\text{Don}})_1 + (E_{\text{Don}})_2 = \frac{RT}{F}\ln\frac{r_2}{r_1} \tag{3.119}$$

where $r_2$ and $r_1$ are the respective Donnan distribution ratios at edges 2 and 1.

In terms of the quantities defined earlier, a simple relationship can be derived for the current density carried by the $i$th ion.[1] For cations, the result is

$$J_i^+ = -Fk^+u_i[(C_i^+)_2 r_2\xi - (C_i^+)_1 r_1] \tag{3.120}$$

where

$$k^+ = \frac{RT}{d}\frac{\ln(K\xi)}{\ln K}\frac{\bar{C}_2^+ - \bar{C}_1^+ - 0.5\omega X\ln\xi}{\bar{C}_2^+\xi - \bar{C}_1^+} \tag{3.121}$$

The corresponding equation for anions is

$$J_i^- = Fk^-u_i\left[\frac{(C_i^-)_2}{r_2\xi} - \frac{(C_i^-)_1}{r_1}\right] \tag{3.122}$$

where

$$k^- = \frac{RT}{d}\frac{\ln(K/\xi)}{\ln K}\frac{\bar{C}_2^- - \bar{C}_1^- - 0.5\omega X\ln\xi}{\bar{C}_2^-/\xi - \bar{C}_1^-} \tag{3.123}$$

From (3.120) and (3.122) we note the dependence of ionic current on the mobility and the overall difference in electrochemical (chemical plus electrical) potential. The relationship is not strictly linear since the coefficient $k$ itself depends on ionic concentrations and membrane charge. If the total current $J = \Sigma J_i^+ + \Sigma J_i^-$ is formed from (3.120) and (3.122), then (3.108) follows from setting $J = 0$.

For biological conditions, the total concentrations of the solutions bathing the membrane are usually equal ($C_1 = C_2$). The Donnan distribution ratios are consequently equal, so that the total membrane potential reduces to the diffusion potential of the membrane itself (that is, $E_{\text{diff}} = E_m$). Under these conditions

$$J_i^+ = -\frac{FRT}{d}\frac{u_i^+\ln\xi}{\xi - 1}[(C_i^+)_2\xi - (C_i^+)_1]r \tag{3.124a}$$

$$J_i^- = \frac{FRT}{d}\frac{u_i^-\ln\xi}{\xi - 1}[(C_i^-)_2 - \xi(C_i^-)_1]r^{-1} \tag{3.124b}$$

[1] Teorell, *ibid.*, pp. 320–321.

noting that $k^+ = RT \ln \xi/[(\xi - 1)d]$ and $k^- = \xi k^+$. It is easily verified that for the condition $J = \Sigma J_i^+ + \Sigma J_i^- = 0$ the simple expression

$$\xi = \frac{U_1 + W_2}{U_2 + W_1} \tag{3.125}$$

results. This can also be obtained directly from (3.108) for the condition $C_1 = C_2$.

If, in addition to the condition $C_1 = C_2$, all ions are univalent, then the resultant potential distribution is linear,[1] so that $d\Phi/dx$ is a constant. The condition of a constant field forms the basis of a membrane model which has the advantage of more simply relating the concentration of constituents to the net membrane potential. This model was developed by Goldman[2] and is considered in the next section.

The quantities in (3.125) can also be related (rather simply) to concentrations in the solution through (3.112) to (3.115); specifically,

$$U_1 = r \sum_i u_i(C_i^+)_1 \qquad W_1 = \frac{1}{r} \sum_i u_i(C_i^-)_1$$

$$U_2 = r \sum_i u_i(C_i^+)_2 \qquad W_2 = \frac{1}{r} \sum_i u_i(C_i^-)_2 \tag{3.126}$$

It is interesting to note that for membranes with a high positive charge density, since $r \to 0$, the determining factor for $\xi$ in (3.125) is $W_2/W_1$. Consequently, it is the highly permeable anions that determine the conductance of the membrane. For the case of negative membranes, the situation is reversed and the conductance is determined by $U_1/U_2$, which depends on the cation flux. [Actually a property of an ideal ion exchanger is its impermeability to the co-ion (ion of same sign as fixed membrane charge).]

Figure 3.6 illustrates application of the general theory to a nonbiological positive membrane in the steady state. Equations for the concentration and potential profile may be found in Teorell.[3] In this figure, although it appears superficially that the $Cl^-$ current exceeds the $H^+$ (in view of the electrical gradient and greater $Cl^-$ concentration), $u_{H^+}/u_{Cl^-} = 5.3$, so that the net current is zero. The approximately linear $V(x)$ is rather exceptional.

[1] A. Finkelstein and A. Mauro, Equivalent Circuits as Related to Ionic Systems, *Biophys. J.*, **3**:233 (1963).

[2] D. E. Goldman, Potential, Impedance, and Rectification in Membranes, *J. Gen. Physiol.*, **27**:37 (1943).

[3] T. Teorell, Zur quantitativen Behandlung der Membranpermeabilität, *Z. Elektrochem.*, **55**:460 (1951).

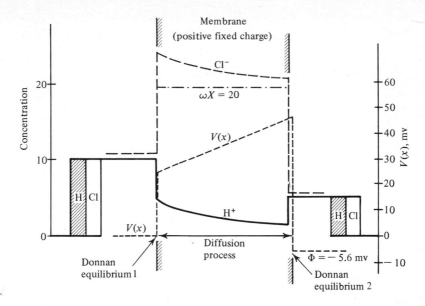

**Fig. 3.6** Application of the fixed-charge theory to a single electrolyte ($u_{H^+}/u_{Cl^-} = 5.3$). Shown are profiles of both ion concentration and electric potential. [*After T. Teorell, Progr. Biophys., Pergamon Press Ltd.,* **3**:305 (1953).]

More recent theoretical work has endeavored to remove some of the assumptions made in the Teorell-Meyer-Sievers fixed-charge membrane. In particular, Conti and Eisenman[1] obtained solutions for fixed-charge membranes whose concentration is permitted to be a function of $x$ (but where the site properties are assumed constant). In this situation, the mobilities and standard chemical potential of each ion species (only two monovalent species were considered) can no longer be assumed constant as in the case of uniformly charged membranes. By assuming instead that $d/dx(u_2/u_1) = 0$ and $d/dx(\mu_1^\circ - \mu_2^\circ) = 0$ and that the membrane is solely permeable to counterions, Conti and Eisenman were able to deduce all properties of the steady-state system of interest.[2]

The above work includes an extension to a mosaic membrane model which consists of insulated separate pathways, each of which is either solely anion-permeable or cation-permeable. Such a model is more satisfactory since biological membranes are generally permeable to both anions and cations. A review of studies on ion exchange membranes

[1] F. Conti and G. Eisenman, The Steady State Properties of Ion Exchange Membranes with Fixed Sites, *Biophys. J.,* **5**:511 (1965).
[2] *Ibid.*

which might be applicable to biological membranes is given by Eisenman, Sandblom, and Walker.[1]

## 3.10 CONSTANT-FIELD-MEMBRANE MODEL

We consider a membrane that is planar, uniform and homogeneous, and neutral $(X = 0)$. The basic assumption is that the field within the membrane is constant. This approximation is fairly good for biological membranes and is exact if the permeable ions are univalent and if the total ionic concentrations on each side of the membrane are equal.[2] A consequence of this hypothesis is that the pertinent Nernst-Planck equation can be integrated directly.

Thus, assuming a constant field, we replace $d\Phi/dx$ by $V/d$, where the potential at $x = 0$ is chosen to be zero and the magnitude at $x = d$ is assigned the value $V$. Then, for ions of biological interest (namely, potassium, sodium, and chloride), (3.29), applied within the membrane, becomes

$$-J_K = u_K RT \frac{d\bar{C}_K}{dx} + \frac{u_K \bar{C}_K F V}{d} \tag{3.127a}$$

$$-J_{Na} = u_{Na} RT \frac{d\bar{C}_{Na}}{dx} + \frac{u_{Na} \bar{C}_{Na} F V}{d} \tag{3.127b}$$

$$-J_{Cl} = -u_{Cl} RT \frac{d\bar{C}_{Cl}}{dx} + \frac{u_{Cl} \bar{C}_{Cl} F V}{d} \tag{3.127c}$$

Equations (3.127) are ordinary first-order separable differential equations, where the dependent variable is the concentration and the independent variable is $x$. The quantities $J_K$, $J_{Na}$, and $J_{Cl}$ are all independent of $x$ by virtue of (3.63). Solving for $\bar{C}_K$ in (3.127a), we obtain

$$\bar{C}_K(x) = A e^{-FVx/RTd} - \frac{J_K}{RTu_K} \frac{RTd}{FV} \tag{3.128}$$

Note that this model yields an exponential concentration variation for each ion species. The integration constant can be evaluated in terms of the concentration (within the membrane) at $x = 0$; thus

$$\bar{C}_K(x = 0) = (\bar{C}_K)_0 = -\frac{J_K d}{u_K F V} + A \tag{3.129}$$

[1] G. Eisenman, J. P. Sandblom, and J. L. Walker, Jr., Membrane Structure and Ion Permeation, *Science*, **155**:965 (1967).

[2] This result was noted in (3.106) for $C_d \rightarrow C_0$. Since $X = 0$, $r_1 = r_2 = 1$ in (3.117); consequently, the condition $\bar{C}_d = \bar{C}_0$ is ensured by equal bulk concentrations. A constant field also arises for univalent ions with equal bulk concentrations even if $X \neq 0$. The proof for this is given in Finkelstein and Mauro, *op. cit.*, p. 215.

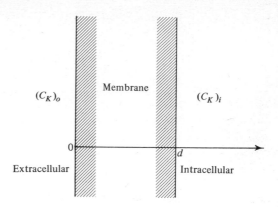

**Fig. 3.7**  Constant-field-membrane model.

Substituting for $A$ in (3.128) the value in (3.129) results in

$$\bar{C}_K(x) = \frac{J_K d}{u_K F V}\left(e^{-FVx/RTd} - 1\right) + (\bar{C}_K)_0 e^{-FVx/RTd} \tag{3.130}$$

If we also define

$$\bar{C}_K(x = d) = (\bar{C}_K)_d = \frac{J_K d}{u_K F V}\left(e^{-FV/RT} - 1\right)$$
$$+ (\bar{C}_K)_0 e^{-FV/RT} \tag{3.131}$$

then, solving for $J_K$, we have

$$J_K = \frac{(\bar{C}_K)_0 e^{-FV/RT} - (\bar{C}_K)_d}{1 - e^{-FV/RT}}\frac{F u_K V}{d} \tag{3.132}$$

Now we shall assume that the concentration $(\bar{C}_K)_0$ at the outer edge of the membrane is proportional to the concentration $(C_K)_o$ in the external medium. Similarly, the concentration $(\bar{C}_K)_d$ at the inner edge of the membrane may be assumed proportional to the intracellular concentration $(C_K)_i$ (see Fig. 3.7). In this notation, $V$ is then the potential of the inside relative to the outside. (That is, $V = \Phi_i - \Phi_o$.) We can write this relationship as

$$(\bar{C}_K)_0 = \beta_K (C_K)_o \tag{3.133a}$$
$$(\bar{C}_K)_d = \beta_K (C_K)_i \tag{3.133b}$$

where $\beta_K$ is the *partition coefficient*, assumed the same at both interfaces. We now define the *permeability coefficient* $P_K$ as

$$P_K = \frac{u_K \beta_K R T}{F d} = \frac{D_K \beta_K}{d} \tag{3.134}$$

With these changes (3.132) becomes

$$J_K = \frac{P_K F^2 V}{RT} \frac{(C_K)_o e^{-FV/RT} - (C_K)_i}{1 - e^{-FV/RT}} \tag{3.135}$$

In general, it is easy to show that for the $m$th ion species the current density $J_m$ is

$$J_m = \frac{P_m F^2 V |z_m|}{RT} \frac{(C_m)_o e^{-F z_m V/RT} - (C_m)_i}{1 - e^{-F z_m V/RT}} \tag{3.136}$$

For the remaining current components of specific interest, we have

$$J_{Na} = \frac{P_{Na} F^2 V}{RT} \frac{(C_{Na})_o e^{-FV/RT} - (C_{Na})_i}{1 - e^{-FV/RT}} \tag{3.137}$$

$$J_{Cl} = \frac{P_{Cl} F^2 V}{RT} \frac{(C_{Cl})_i e^{-FV/RT} - (C_{Cl})_o}{1 - e^{-FV/RT}} \tag{3.138}$$

Under static equilibrium conditions, no change in composition can occur; that is, all concentrations remain the same. In this case $J_K = J_{Na} = J_{Cl} = 0$ and we have a Donnan equilibrium. We note that (3.135), (3.137), and (3.138) yield the appropriate Nernst potentials in this case.

Under quasi-static conditions, such as assumed for the Planck and fixed-charge membrane, the composition in the extra- and intracellular fluids is assumed constant. If we impose the requirement that there be no *net* electric flow of current since no external electric circuit is provided, then we require

$$J_K + J_{Na} + J_{Cl} = 0 \tag{3.139}$$

Solving for $V$ by substituting (3.135), (3.137), and (3.138) into (3.139), we obtain

$$V = \frac{RT}{F} \left[ \ln \frac{P_K (C_K)_o + P_{Na}(C_{Na})_o + P_{Cl}(C_{Cl})_i}{P_K (C_K)_i + P_{Na}(C_{Na})_i + P_{Cl}(C_{Cl})_o} \right] \tag{3.140}$$

Equation (3.140) is the result utilized by Hodgkin and Katz,[1] and serves as a basis for evaluating the resting-membrane potential, given the concentration of sodium, potassium, and chloride in the intra- and extracellular media. The permeability coefficients are inferred from potential measurements and are assumed to be independent of ion concentration. Discussion of this equation, based upon measurements on physiological membranes, is given in the next chapter.

The constant-field model has been extensively utilized in the characterization of biological membranes. It has an advantage in that it

[1] A. L. Hodgkin and B. Katz, The Effect of Sodium Ions on the Electrical Activity of the Giant Axon of the Squid, *J. Physiol.*, **108**:37 (1949).

provides relatively simple relationships. By assigning appropriate values to the permeability, one can take account of the selective properties of the membrane. In addition, the membrane-solution interfacial phenomena can be considered by means of the partition coefficient, although the linear relationship of (3.133) is empirical. (A more rational approach is given in the fixed-charge analysis, where a Donnan equilibrium is assumed. This difference in characterizing the membrane-solution interface is probably the major difference between the two treatments.) By assigning appropriate relative values of permeability for the resting membrane and another set for the peak of the action current, the dependence of transmembrane potential on concentration has a particularly simple form in each case. In an evaluation presented in Chap. 4, we shall see that the results are at least qualitatively correct. An important criticism of the constant-field model is that the permeability coefficients are not obtainable from fundamental microscopic membrane properties, but rather are inferred from the overall behavior. Other shortcomings in its application to biological membranes are discussed in Chap. 4.

### 3.11 MEMBRANE CONDUCTIVITY AND RECTIFICATION

The electric conductance of a membrane is considered to be of some importance as a measure of membrane permeability. The current produced by applying an electric field across the membrane cannot be expected to be a linear function of that field since internal membrane changes are also produced, which complicates the relationship. Because the ratio of current to voltage does not have the usual significance, and in view of the anticipated nonlinearity, one defines instead

$$G = -\frac{dJ}{d\overline{V}} \tag{3.141}$$

which is a *dynamic* conductance (or incremental conductance). The total current density, in (3.141), is the sum of the partial ionic currents and may be expressed as

$$J = \sum_i J_i^+ + \sum_i J_i^- \tag{3.142}$$

where $J_i^+$ and $J_i^-$ are the current densities as obtained from, say, (3.29) and represent flow of electric current within the membrane from "left" to "right." If we use the relationship $(\Phi_2 - \Phi_1) = (RT/F) \ln \xi$, then

$$G = -\frac{F}{RT} \frac{d\left(\sum_i J_i^+ + \sum_i J_i^-\right)}{d(\ln \xi)} \tag{3.143}$$

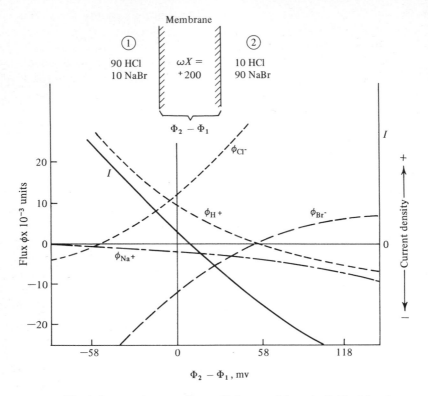

**Fig. 3.8** The influence of externally applied potentials on individual ion flux ($\phi$) and resultant current density ($I$). ($\Phi_1 - \Phi_2$) is the potential drop across the membrane. The composition of solutions and fixed-charge membrane is shown in the inset. [*After T. Teorell, Progr. Biophys., Pergamon Press Ltd., 3:305 (1953).*]

We show, in Fig. 3.8, the influence of membrane voltage on the individual current components and the total current, where the computation has been carried out by using (3.120) and (3.122) for the specific case that $C_1 = C_2$. The case of *free diffusion* is represented by the condition $J = 0$, whereupon ($\Phi_2 - \Phi_1$) = $E_{\text{diff}}$, as noted previously. As predicted, the curves of Fig. 3.8 show a nonlinear relationship. By substituting into (3.143) the appropriate quantities as given by (3.120) and (3.122), the indicated differentiation can be carried out, and an explicit expression for $G$ obtained. Then, noting the restriction to the case where the concentrations in the surrounding bulks are the same ($C_1 = C_2$), one obtains

$$G_D = \frac{F^2}{d} \frac{(U_2 + W_1)(U_1 + W_2)}{(U_1 + W_2) - (U_2 + W_1)} \ln \xi_D \qquad I = 0 \qquad (3.144)$$

and

$$G = \frac{F^2}{d} \frac{(U_2 + W_1)\xi}{\xi - 1} \left[ \frac{1 - \xi_D}{1 - \xi} \ln \xi + \left(1 - \frac{\xi_D}{\xi}\right) \right] \qquad I \neq 0 \quad (3.145)$$

In these equations we have

$$\xi_D = \frac{U_1 + W_2}{U_2 + W_1} = e^{F E_{\text{diff}}/RT}$$

and

$$\xi = e^{F(\Phi_2 - \Phi_1)/RT}$$

corresponding, essentially, to the earlier usage except that $\xi$ is now considered a function of $(\Phi_2 - \Phi_1)$ while $\xi_D$ is the specific value of $\xi$ when $I = 0$. The quantities $U$ and $W$ are defined as before, and involve ionic concentrations within the membrane.

A study of these results shows that for "neutral" (uncharged) membranes, an increase of any partial ionic flux is usually accompanied by a rise in total conductance. For charged membranes, such a correlated effect can no longer be predicted, and the converse may actually be true in a given situation. Thus, for equal concentrations in the bounding solutions, increasing the ion concentration of a positive-ion membrane increases the conductance while decreasing the positive-ion flux.

When $\xi = +\infty$ ($C_1 = C_2$), a limiting value of conductance is obtained corresponding to current flow from right to left within the membrane. The limiting conductance for a reversal of current corresponds to $\xi = 0$ ($E_{\text{diff}} = -\infty$). Reference to (3.145) yields

$$G_1(\xi = \infty) = \frac{F^2}{d}(U_2 + W_1) \qquad G_2(\xi = 0) = \frac{F^2}{d}(U_1 + W_2) \quad (3.146)$$

We note that the conductance depends on the direction of current flow, and the ratio

$$\frac{G_1}{G_2} = \frac{U_1 + W_2}{U_2 + W_1} = \xi_D \qquad\qquad (3.147)$$

is a measure of the *rectification* property of the membrane. The ratio in (3.147) is called the *rectification ratio*.

The above result is shown here as arising from the extended fixed-charge theory of Teorell. An identical result is also obtained by Goldman, based on the constant-field theory (although in this case the results are expressed only in terms of concentrations within the membrane, in the absence of a membrane-solution relationship). That the two results are similar is a consequence of the fact, already noted, that when $C_1 = C_2$,

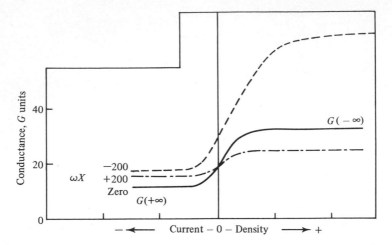

**Fig. 3.9** Conductance vs. $I$ for membrane with positive and negative fixed charges and for neutral membrane. (Bulk solutions have the same composition as in Fig. 3.8.) $G(-\infty)$ and $G(+\infty)$ denote the limiting conductance at very high negative and positive membrane potentials, respectively. [*After T. Teorell, Progr. Biophys., Pergamon Press Ltd.,* **3**:305 (1953).]

and for univalent ions, the potential variation in a charged (ionic) membrane is linear. A plot of conductance vs. total current is given in Fig. 3.9, and the effect of reversal of direction of current is quite clear.

## 3.12  MEMBRANE IMPEDANCE CHARACTERISTICS

In this and the following section, we shall consider the response of a membrane to time-varying stimuli of small magnitude. It might seem at first that this subject belongs more appropriately in the next chapter, where the dynamic characteristics of biological membranes are considered. It turns out, however, that if *subthreshold* time-varying signals are applied to a membrane, the behavior may be approximated by *linear* equivalent electric networks.[1] Nonlinear behavior which is initiated by transthreshold stimuli *is* reserved for the next chapter.

Experimental investigations of the impedance properties of membrane often utilize a section of nerve or muscle fiber or suspension of cylindrical or spherical cells. Consequently, it may be necessary to include in the model the electrical characteristics of the internal and

[1] O. H. Schmitt, Dynamic Negative Admittance Components in Statically Stable Membranes, in T. Shedlovsky (ed.), "Electrochemistry in Biology and Medicine," John Wiley & Sons, Inc., New York, 1955.

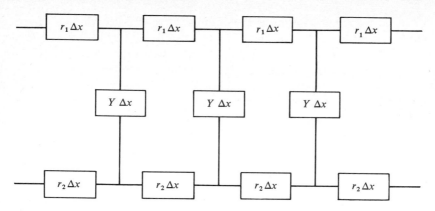

**Fig. 3.10**  Core-conductor model of nerve or muscle fiber.  Resistances per unit length in axoplasm and interstitial medium are $r_1$ and $r_2$, respectively.  Membrane conductance per unit length is $Y$.

external media.  Thus, for essentially axial current flow along a long cylindrical fiber a reasonable equivalent circuit is that depicted in Fig. 3.10, where $r_1$ is the resistance of the external medium per unit length,[1] $r_2$ is the internal resistance per unit length, and $Y$ is the complex admittance for a length of unity.  In this model, $x$ is the axial distance, and the iterative structure based on segments $\Delta x$ in extent is shown.  The resultant network is the classical *core-conductor model*, and its properties will be considered in greater detail in the next section.  In this and other models, the internal and external media are represented by pure resistances $r_2$ and $r_1$, an approximation well justified in a frequency range up to, perhaps, several hundred kilohertz.[2]

An effective specific impedance of a suspension of cells of spherical, ellipsoidal, or cylindrical shape can be calculated.  This result is related to the internal and external resistivities $\rho_2$ and $\rho_1$, respectively, and the membrane admittance per unit area, $y$, as well as the shape and particle density.[3]  For spherical cells under time harmonic conditions the result is

$$z = \rho_1 \frac{(1 - \beta)\rho_1 + (2 + \beta)(\rho_2 + 1/ya)}{(1 + 2\beta)\rho_1 + 2(1 - \beta)(\rho_2 + 1/ya)} \tag{3.148}$$

[1] For an external medium whose extent is large compared with the axoplasm, little error results from setting $r_1 = 0$.

[2] S. A. Briller, et al., The Electrical Interaction between Artificial Pacemakers and Patients with Application to ECG, *Am. Heart J.*, **71**:656 (1966).

[3] K. S. Cole and H. J. Curtis, Bioelectricity: Electric Physiology, in O. Glasser (ed.), "Medical Physics," vol. II, The Year Book Medical Publishers, Inc., Chicago, 1950.

where $\beta = 4\pi a^3/3A^3$ and $a$ is the radius, while $A$ is the distance between spherical centers. In (3.148), $z$ is a complex phasor if $y$ is complex. Measurement of $z(\omega)$ and independent measurement of $\rho_1$ and $\rho_2$ permit $y(\omega)$, the membrane admittance itself, to be determined.

Equation (3.148) has been applied[1] to measurements of the conductivity of suspensions of spherical bacterial cells where procedures are available for the determination of the corresponding $\beta$ and $\rho_1$. From the values of $z$, $\beta$, and $\rho_1$, it is then possible to calculate the effective internal resistivity ($\rho_2 + 1/ya$). It turns out that the cell wall is almost entirely responsible for the cell conductivity (of *E. coli* and *M. lysodeikticus*) so that the value of $y$ is essentially determined by this procedure.

For many cell and nerve axon membranes, $y$ is very well approximated by a parallel combination of transverse resistance $r_m$ and capacitance $C_m$ (per unit area). Then the conductance $G$ and susceptance $B$ are given by

$$G = \frac{1}{r_m} \qquad B = \omega C_m \tag{3.149}$$

The transmembrane impedance $Z$ is then

$$Z = (G + jB)^{-1} = R_s + jX_s \tag{3.150}$$

Substituting (3.149) into (3.150) and solving for $R_s$ and $X_s$ gives

$$R_s(\omega) = \frac{r_m}{1 + \omega^2\tau^2} \qquad X_s(\omega) = \frac{-\omega\tau r_m}{1 + \omega^2\tau^2} \tag{3.151}$$

where the time constant $\tau$ is given by

$$\tau = r_m C_m \tag{3.152}$$

The locus of $R_s$ versus $-X_s$ as a function of frequency is a circle of radius $r_m/2$ and center at $(0, r_m/2)$, as illustrated in Fig. 3.11. The angular frequency for which the real and imaginary parts of the impedance are equal corresponds to the reciprocal of the membrane time constant.

For a suspension of spherical cells whose membrane can be approximated by a pure capacitance (3.148) reduces to

$$z = \rho_\infty + \frac{\rho_0 - \rho_\infty}{1 + j\omega T} = \left(\rho_\infty + \frac{\rho_0 - \rho_\infty}{1 + \omega^2 T^2}\right) - j\frac{(\rho_0 - \rho_\infty)\omega T}{1 + \omega^2 T^2} \tag{3.153}$$

[1] E. L. Carstensen, H. A. Cox, Jr., W. B. Mercer, and L. A. Natale, Passive Electrical Properties of Microorganisms I. Conductivity, *Biophys. J.*, **5**:289 (1965).

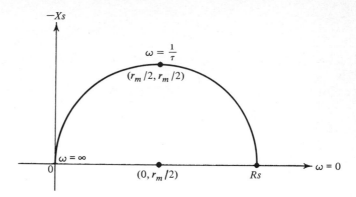

**Fig. 3.11**   Impedance locus for an idealized $RC$ membrane.

where $\rho_\infty$ is the effective resistivity of the suspension medium when $\omega \to \infty$, while $\rho_0$ is the resistivity at zero frequency.   These quantities, as determined from (3.148), are given by

$$\rho_\infty = \rho_1 \frac{(1 - \beta)\rho_1 + (2 + \beta)\rho_2}{(1 + 2\beta)\rho_1 + 2(1 - \beta)\rho_2} \tag{3.154}$$

$$\rho_0 = \frac{2 + \beta}{2(1 - \beta)} \tag{3.155}$$

while the time constant $T$ is

$$T = aC_m \left[ \frac{(1 + 2\beta)\rho_1}{2(1 - \beta)} + \rho_2 \right] \tag{3.156}$$

One notes that the form of $z$ in (3.153) corresponds to $Z$ in (3.151), so that the impedance locus will be circular with diameter $(\rho_\infty - \rho_0)$, but with an origin shifted along the resistance axis by the value $\rho_\infty$ (that is, the origin is at $\{[\rho_\infty + (\rho_0 - \rho_\infty)/2], 0\}$).

An impedance locus obtained by measurements on a frog egg is given in Fig. 3.12.[1]   The deviation of $y$ from a pure capacitance has resulted in the locus being circular, but with its center displaced vertically downward.   Such an effect can be accounted for by taking the membrane admittance as $(j\omega)^{\theta/90°}y_0$, where $y_0$ is a constant and $\theta$ is a constant phase angle ($\theta \leq 90°$).   This corresponds physically to a frequency-dependent parallel resistance and capacitance, where

$$(j\omega)^{\theta/90°}y_0 = y_0\omega^{\theta/90°} \cos \theta + jy_0\omega^{\theta/90°} \sin \theta \tag{3.157}$$

[1] K. S. Cole and R. M. Guttman, Electrical Impedance of Frog Egg, *J. Gen. Physiol.*, **25**:765 (1942).

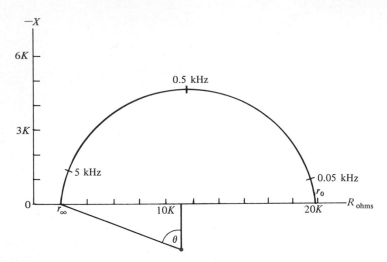

**Fig. 3.12** Impedance locus for the single frog egg. [*After K. S. Cole and R. M. Guttman. Reprinted by permission of the Rockefeller University Press from J. Gen. Physiol.*, **25**:765 (1942).]

That the locus is as stated can be confirmed by first noting that (3.153) can be written as

$$z = \rho_\infty + \frac{\rho_0 - \rho_\infty}{1 + j\omega C_m \gamma} \tag{3.158}$$

where $\gamma = T/C_m$ and is a constant. For a membrane described by (3.157), we can apply (3.158) by simply replacing $j\omega C_m$ by $(j\omega)^{\theta/90°}y_0$. If we let $\gamma y_0 = T_1^{\theta/90°}$, a constant, then

$$z = \rho_\infty + \frac{\rho_0 - \rho_\infty}{1 + (j\omega T_1)^{\theta/90°}} \tag{3.159}$$

The impedance locus of (3.159) is plotted in Fig. 3.13. With reference to this figure $u = z^* - \rho_\infty$, and $v = \rho_0 - z^*$, where $u$ and $v$ are complex and considered as vectors and $z^*$ is the complex conjugate of $z$ [replace $j$ by $-j$ in (3.159)]. Then we have

$$u = \frac{\rho_0 - \rho_\infty}{1 + (-j\omega T_1)^{\theta/90°}} \qquad v = (-j\omega T_1)^{\theta/90°}u$$

so that the angle between $u$ and $v$, $\phi$ (see Fig. 3.13), is the constant $\theta$. From this fact, it immediately follows that the locus of $z$ is a circle and that the central angle subtended by $\rho_0 - \rho_\infty$ is $(2\theta)$. For $\theta = 90°$ the previous result is obtained. A discussion of the dielectric property of

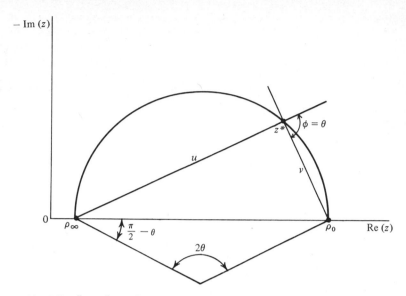

**Fig. 3.13** Impedance locus for a suspension of spherical cells in a resistive medium where the cell membrane is a constant phase admittance.

materials with behavior that results in (3.157) is given by Cole and Cole.[1] It should be noted that the form of (3.159) can arise from a resistance $\rho_\infty$ in series with the parallel combination of resistance $(\rho_0 - \rho_\infty)$ and constant phase impedance of the form (3.157).

Electrical impedance characteristics of cells and tissues are given in Table 3.5. The results for the squid axon are of particular interest since it has been almost exclusively the source of experimental data for biological membrane studies. We note that the membrane has a capacitance of 1 $\mu$f/cm². A number of investigations have shown that this value is essentially independent of frequency and excitation.[2] If one considers the physical membrane as a capacitor,[3] then the thickness in angstroms required to yield 1 $\mu$f/cm² is $10\kappa$, where $\kappa$ is the relative permittivity. On this basis for a membrane of around 100 Å, as determined from electron micrographs, $\kappa = 10$ is required. This may seem to

---

[1] K. S. Cole and R. H. Cole, Dispersion and Absorption in Dielectrics, *J. Chem. Phys.*, **9**:341 (1941).

[2] H. Schwan and K. S. Cole, Bioelectricity, in O. Glasser (ed.), "Medical Physics," vol. III, The Year Book Medical Publishers, Inc., Chicago, 1960.

[3] For a parallel-plane capacitor of area $A$, plate separation $d$, and permittivity $\kappa$, and assuming a uniform field, we have $C = \kappa\epsilon_0 A/d$, where $\epsilon_0$ is the free-space dielectric constant.

**Table 3.5   Electrical impedance characteristics of cells and tissues**

| Biological material | $C_m$ | $\theta_m$ | $\rho_m$ | $\rho_2$ | $T^{-1}$ |
|---|---|---|---|---|---|
| Plant cells: | | | | | |
|   Yeast | 0.6 | 87 | $\infty$ | 460 | 2,000 |
|   (*Nitella flexilis*) | | | | | |
|     Resting | 0.94 | 80 | $2.5 \times 10^5$ | 87 | |
|     Excited | 0.80 | 80 | $5 \times 10^2$ | 87 | 0.6 |
| Marine egg cells | | | | | |
|   (*Arbacia*) | 1.1 | 90 | 100 | 210 | $2.5 \times 10^4$ |
| Blood cells | | | | | |
|   (Human red blood cell) | 0.8 | 90 | $\infty$ | | |
| Muscle | | | | | |
|   (*Rana pipiens*) | 1.5 | 70 | 40 | 250 | 15 |
| Nerve: | | | | | |
|   Squid: | | | | | |
|     Resting | 1.1 | 75 | $10^3$ | 30 | |
|     Excited | 1.1 | 75 | 25 | 90 | 530 |
|   Frog sciatic | 0.55 | 40 | $\infty$ | 560 | 2,300 |
|   Cat sciatic | 0.65 | 40 | $\infty$ | 720 | 1,000 |

Symbols: $C_m$ = membrane capacitance, $\mu$f/cm$^2$

$\theta_m$ = membrane phase angle, deg

$\rho_m$ = membrane resistance, ohm-cm$^2$

$\rho_2$ = internal specific resistivity, ohm-cm

contradict a reasonable value of $\kappa = 3$ based on the lipid content of the membrane; however, recent measurements of the effective dielectric permittivity of hydrated proteins give values for $\kappa$ between 20 and 30.[1] A physical basis for membrane capacitance, other than that of a simple dielectric between conducting plates,[2] may be another factor responsible for the observations.

In the case of muscle, values of capacitance of 5 to 8 $\mu$f/cm$^2$ appear to violate the picture given above.   The possibility of extensive membrane folding is ruled out, based on electromicrographs.   A careful study by Falk and Fatt[3] has resulted in a modified membrane model involving two time constants.   An electrical representation of the shunt admittance per unit length of fiber, $Y$, is shown in Fig. 3.14.   In addition to

[1] H. Schwan, Electrical Properties of Tissue and Cell Suspensions, *Advan. Biol. Med. Phys.*, **5** (1957).

[2] A. Mauro, Space Charge Regions in Fixed Charge Membranes and the Associated Property of Capacitance, *Biophys. J.*, **2**:179 (1962).

[3] G. Falk and P. Fatt, Linear Electrical Properties of Striated Muscle Fibers Observed with Intracellular Electrodes, *Proc. Roy. Soc.*, **B160**:69 (1964).

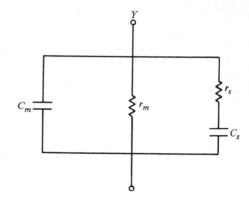

**Fig. 3.14** Electric-circuit representation for the shunt admittance per unit length, $Y$, of a muscle fiber.

the parallel capacitance and resistance $C_m$ and $r_m$, there are introduced the series elements $r_s$ and $C_s$. The admittance is given as

$$Y = G + jB = \frac{1}{r_m} + \frac{\omega^2 C_s{}^2 r_s}{\omega^2 C_s{}^2 r_s{}^2 + 1} + j\left(\omega C_m + \frac{\omega C_s}{\omega^2 C_s{}^2 r_s{}^2 + 1}\right)$$

$$(3.160)$$

For an ideal two-time-constant network, such as illustrated in Fig. 3.15, where the capacitances are widely separated in value, then two non-overlapping circular loci result. Thus, in the low range of frequencies, one branch $(R_b + j/\omega C_2)$ remains a virtual open circuit, so that a typical single-time-constant behavior due to the second branch

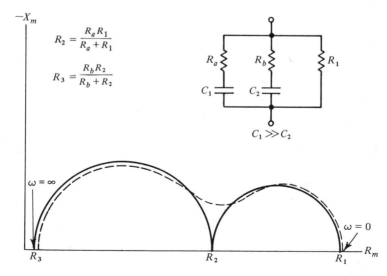

**Fig. 3.15** Theoretical impedance locus of a two-time-constant network. Solid curve for $C_1 \gg C_2$; dotted locus for $C_1 > C_2$.

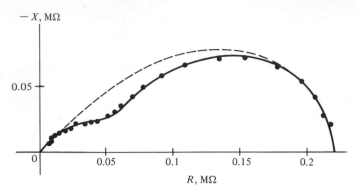

**Fig. 3.16**  Impedance locus for frog sartorius muscle fiber over the frequency range 1 to 10,000 Hz.  Dots are corrected experimental points.  The theoretical curve for the two-constant model is solid, while the single-constant curve is shown slashed.  $R + jX = Z$, where $Z$ is the characteristic impedance.  [*After G. Falk and P. Fatt, Proc. Roy. Soc.,* **B160**:69 (1964).]

results.   As the frequency is further increased, the latter branch behaves like a pure resistance $R_2$ (negligible capacitive reactance), while in the other branch the admittance is no longer zero.   Again the behavior is characterized by that of a single-time-constant circuit.   For $R_b = 0$ the circuit of Fig. 3.15 reduces to that shown in Fig. 3.14 and has the property that $Y(\omega \rightarrow \infty) = 0$.

When the capacitances are not so widely separated, overlap occurs and a locus such as the dotted curve in Fig. 3.15 results.   The latter is typical of results obtained from muscle measurements which yield the two dispersions as shown.   The ratio of transmembrane voltage to current, the so-called characteristic impedance $Z = R + jX$, is the quantity normally measured.   The impedance plot of $Z$ corresponds to the dotted curve in Fig. 3.15.

A characteristic-impedance locus of measurements on a frog sartorius muscle fiber is given in Fig. 3.16.[1]   The dots are the measured values (corrected for stray capacitances), and it is seen that the solid theoretical curve for the two-constant model is a much better fit than the dashed curve representing the single-constant model.   Averaged values for the electrical parameters of the frog sartorius muscle fiber (approximate radius 50 microns), as found by Falk and Fatt,[2] are

$$C_m = 2.6 \ \mu f/cm^2 \qquad r_m = 3,100 \ ohm\text{-}cm^2$$
$$C_s = 4.1 \ \mu f/cm^2 \qquad r_s = 330 \ ohm\text{-}cm^2$$

[1] *Ibid.*
[2] *Ibid.*

Results obtained for crayfish fibers are $C_m = 3.9\ \mu f/cm^2$; $C_s = 17\ ^2\mu f/cm^2$; $r_s = 35$ ohm-cm$^2$; $r_m = 680$ ohm-cm$^2$.

### 3.13 ELECTROTONUS

In this section, we shall consider in some detail the electrical response of a cylindrical nerve fiber to a subthreshold stimulus. The problem is defined in a mathematical sense by the electric circuit of Fig. 3.10, the core-conductor model, which we shall accept as fully valid at this point. (A further discussion of this subject will be given in Chap. 5.) The material of this section is based heavily on the paper of Davis and Lorente de Nó;[1] a similar treatment by Hodgkin and Rushton[2] is also available. The results of this section will prove useful not only for obtaining a better understanding of passive nerve properties but also for understanding their role in the propagated action potential, to be discussed in the next chapter.

In Fig. 3.17, the nerve trunk is shown, along with a pair of electrodes for providing polarizing (stimulating) current and a pair for recording voltage. The nerve trunk is assumed to be cylindrical and infinite in extent and to contain a large number of cylindrical fibers uniformly spaced within the sheath and each of the same diameter. The interstitial space, like the intracellular space, consists of many confined axial paths, and indeed the cross-sectional areas of each are of the same order of magnitude. From such a set of conditions it is reasonable to suppose that the current is essentially axial in the interstitial fluid and the axoplasm. In view of the assumed uniformity the problem is one-dimensional. That is, since the physical environment of each fiber is

[1] L. Davis, Jr., and R. Lorente de Nó, Contribution to the Mathematical Theory of the Electrotonus, in "A Study of Nerve Physiology," chap. IX, Rockefeller Institute for Medical Research, New York, 1947.

[2] A. L. Hodgkin and W. A. Rushton, The Electrical Constants of a Crustacean Nerve Fiber, *Proc. Roy. Soc.*, **B133**:444 (1946).

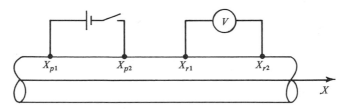

**Fig. 3.17** Geometry for polarizing (stimulating) and recording electrodes for a nerve trunk.

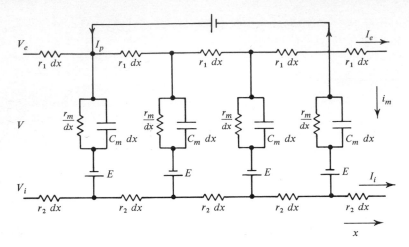

**Fig. 3.18**   Core-conductor network.  $I_e$ and $I_i$ are the longitudinal external and internal currents, respectively, and are positive in the direction of increasing $x$, i.e., left to right.  The transmembrane current per unit length is positive from outside to inside, while the transmembrane voltage is $V = V_e - V_i$.

essentially identical and uniform, only the axial variations of a single representative fiber need be considered.  The behavior of each fiber resembles that of a coaxial cable with leakage through the insulation (membrane) separating the "inner conductor" and outer "sheath." The lumped-parameter version shown in Fig. 3.18 constitutes an elaboration of the core-conductor model as depicted in Fig. 3.10.  Here, $r_1$ and $r_2$ are the external and internal resistances per unit length, while $r_m$ and $C_m$ are the shunt resistance and capacitance for a unit length which is appropriate for the *nerve* model,[1] as noted previously.[2]  In view of the passive nature of the network, the energy source is exclusively that due to the external battery (which causes the current $I_p$), as shown.

Since $E$ (which represents the resting-membrane potential) is assumed to be constant and since time-varying components alone are of interest, $E$ may be set equal to zero.[3]  If we let $V_e$ be the external

---

[1] L. Hermann, Allgemeine Muskelphysik; Allgemeine Nervenphysiologie; in L. Hermann (ed.), "Handbuch der Physiologie."

[2] For a *muscle* fiber the shunt network would require an additional shunt path consisting of $R_s$ and $C_s$ in series, as noted in Sec. 3.12.  The mathematical analysis must be modified in this case.  Some details are presented later in this section.

[3] Application of superposition reveals that if the emf of the resting membrane, denoted as $E$ in Fig. 3.18, is constant, then its contribution to either the transmembrane or external potential is also constant.  The transient solution, which is of interest here, constitutes a component of the total solution, obtainable by superposition, with $E$ set equal to zero.  The simplified model was actually introduced in Fig. 3.10.

potential, $V_i$ the internal potential, $I_e$ the net axial external current, $I_i$ the net axial internal current, and $i_m$ the membrane current per unit length from the external to the internal medium, then the following "Ohm's law" types of relationships must exist:

$$\frac{\partial V_e}{\partial x} = -I_e r_1 \tag{3.161}$$

$$\frac{\partial V_i}{\partial x} = -I_i r_2 \tag{3.162}$$

$$\frac{\partial I_i}{\partial x} = i_m \tag{3.163}$$

In the above expressions, $I_e$ and $I_i$ are positive when flowing in the direction of increasing $x$, that is, from left to right in Fig. 3.18. We designate $I_e + I_i = I$, where, in the absence of an applied (polarizing) current, $I$ must be a constant because any increase, say, in $I_i$, due to an inflow of $i_m$ causes $I_e$ to decrease by exactly the same amount, and vice versa. If we let $i_p$ be the polarizing (applied) current per unit length, then

$$\frac{\partial I}{\partial x} = i_p \tag{3.164}$$

Note that the exciting electrodes must be axially symmetric to justify the assumption that for all quantities $\partial/\partial\phi = 0$. We define $V = V_e - V_i$, so that, from (3.162) and (3.161), we have

$$\frac{\partial V}{\partial x} = \frac{\partial V_e}{\partial x} - \frac{\partial V_i}{\partial x} = -I_e r_1 + I_i r_2 = -(r_1 + r_2)I_e + I r_2 \tag{3.165}$$

$$\frac{\partial V}{\partial x} = \frac{r_1 + r_2}{r_1}\frac{\partial V_e}{\partial x} + I r_2 \tag{3.166}$$

Solving for $\partial V_e/\partial x$ gives

$$\frac{\partial V_e}{\partial x} = \frac{r_1}{r_1 + r_2}\frac{\partial V}{\partial x} - \frac{r_1 r_2}{r_1 + r_2}I \tag{3.167}$$

We choose as zero reference the potential at infinity

$$[V_e(+\infty,t) = V_i(+\infty,t) = 0]$$

hence if (3.167) is integrated from $x$ to infinity, we get

$$V_e(\infty,t) - V_e(x,t) = \frac{r_1}{r_1 + r_2}[V(\infty,t) - V(x,t)]$$

$$-\frac{r_1 r_2}{r_1 + r_2}\int_x^\infty I(x)\,dx \tag{3.168}$$

Let us consider that the applied current is a constant $I_0$. Then $I(x) = I_0$ if $X_{p1} < x < X_{p2}$ (see Fig. 3.17), and $I(x) = 0$ elsewhere. Consequently,

$$V_e(x,t) = \frac{r_1}{r_1 + r_2} V(x,t)$$

$$+ \begin{cases} 0 & x \geq X_{p2} \\ \dfrac{I_0 r_1 r_2}{r_1 + r_2} (X_{p2} - x) & X_{p1} \leq x \leq X_{p2} \\ \dfrac{I_0 r_1 r_2}{r_1 + r_2} (X_{p2} - X_{p1}) & x \leq X_{p1} \end{cases} \quad (3.169)$$

In order to use (3.169), we first find an expression for $V(x,t)$. If (3.165) is differentiated with respect to $x$, one obtains

$$\frac{\partial^2 V}{\partial x^2} = -r_1 \frac{\partial I_e}{\partial x} + r_2 \frac{\partial I_i}{\partial x} = (r_1 + r_2) \frac{\partial I_i}{\partial x} - r_1 \frac{\partial I}{\partial x} \quad (3.170)$$

Using (3.163) and (3.164) results in

$$\frac{\partial^2 V}{\partial x^2} = (r_1 + r_2) i_m - r_1 i_p \quad (3.171)$$

The relationship between $i_m$ and $V$ is found from the shunt circuit in Fig. 3.18 and is

$$i_m = \frac{V}{r_m} + C_m \frac{\partial V}{\partial t} \quad (3.172)$$

so that (3.171) becomes

$$\lambda^2 \frac{\partial^2 V}{\partial x^2} - \tau \frac{\partial V}{\partial t} - V = -r_1 \lambda^2 i_p \quad (3.173)$$

where $\lambda = \sqrt{r_m/(r_1 + r_2)}$ and $\tau = r_m C_m$. This differential equation is solved subject to the initial condition that $V(x,0) = 0$, which corresponds to a resting nerve. If we make the substitution

$$V = v e^{-t/\tau} \quad (3.174)$$

then (3.173) transforms into the standard form of the diffusion equation, namely,

$$\frac{\lambda^2}{\tau} \frac{\partial^2 v}{\partial x^2} - \frac{\partial v}{\partial t} = -q(x,t) \quad (3.175)$$

where

$$q(x,t) = \frac{1}{\tau} e^{t/\tau} (r_1 \lambda^2 i_p) \quad (3.176)$$

The solution to the homogeneous equation, which results when $q$ is set equal to zero in (3.175), is readily found by using the method of separation of variables. Thus, if one assumes that $v = v_x(x)v_t(t)$, then

$$\frac{d^2 v_x}{dx^2} = \Gamma^2 v_x \qquad \frac{dv_t}{dt} = \frac{\Gamma^2 \lambda^2}{\tau} v_t$$

where $\Gamma$ is the separation constant. The form of solution that satisfies the boundary conditions is

$$v_x = A(\Gamma)e^{i\Gamma x} \qquad v_t = B(\Gamma)e^{-\Gamma^2 \lambda^2 t/\tau}$$

and a general solution is

$$v = \int_{-\infty}^{\infty} C(\Gamma)e^{i\Gamma x - \Gamma^2 \lambda^2 t/\tau} \, d\Gamma \tag{3.177}$$

For the case where the source is a two-dimensional delta function[1] of space and time located at $t = 0$, $x = 0$, the initial condition can be expressed as

$$v_g(x,0) = \delta(x)$$

The function $C(\Gamma)$ can now be established from (3.177) by an inverse Fourier transform, and comes out

$$C(\Gamma) = \frac{1}{2\pi} \int_{-\infty}^{\infty} \delta(x)e^{-i\Gamma x} \, dx = \frac{1}{2\pi} \tag{3.178}$$

Substituting this value into (3.177) and performing the integration with respect to $\Gamma$ yields

$$v_g(x,t) = \frac{1}{2\sqrt{\lambda^2 \pi t/\tau}} e^{-x^2 \tau/4\lambda^2 t}$$

where $v_g$ designates the solution to the diffusion equation with a delta-function source. That is,

$$\frac{\lambda^2}{\tau} \frac{\partial^2 v_g}{\partial x^2} - \frac{\partial v_g}{\partial t} = \delta(x)\,\delta(t) \tag{3.179}$$

The solution to (3.175) is readily found by considering $q(x,t)$ as an infinite sum of delta functions of the corresponding $x$ and $t$, so that $v(x,t)$ consists of the superposition of the solutions for each such element. As a result, we obtain (replace $\lambda^2/\tau$ by $k$)

$$v = \int_{-\infty}^{t} \int_{-\infty}^{\infty} \frac{q(\xi,\sigma)}{2\sqrt{\pi k(t - \sigma)}} e^{-[(x-\xi)^2/4k(t-\sigma)]} \, d\xi \, d\sigma \tag{3.180}$$

---

[1] The delta function $\delta(x)$ has the property that $\delta(0) = \infty$, $\delta(x) = 0$, $x \neq 0$, $\int_{-\infty}^{\infty} \delta(x) \, dx = 1$.

Substituting the expression for $q(\xi,\sigma)$ and using (3.174) to convert back to the potential $V$ yields

$$V = \frac{1}{2\lambda\sqrt{\pi\tau}} \int_{-\infty}^{t} \int_{-\infty}^{\infty} \lambda^2 r_1 i_p(\xi,\sigma) e^{-\{[(t-\sigma)/\tau]-[\tau(x-\xi)^2/4\lambda^2(t-\sigma)]\}} \frac{d\xi\, d\sigma}{\sqrt{t-\sigma}}$$

(3.181)

If, at $t = 0$, a current $I_0$ is caused to flow into the electrode at $X_{p1}$ of differential width $\delta$ and return from the electrode at $X_{p2}$ (also of differential width $\delta$), then $i_p(x,t)$ has the character

$$i_p(x,t) = \begin{cases} -\dfrac{I_0}{\delta} & t > 0,\ X_{p2} < x < X_{p2} + \delta \\[2mm] \dfrac{I_0}{\delta} & t > 0,\ X_{p1} < x < X_{p1} + \delta \\[2mm] 0 & \text{elsewhere} \end{cases}$$

(3.182)

The current as a function of time, specified by (3.182), is a step.[1]  Substituting (3.182) into (3.181) gives

$$-V(x,t) = \tfrac{1}{2} r_1 \lambda I_0 [2F(x - X_{p1},\, t) - 2F(x - X_{p2},\, t)]$$ (3.183)

where

$$F(x,t) = \frac{1}{2\sqrt{\pi\tau}} \int_0^t e^{-[\tau x^2/4\lambda^2(t-\sigma)]} e^{-[(t-\sigma)/\tau]} \frac{d\sigma}{\sqrt{t-\sigma}}$$ (3.184)

After some algebraic operations, the integrals in (3.184) can be evaluated and are

$$F(x,t) = \tfrac{1}{4} \left\{ e^{-(|x|/\lambda)} \left[ 1 - \text{erf}\left( \frac{|x|}{2\lambda}\sqrt{\frac{\tau}{t}} - \sqrt{\frac{t}{\tau}} \right) \right] \right.$$
$$\left. - e^{|x|/\lambda} \left[ 1 - \text{erf}\left( \frac{|x|}{2\lambda}\sqrt{\frac{\tau}{t}} + \sqrt{\frac{t}{\tau}} \right) \right] \right\}$$ (3.185)

which is expressed in terms of the error function erf $(u) = 2/\sqrt{\pi} \int_0^u e^{-x^2} dx$. While this result depends on the one-dimensional formulation implicit in Fig. 3.18 and Eqs. (3.161) to (3.164), it can be shown to be well justified on the basis of a formal three-dimensional analysis.[2]  Note that length and time appear in the form $x/\lambda$ and $t/\tau$ so that $\lambda$ and $\tau$ behave as normalizing spatial and temporal factors.

Based on (3.183) and (3.185), the electrotonic membrane potential may be determined for a variety of conditions.  As an illustration, the potential distribution for the case where $X_{p1} = -\infty$ and $X_{p2} = 0$ is

[1] The unit step $u(t)$ is defined as $u(t) = 0$ for $t < 0$; $u(t) = 1$ for $t \geq 0$.
[2] D. Hellerstein, Passive Membrane Potentials, *Biophys. J.*, **8**:358 (1968).

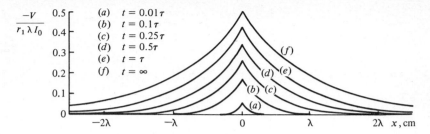

Fig. 3.19  Electrotonic membrane potential at successive instants of time due to step-current input at the origin. (*After L. Davis, Jr., and R. Lorente de Nó, p. 442, Rockefeller Institute, Monograph 131, 1947.*)

plotted in Fig. 3.19 at several instants of time $t$.  An additional example is given in Fig. 3.20, which shows the external potential as a function of time at $x = 0.4\lambda$ for several current pulse durations $t_0$.  The latter curves are readily found by superposing the potentials produced by two step functions, where the second is delayed by $t_0$ and is of opposite sign relative to the first.

From the above examples, one obtains an interpretation of the *space constant* $\lambda$ and *time constant* $\tau$.  In Fig. 3.19, it is seen that the space constant is a measure of the spatial extent of nerve that is affected by current injected at one electrode.  Figure 3.20 shows that the time constant is a measure of the time required to reach steady-state conditions.

An important application of electrotonic theory is in the experimental determination of nerve parameters $r_1$, $r_2$, $\lambda$, and $\tau$.  The time unit $\tau$ can be found from the temporal curve of external electrotonic potential at the polarizing electrode.  If $x$ is set equal to zero, then

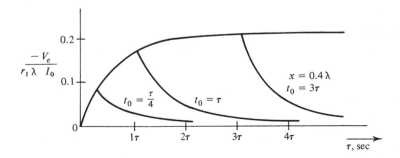

Fig. 3.20  External electrotonic potentials at $x = 0.4\lambda$ for several polarizing pulse durations. (*After L. Davis, Jr., and R. Lorente de Nó, p. 442, Rockefeller Institute, Monograph 131, 1947.*)

(3.183), with $X_{p1} = -\infty$ and $X_{p2} = 0$, leads to

$$V(0,t) = \tfrac{1}{2}r_1\lambda I_0 \operatorname{erf}\sqrt{\frac{t}{\tau}} \qquad (3.186)$$

By making use of (3.169), the following relationship is obtained for $V_e$:

$$V_e(0,t) = \frac{r_1}{r_1 + r_2} V(0,t) \qquad (3.187)$$

and it is seen that $\tau$ equals the time required by the external potential $V_e$ to reach 84 percent of its steady-state value at the polarizing electrode.

The characteristic length $\lambda$ can be determined from steady-state measurements of the external potential along the axis $x > 0$. Thus, for $t \to \infty$ [since erf $(u) = \pm 1$ as $u \to \pm \infty$] one obtains from (3.183) and (3.185) that

$$V(x,\infty) = \tfrac{1}{2}r_1\lambda I_0 e^{-x/\lambda} \qquad (3.188)$$

Application of (3.169) then yields

$$V_e(x,\infty) = \frac{1}{2}\frac{r_1{}^2}{r_1 + r_2} \lambda I_0 e^{-(x/\lambda)} \qquad (3.189)$$

A plot of log $V_e$ versus $x$ should be a straight line with slope $-\log e/\lambda$, from which $\lambda$ itself may be obtained. ($\lambda$ is also the interval along the $x$ axis corresponding to a change in log $V_e$ of log $e$; with reference to the origin, it is the $x$ value at which $V_e$ has diminished to $1/e = 1/2.718$ of its original value.)

The potential in the interpolar region, with $X_{p1} = -\infty$ and $X_{p2} = 0$, can be found from (3.169) to be

$$V_e(x,\infty) = \frac{r_1}{r_1 + r_2} V(x,\infty) - \frac{I_0 r_1 r_2 |x|}{r_1 + r_2} \qquad (3.190)$$

If we measure $V_e$ at two points $x_1$ and $x_2$ which are sufficiently large that $V(x_1,\infty) = V(x_2,\infty) = 0$, and if, furthermore, $x_2 = x_1 - 1$, then using (3.190)

$$V_e(x_1,\infty) - V_e(x_1 - 1,\ \infty) = I_0\frac{r_1 r_2}{r_1 + r_2} \qquad (3.191)$$

Alternatively, if $x_2 = x_1 - \Delta x$, then for $x_2$ sufficiently large and with $\Delta x \to 0$ we get

$$\frac{\partial V_e}{\partial x}\bigg|_{x=x_2} = I_0\frac{r_1 r_2}{r_1 + r_2} \qquad (3.192)$$

At the polarizing electrode, $x = 0$, (3.189) reduces to

$$\frac{2V_e(0,\infty)}{\lambda I_0} = \frac{r_1{}^2}{r_1 + r_2} \qquad (3.193)$$

Thus having found $\lambda$ in the manner described above, we see that (3.191) and (3.193) constitute two equations from which the two unknowns, $r_1$ and $r_2$, can be determined.

An independent measurement of $r_1r_2/(r_1 + r_2)$ is also possible by placing the nerve in a nonconducting medium (sucrose, for example), in which case the net axial resistance can be determined by conventional means. For the length of the nerve, $l$, the parallel combination of $lr_1$ and $lr_2$ constitutes the resistance that is measured, that is, $l[r_1r_2/(r_1 + r_2)]$.

For sinusoidal excitation, the steady-state solutions to (3.171) can be found quite readily since all voltages and currents must have the same harmonic time dependence. Considering such quantities to be complex phasors leads to the following equation where $Y$ is the membrane admittance:

$$\frac{\partial^2 V}{\partial x^2} - (r_1 + r_2)YV = -r_1 i_p \tag{3.194}$$

The general solution is

$$V(x) = A_1 e^{-\gamma x} + A_2 e^{\gamma x} \tag{3.195}$$

where

$$\gamma = \sqrt{(r_1 + r_2)Y} \tag{3.196}$$

This relationship is the basis for the experimental determination of $Y$ in the muscle-fiber experiments of Falk and Fatt mentioned in the previous section. In these experiments it is convenient to separate the electrode which is injecting current from that which measures the voltage. Consequently, it is a transfer impedance that is computed by dividing the voltage at a small distance $x$ from the origin by the input current at the origin. Designating the latter as $I_0$, we have, from (3.195), with $A_2 = 0$ since $x \geq 0$,

$$Z = \frac{V(x)}{I_0} = Z_0 e^{-\gamma x} \tag{3.197}$$

and $Z_0$ is the desired characteristic impedance.

When the membrane can be represented as a parallel resistance and capacitance, (3.196) can be written as

$$\gamma = \sqrt{(r_1 + r_2)\left(\frac{1}{r_m} + j\omega C_m\right)} \tag{3.198}$$

For $x \geq 0$, $A_2$ must be set equal to zero in (3.195) to avoid a physically unacceptable infinite transmembrane potential. The resultant expression can be written as

$$V(x,t) = A e^{-\alpha x} \cos(\omega t - \beta x) \tag{3.199}$$

One notes an exponential decrement characterized by $\alpha$, and a linear phase shift $\beta$, in the behavior of the electrotonic spread. By solving for the real and imaginary parts of $\gamma$ in (3.198) the attenuation and phase constants can be related to the electrical nerve parameters. The result is

$$\alpha = \lambda^{-1} \sqrt{\tfrac{1}{2}(\sqrt{1 + \omega^2\tau^2} + 1)}$$

$$= \sqrt{\frac{1}{2}\left[\sqrt{(r_1 + r_2)^2 \left(\frac{1}{r_m^2} + \omega^2 C_m^2\right)} + \frac{r_1 + r_2}{r_m}\right]} \quad (3.200)$$

$$\beta = \lambda^{-1} \sqrt{\tfrac{1}{2}(\sqrt{1 + \omega^2\tau^2} - 1)}$$

$$= \sqrt{\frac{1}{2}\left[\sqrt{(r_1 + r_2)^2 \left(\frac{1}{r_m^2} + \omega^2 C_m^2\right)} - \frac{r_1 + r_2}{r_m}\right]} \quad (3.201)$$

From (3.200) and (3.201), given measurements of $\alpha$ and $\beta$ and the loop resistance $(r_1 + r_2)$, it is possible to determine the membrane conductance $1/r_m$ and susceptance $\omega C_m$. This method is described by Tasaki and Hagiwara[1] as giving very good results. It is interesting to note that the attenuation constant $\alpha$ increases as a function of frequency; for $\omega = 0$ the attenuation constant value reduces to $\lambda^{-1}$, as expected. For a muscle fiber $\tau \approx 30$ msec, so that, according to (3.200), the effective space constant at 15 Hz is one-half that under dc conditions, while at 1.06 kHz the space constant diminishes to one-tenth.

The behavior of the two-time-constant model is contained in the analyses of the transform impedance of the circuit in Fig. 3.14, which is

$$Z(s) = \left(\frac{1}{r_m} + sC_m + \frac{sC_s}{1 + sC_sr_s}\right)^{-1} \quad (3.202)$$

$$Z(s) = \frac{sr_mr_sC_s + r_m}{s^2C_sC_mr_sr_m + s(r_mC_m + r_sC_s + r_mC_s) + 1} \quad (3.203)$$

The roots $m_{1,2}$ of the quadratic in the denominator of (3.203) are

$$m_{1,2} = \frac{-(C_s + C_m)r_m - r_sC_s \pm \sqrt{(C_s + C_m)^2r_m^2 + r_sC_s(r_sC_s + 2r_mC_s - 2r_mC_m)}}{2C_sC_mr_sr_m}$$

$$= \frac{-[(C_s + C_m)r_m + r_sC_s] \pm (C_s + C_m)r_m\sqrt{1 + \dfrac{r_sC_s[r_sC_s + 2r_m(C_s - C_m)]}{r_m^2(C_s + C_m)^2}}}{2C_sC_mr_sr_m} \quad (3.204)$$

[1] I. Tasaki and S. Hagiwara, Capacity of Muscle Fiber Membrane, *Am. J. Physiol.*, **188**:423 (1957).

We define the time constants $T_1$ and $T_2$ as

$$T_1 = r_s \frac{C_s C_m}{C_s + C_m} \tag{3.205}$$

$$T_2 = r_m(C_s + C_m) \tag{3.206}$$

and note from the muscle data given earlier that $T_2 \gg T_1$. Using this data, we see that the quotient under the radical in (3.204) is small compared with unity, so that, using a Taylor expansion for the square root, we get

$$m_1 = \frac{-2(C_s + C_m)r_m}{2C_s C_m r_s r_m} = -\frac{1}{T_1} \tag{3.207}$$

$$m_2 = \frac{-2r_m r_s C_s (C_s + C_m) + r_s C_s[r_s C_s + 2r_m(C_s - C_m)]}{2(C_s + C_m)r_m{}^2 r_s C_s C_m}$$

$$= -\frac{1}{T_2} \tag{3.208}$$

This result permits one to interpret the shunt network as (approximately) a series combination of two parallel $RC$ networks, one with a short time constant $T_1$, and the other with a long time constant $T_2$.

For muscle fibers which are represented by the two-time-constant model the dc method (step) is inadequate for the determination of the short time constant $T_1$. This is because the Fourier components of the step emphasize the lower frequencies (the energy density is inversely proportional to the square of the frequency). At the same time in the low-frequency range the impedance locus is approximately circular (see Fig. 3.16) and is characterized by the long-time-constant component as if it existed alone.[1] A comparison of the response of the core-conductor model utilizing a two-time-constant network and a single- (long-) time-constant one, and where the polarizing current is a step, is given by Pugsley,[2] based upon data from Falk and Fatt, and is reproduced in

[1] The voltage across $Z(s)$, defined in (3.203), due to a step current should be a rough gage of the response of the core-conductor model to a polarizing step current when $Z(s)$ is the membrane impedance. The aforementioned voltage is then the inverse transform of $(1/s)Z(s)$, where

$$\frac{1}{s} Z(s) = \frac{1}{C_s C_m r_s} \frac{s r_s C_s + 1}{s(s + 1/T_1)(s + 1/T_2)}$$

and the result (for $T_2 \gg T_1$) is

$$v(t) = r_m \left(1 - \frac{C_s}{C_m} \frac{T_1}{T_2} e^{-t/T_1} - e^{-t/T_2}\right)$$

The response is clearly dominated by the long time constant $T_2$, except for $t \ll T_1$.

[2] I. D. Pugsley, Microelectrode Measurements of Membrane Resistance and Capacitance of the Sartorius Muscle of the Toad, *Australian J. Exptl. Biol.*, **41**:615 (1963).

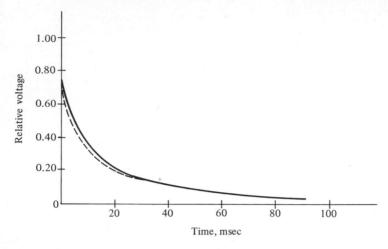

**Fig. 3.21** Comparison of response to a step for a single- and a two-time-constant network. $C_m = 2.4$ $\mu f/cm^2$, $r_m = 3,700$ ohm-cm$^2$, $C_s = 5.8$ $\mu f/cm^2$, $r_s = 370$ ohm-cm$^2$. The single time constant is formed by setting $r_s = 0$ and its response is shown by the solid line. [*After I. D. Pugsley, Australian J. Exptl. Biol.*, **41**:615 (1963).]

Fig. 3.21. It is seen from this that if dc techniques are used it is virtually impossible to distinguish between the two models. But for ac impedance studies a clear discrimination is possible, as noted by Falk and Fatt,[1] particularly if phase information is considered.

In contrast to the dc method, the technique employing the "foot of the action potential"[2] appears to emphasize a short time constant. This is apparently due to the relatively short interval during which the phenomena can be characterized as passive and the fact that the driving force during this time is more nearly a ramp than a step. According to Falk and Fatt,[3] the capacitance determined by dc methods corresponds to an effective frequency of 8 Hz, while that utilizing the foot of the action potential involves an equivalent frequency of 800 Hz.

Interest in electrotonus stems not only from its application to the measurement of nerve parameters, as noted above, but from the role that it plays in the electrical propagation of nervous activity. This point will be discussed in the next chapter when the propagated action potential is considered.

[1] G. Falk and P. Fatt, Linear Electrical Properties of Striated Muscle Fibers Observed with Intracellular Electrodes, *Proc. Roy. Soc.*, **B160**:69 (1964).

[2] This method is discussed in Chap. 4.

[3] *Loc. cit.*

## BIBLIOGRAPHY

Backris, J., and B. E. Conway: "Modern Aspects of Electrochemistry," Butterworth Scientific Publications, London, 1954.

Biological Membranes: Recent Progress, *Ann. N.Y. Acad. Sci.*, art. 2, **137**:403–1048 (July, 1966).

Clarke, H. T. (ed.): "Ion Transport across Membranes," Academic Press Inc., New York, 1954.

Cole, K.S.: "Membranes, Ions, and Impulses," University of California Press, Berkeley, Calif., 1968.

Davson, H.: "A Textbook of General Physiology," 3d ed., Little, Brown and Company, Boston, 1964.

Harris, E. J.: "Transport and Accumulation in Biological Systems," 2d ed., Butterworth Scientific Publications, London, 1960.

Johnson, F. H., H. Eyring, and M. J. Polissar: "The Kinetic Basis of Molecular Biology," John Wiley & Sons, Inc., New York, 1954.

MacInnes, D. A.: "The Principles of Electrochemistry," Dover Publications, Inc., New York, 1961.

Ruch, T. C., and H. D. Patton (eds.): "Physiology and Biophysics," W. B. Saunders Company, Philadelphia, 1965.

Shanes, A. M. (ed.): "Biophysics of Physiological and Pharmacological Actions," American Association for Advancement of Science, Washington, D.C., 1961.

Shedlovsky, T. (ed.): "Electrochemistry in Biology and Medicine," John Wiley & Sons, Inc., New York, 1955.

Stacy, R., et al.: "Essentials of Biological and Medical Physics," McGraw-Hill Book Company, New York, 1955.

Van Winkle, Q.: Study of Bioelectric Energy Sources, *Tech. Doc. Rept. ASD-TDR-62-377, Wright-Patterson Air Force Base, Ohio*, June, 1962.

Whitelock, O. (ed.): Second Conference on Physicochemical Mechanism of Nerve Activity, *Ann. N.Y. Acad. Sci.*, **81**:215–510 (August, 1959).

# 4
# Membrane Action Potentials

## 4.1 INTRODUCTION

We have stated earlier that many living cells are excitable. That is, when the cell is stimulated, an active electrophysiological process is initiated. Thus, if a sufficiently strong impulse is applied at point $S$ on a nerve, the recording of transmembrane potential at $R$ is as illustrated in Fig. 4.1. We note that the "stimulus artifact" is followed (after a short time interval which depends on the separation of $S$ and $R$) by an electrical event known as an *action potential*.

If the transmembrane potential were plotted as a function of (axial) distance at a fixed instant of time, a curve similar in shape to Fig. 4.1 would result. The space-time behavior of the action potential thus corresponds to a propagated wave. Propagation velocities for nerve fibers lie in the range of 0.1 to 100 m/sec.

Parameters which characterize the action potential are resting potential, threshold for activation, overshoot (peak value of transmembrane potential), undershoot (negative peak value following overshoot),

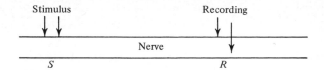

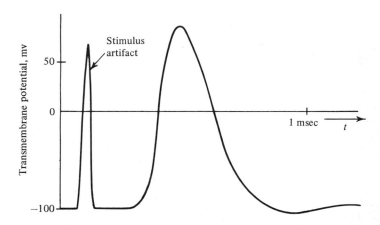

**Fig. 4.1** Transmembrane potential as a function of time measured at the recording site following the application of a brief suprathreshold stimulus.

duration of "spike," rise time of initial phase, and propagation velocity. In this chapter, we shall be interested in relating the above parameters to the underlying nerve structure and electrochemistry. We shall look for quantitative explanations, but as we shall learn, many of the phenomena can be described only in empirical terms.

The resting-membrane potential for most excitable cells lies in the range of 50 to 100 mv with the inside negative with respect to the outside. Some typical values are given in Table 3.3. In that table, it may be noted that the observed resting potential is roughly equal to the potassium Nernst potential. If the assumption is made that the resultant potential is determined by the major ions of potassium, sodium, and chloride and that the potassium permeability at rest greatly exceeds that for sodium, then (3.140) can be made to give a reasonably good fit to data obtained from varied internal and external composition. This statement is documented in Sec. 4.2, where a variety of experimental and theoretical results are compared.

The peak of the action potential has a magnitude in the range 20 to 80 mv with the inside positive with respect to the outside. The value obtained is, in this case, very close to the sodium Nernst potential. Such a result can be made consistent with the constant-field model if it is assumed that the sodium permeability now greatly exceeds that of potassium or chloride; details are provided in the next section.

Excitation of the nerve thus results in an enhancement of the sodium permeability, so that a local influx of sodium ions results and a reversal in sign of transmembrane potential takes place. This is followed by a relatively greater outflow of potassium current and a recovery to the original resting condition. As a consequence of the sodium and potassium ion flow across the active region of the membrane, electric currents must flow in the surrounding media. In particular these currents cause ohmic transmembrane potentials in the adjacent passive regions of the membrane which, as the current builds up, can reach threshold for an action potential. In this manner the activity spreads (propagates) into the nearby regions. Note that prior to threshold being reached, the electrical behavior of the membrane is characterized by linear, passive phenomena; i.e., the initial phase is electrotonic.

## 4.2 SODIUM THEORY

The *sodium theory*, outlined in the previous section, seeks to explain the resting potential and action potential peaks on the basis of the constant-field model and Eq. (3.140). On an empirical basis, values of sodium, potassium, and chloride permeability are chosen to fit the measured potentials under resting and active conditions. Verification of the hypothetical model has been sought by varying the concentration of the ions that enter (3.140), utilizing the aforementioned permeabilities, and comparing the computed values with those that are measured.

Many of the measurements of the above type have been performed on the squid giant axon. The reason for such extensive use of this axon lies in its large size; it is possible to secure specimens which are several centimeters in length and 300 to 800 microns in diameter. In measurements made during the 1940s and 1950s, the axon was suspended in a chamber into which solutions of different composition could be introduced. An axial electrode was placed within the axon, and the transmembrane potential with respect to a reference electrode in the external medium recorded as a function of the external ion concentration.

In one related set of experiments, Adrian[1] noted that the resting potential was essentially determined by the potassium concentration. That is, if the external potassium concentration is varied and the resulting

[1] R. H. Adrian, The Effect of Internal and External Potassium Concentration on the Membrane Potential of Frog Muscle, *J. Physiol.* (*London*), **133**:631 (1956).

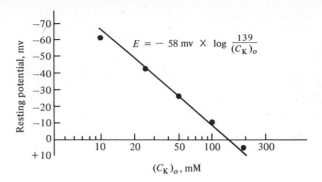

$$E = -58 \text{ mv} \times \log \frac{139}{(C_K)_o}$$

**Fig. 4.2** Effect of external potassium concentration $(C_K)_o$ on resting-membrane potential (sartorius muscle, sulfate solution). Internal potassium concentration $(C_K)_i$ is 139 mM/liter. Straight line in the figure follows the theoretical equation. [*From R. H. Adrian, J. Physiol.* (*London*), **133**:631 (1956).]

resting potentials plotted, as in Fig. 4.2, the results fall very close to the Nernst potential

$$V_K = 58 \log \frac{(C_K)_o}{139}$$

where $(C_K)_i = 139$ mM/liter.

According to the sodium theory, the activation of a cell involves a several-hundred-fold increase in sodium permeability and a resultant influx of sodium into the cell interior. This process stabilizes when the transmembrane potential reaches the sodium Nernst potential (at the peak of the action potential). At this point, *inactivation* takes place whereby the sodium permeability decreases to its resting value. In the meantime, since the potassium concentration gradient is no longer equilibrated by a suitably high (inward) electric field, an outflow of potassium occurs. At the end of the action potential, a net transfer of $Na^+$ inward and $K^+$ outward has occured. [We shall discuss later the requirement for a metabolic process (sodium-potassium pump) to return the system to its original state.] As a check on this aspect of the theory, radioactive sodium and potassium were used and the net gain of sodium and loss of potassium by a nerve fiber were measured. A close fit of electrical and chemical observations was found.[1]

A further check on this hypothesis was given by Hodgkin and

[1] R. D. Keynes and P. R. Lewis, Resting Exchange of Radioactive Potassium in Crab Nerve, *J. Physiol.*, **113**:73 (1951).

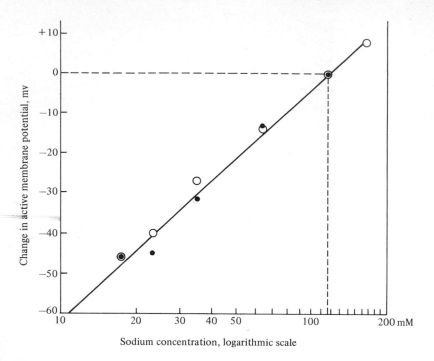

**Fig. 4.3** Average change in active membrane potential (crest of the spike) resulting from alterations in the external sodium concentration. The straight line is from the equation $\Delta V = 58 \log (C_{Na})_{test}/(C_{Na})_{Ringer}$. [*From W. L. Nastuk and A. L. Hodgkin, J. Cellular Comp. Physiol.,* **35**:39, fig. 12, (1950).]

Katz,[1] who replaced the sodium in the external solution by choline, an inert cation. They found that the action potential was abolished under these conditions.

By varying the concentration of the external sodium, Nastuk and Hodgkin[2] were able to check on the quantitative results predicted by (3.140) assuming $P_K$ and $P_{Cl}$ negligible compared with $P_{Na}$. Figure 4.3 shows that fairly satisfactory results were obtained.

Hodgkin and Katz[3] applied (3.140) to measurements performed by Curtis and Cole[4] and found that under resting conditions

$$P_K : P_{Na} : P_{Cl} = 1 : 0.04 : 0.45$$

[1] A. L. Hodgkin and B. Katz, The Effect of Sodium Ions on the Electrical Activity of the Giant Axon of the Squid, *J. Physiol.,* **108**:37 (1949).

[2] W. L. Nastuk and A. L. Hodgkin, The Electrical Activity of Single Muscle Fibers, *J. Cellular Comp. Physiol.,* **35**:39 (1950).

[3] *Loc. cit.*

[4] H. J. Curtis and K. S. Cole, Membrane Resting and Action Potentials from the Squid Giant Axon, *J. Cellular Comp. Physiol.,* **19**:135 (1942).

fit the data well for the physiological range of potassium concentrations. For high potassium concentrations a better fit was obtained with $P_K:P_{Na}:P_{Cl} = 1:0.025:0.3$. It also fit their own experiments, which involved test solutions of various concentrations of potassium, sodium, and chloride. These data are shown in Table 4.1.

A fair satisfaction of the constant-field equation was obtained for the peak of the action potential, utilizing a permeability ratio of $P_K:P_{Na}:P_{Cl} = 1:20:0.45$. The predicted and measured results for vari-

**Table 4.1  Comparison of observed changes in membrane potential with changes calculated from constant-field theory***

| State of nerve | Composition of test solution | | | | Change in membrane potential when test solution was substituted for seawater, or artificial seawater | |
| | K, mM | Na, mM | Cl, mM | Type of seawater† | Observed mv | Calculated‡ mv |
| --- | --- | --- | --- | --- | --- | --- |
| Resting | 0 | 465 | 587 | A | +3 | +5 |
| | 15 | 450 | 587 | A | −2 | −2 |
| | 20 | 445 | 587 | A | −4 | −4 |
| | 7 | 324 | 384 | B | 0 | +1 |
| | 5 | 227 | 270 | B | +2 | +2 |
| | 3 | 152 | 180 | B | +2 | +2 |
| | 2 | 91 | 108 | B | +4 | +3 |
| | 10 | 573 | 658 | B | +1 | 0 |
| | 10 | 711 | 796 | B | −2 | 0 |
| Active (peak of spike) | 0 | 465 | 587 | A | −1 | −1 |
| | 15 | 450 | 587 | A | +1 | 0 |
| | 20 | 445 | 587 | A | +5 | +1 |
| | 7 | 324 | 384 | B | +9 | +8 |
| | 5 | 227 | 270 | B | +21 | +16 |
| | 3 | 152 | 180 | B | +44 | +25 |
| | 2 | 91 | 108 | B | +59 | +38 |
| | 10 | 573 | 658 | B | −3 | −5 |
| | 10 | 711 | 796 | B | −9 | −10 |

* A. L. Hodgkin and B. Katz, The Effect of Sodium Ions on the Electrical Activity of the Giant Axon of the Squid, *J. Physiol.*, **108**:37 (1949).

† Type *A* (artificial seawater): 10 mM K, 455 mM Na, 587 mM Cl
  Type *B* (seawater): 10 mM K, 455 mM Na, 540 mM Cl
  Internal ionic composition: 345 mM K, 72 mM Na, 61 mM Cl

‡ Permeability coefficients used for calculation $(P_K:P_{Na}:P_{Cl})$:resting, 1:0.04:0.45; active, 1:20:0.45

**Table 4.2   Sodium and potassium movements across resting membrane**

|              | Fluxes, $pM/cm^2$ sec | Concentrations, $mM/kg$ $H_2O$ |
|--------------|-----------------------|--------------------------------|
| Na influx    | 3.5                   | $[Na]_o = 120$                 |
| Na efflux    | 3.5                   | $[Na]_i = 9.2$                 |
| K influx     | 5.4                   | $[K]_o = 2.5$                  |
| K efflux     | 8.8                   | $[K]_i = 140$                  |

ous combinations of potassium, chloride, and sodium ion concentrations are shown in Table 4.1.

Keynes[1] found a permeability ratio $P_{Na}/P_K$ equal to 0.077 in radioactive tracer measurements on *Sepia* axons, which is in the same general range as the ratio of 0.04 of Hodgkin and Katz.

The constant-field theory predicts a resting potential which is less than the resting potential given by the Nernst equation for potassium. This prediction is substantiated experimentally, as seen in Table 3.3. Consequently, the potassium diffusion force exceeds the electrical gradient, and a net outward flow of ions is predicted. This conclusion may also be drawn from (3.135) under the condition that the magnitude of $V$ (which is negative) is less than that which results in zero current (i.e., less than the Nernst potential). A similar analysis shows that an inward sodium and chloride flux is to be expected. Measurements of the $Na^+$ and $K^+$ fluxes, along with their concentrations for isolated frog skeletal muscle fibers, were performed by Hodgkin and Horowicz,[2] and the results are given in Table 4.2. The net sodium flux is approximately zero; a small net potassium flux, however, is observed.

Assuming that only passive (electrochemical) mechanisms are involved and that the constant-field model can be applied, then the component potassium flux can be found from (3.135) and is

$$\text{Potassium efflux} = P_K(C_K)_i \left| \frac{F^2 V}{RT} \frac{1}{1 - e^{-FV/RT}} \right| \tag{4.1}$$

$$\text{Potassium influx} = P_K(C_K)_0 \left| \frac{F^2 V}{RT} \frac{1}{1 - e^{FV/RT}} \right| \tag{4.2}$$

Substituting the values in Table 4.2 into (4.1) gives a value for $P_K$ of $6.4 \times 10^{-7}$ cm/sec, while using (4.2) results in $P_K$ equal to $5.8 \times 10^{-7}$ cm/sec. Thus reasonably consistent results are obtained.

[1] R. D. Keynes, The Ionic Movements during Nervous Activity, *J. Physiol.*, **114**:119 (1951).

[2] A. L. Hodgkin and P. Horowicz, Movements of Na and K in Single Muscle Fibers, *J. Physiol.* (*London*), **145**:405 (1959).

Sodium permeability may be calculated from sodium influx by using (4.2) with sodium parameters. One obtains in this way a value of $7.9 \times 10^{-9}$ cm/sec. The relative sodium-potassium permeability is, consequently, 0.013. This is in the same order of magnitude as that given earlier.

A calculation of sodium permeability from sodium efflux gives a value of $3.9 \times 10^{-6}$ cm/sec, which is completely at variance with the earlier result. That this value would be incorrect could have been anticipated since the outward sodium flux goes against both the concentration and the electrical gradient. Consequently, it cannot be a passive process and would therefore not be described by the constant-field equation. The earlier assumption that the potassium flux is passive is probably not completely correct either; there is some evidence that an active process may couple the efflux of sodium with a potassium influx. The active process is the link between the metabolic activity of the cell and the available electric energy in the action potential.[1] Some additional remarks are given in Sec. 4.9.

For the membrane described in Table 4.2 a transmembrane potential of $-90$ mv was obtained, while the Nernst potential for potassium is $-101$ mv. Thus, as noted earlier, the numerical value of the resting potential does not quite reach that of the potassium potential. If the relative permeability of sodium to potassium of 0.013 is utilized, then the constant-field equation gives

$$V = \frac{RT}{F} \ln \frac{(C_K)_o + b(C_{Na})_o}{(C_K)_i + b(C_{Na})_i} \tag{4.3}$$

where $b = P_{Na}/P_K = 0.013$. [It is assumed in Eq. (4.3) that the chloride ion is in equilibrium.] Substituting values from Table 4.2 into (4.3) gives

$$V = -89 \text{ mv}$$

which is in excellent agreement with the measured value.

## 4.3 PERFUSION EXPERIMENTS

With a technique developed by Baker, Hodgkin, and Shaw,[2] it is possible to squeeze out the axoplasm of the squid fiber (this is done much as one squeezes out toothpaste) without serious damage. The fiber may then be reinflated with solutions of arbitrary composition. This perfusion

[1] J. W. Woodbury, Interrelationships between Ion Transport Mechanisms and Excitatory Events, *Federation Proc.*, **22**:31 (1963).
[2] P. F. Baker, A. L. Hodgkin, and T. I. Shaw, Replacement of the Protoplasm of a Giant Nerve Fiber with Artificial Solutions, *Nature*, **190**:885 (1961).

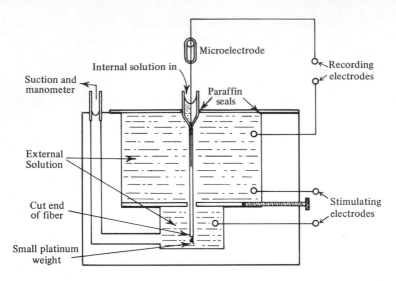

**Fig. 4.4** Apparatus for recording from a perfused axon. [*P. F. Baker, A. L. Hodgkin, and T. I. Shaw, J. Physiol.* (*London*), **164**:330 (1962).]

technique is illustrated in Fig. 4.4, which shows a cannula inserted longitudinally into the axon which carries the perfusate and the axial electrode. With such a preparation, it becomes possible to control the composition of both the intracellular and the extracellular medium. This makes possible a much wider range of experiments to check the validity of (3.140) and the sodium theory. Although not all axoplasm is removed, this fact appears to be unimportant, as determined from tracer studies, by observation of very short time constants in applying step changes in ionic composition, and by direct determination that this region is isopotential with the perfusate.

In one series of experiments, the relative strength of the internal potassium and sodium ion concentrations was varied by replacing isotonic $K_2SO_4$ by mixtures of $K_2SO_4$ and $Na_2SO_4$.† Since the effect of the chloride ion is absent, (4.3) applies. Experimental results for the resting potential and underswing were fit reasonably well by the choice of $P_{Na}/P_K = b = 0.03$ for low values of resting potential and $b = 0.08$ for high values. Other experiments showed that when $(C_K)_i = (C_K)_o$ the resting potential became zero, while the sign reversed for $(C_K)_i < (C_K)_o$, as predicted by (4.3). For the active membrane, a

---

† P. F. Baker, A. L. Hodgkin, and T. I. Shaw, The Effects of Changes in Internal Ionic Concentrations on the Electrical Properties of Perfused Giant Axons, *J. Physiol.* (*London*), **164**:355 (1962).

value of $P_{Na}/P_K = 7$ gives a good fit to the measured data.  A plot of both experimental and computed values is given in Fig. 4.5.

More recent work by Chandler and Meves[1] gives a permeability ratio of $P_{Na}/P_K = 12$ for the peak of the action potential.  Note that

[1] W. K. Chandler and H. Meves, Voltage Clamp Experiments on Internally Perfused Giant Axons, *J. Physiol.* (*London*), **180**:788 (1965).

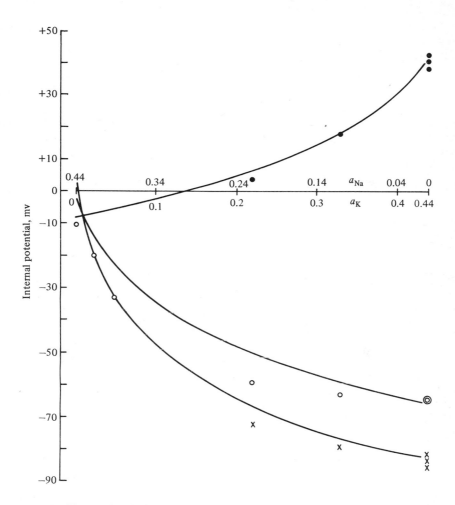

**Fig. 4.5**  Effect of replacing K by Na in an internal solution.  Ordinate: internal potential at the crest of the spike (●), in the resting state (○), and at the bottom of the underswing (×).  Abscissa: activity (moles per liter) of K and Na.  Continuous curves from (4.3) with $b = 7$ for the upper, $b = 0.08$ for the middle, and $b = 0.03$ for the lower curve.  [*P. F. Baker, A. L. Hodgkin, and T. I. Shaw, J. Physiol.* (*London*), **164**:355 (1962).]

**Table 4.3   Measured and computed peak (action) potential as a function of internal composition**

| $(C_K)_i$ | $(C_{Na})_i$ | $V_R$ | $V_A$ | $V_{Na}$ | $V_{Na+0.2K}$ |
|---|---|---|---|---|---|
| 300 | 0 | $-88.5$ | 47.5 | .... | 53.5 |
| 150 | 150 | $-66$ | 26 | 30.4 | 25.8 |
| 100 | 300 | $-59$ | 5 | 4.9 | 3.3 |

under these conditions the sodium Nernst potential represents the approximate (and limiting) magnitude of the action-potential maxima.

In a series of experiments by Chandler and Hodgkin[1] the main internal anion was chosen to be glutamate. For $(C_{Na})_o = 470$ mM, the results shown in Table 4.3 were obtained. In Table 4.3, $V_R$ is the resting potential; $V_A$, the measured action potential peak; $V_{Na}$, the sodium Nernst potential $\{V_{Na} = 60 \ \log [(C_{Na})_o/(C_{Na})_i]\}$; while $(C_K)_i$, $(C_{Na})_i$, etc, are ionic concentrations in millimoles per liter. In computing the sodium Nernst potential, the sodium activities were actually used. In addition, the relation based on (4.3), namely,

$$V_{Na+0.2K} = 60 \log \frac{(C_{Na})_o}{(C_{Na})_i + 0.2(C_K)_i} \tag{4.4}$$

was used for the theoretical analysis. This amounts to choosing $P_{Na}/P_K = 5$, approximately.

This value is conceded to be a little on the low side. Note that for the condition $(C_{Na})_i = 0$, it is necessary to include the effect of the internal potassium ion to prevent an absurd result. [A similar question arises in discussing the resting potential for the case where either $(C_K)_o$ or $(C_K)_i$ is zero.] Chandler and Hodgkin nowhere found $V_A$ to exceed $V_{Na}$, although they considered fluoride as well as glutamate. The agreement between measured and computed action potential peak values, as exemplified by data in Table 4.3, is good.

In an investigation by Baker, Hodgkin, and Shaw[2] the effect of changing the internal potassium concentration from 150 mM $[(C_{Ko}) = 10$ mM] to 600 mM resulted in a change in resting potential from $-$ (40 to 50) mv to $-$ (50 to 60) mv. This contrasts with a 36-mv change required by the Nernst equation for potassium. The discrepancy is ascribed to a potassium saturation which prevents the resting potential from exceeding $-$(50 to 60) mv. (A similar effect is noted when the external concentration is substantially reduced while the internal composition is

[1] W. K. Chandler and A. L. Hodgkin, The Effect of Internal Sodium on the Action Potential in the Presence of Different Internal Anions, *J. Physiol.*, **181**:594 (1965).
[2] *Ibid.*

maintained at normal value.) A suggested explanation for this phenomenon, given by Hodgkin and Chandler,[1] is that the permeability ratios $P_K:P_{Na}:P_{Cl}$ are dependent on the internal ionic concentration. This effect was noted earlier in connection with Fig. 4.4 and Eq. (4.3). Based on their work, Baker et al.[2] concluded that the permeability ratios changed from $P_K:P_{Na}:P_{Cl} = 1:0.05:0.1$ at internal KCl concentrations of 100 to 600 mM to $1:0.035:0.02$ at internal concentrations below 50 mM. According to their hypothesis, a negative-fixed-charge layer exists on the inner membrane surface, and this causes a potential dip following the Debye-Hückel theory that is greater for decreasing internal ion strengths. The overall effect is to concentrate cations near the inside of the membrane, which, according to the constant-field theory, is equivalent to increasing the permeability of the cations relative to the anions.

The details of this hypothesis can be reproduced by applying (3.38) to the inside of the axon, utilizing the plane parallel model as illustrated in Fig. 4.6. The field within the membrane is assumed constant, and the external reference is taken to be zero. We seek to determine the effect of a (positive) charge density $\sigma$ at the plane $x = 0$. This charge imposes a boundary condition given by

$$\epsilon_m \left[\frac{d\Phi}{dx}\right]_{0^-} - \epsilon \left[\frac{d\Phi}{dx}\right]_{0^+} = \sigma \tag{4.5}$$

[1] A. L. Hodgkin and W. K. Chandler, Effects of Changes in Ionic Strength on Inactivation and Threshold in Perfused Nerve Fibers of *Loligo, J. Gen. Physiol.*, **48**:27 (1965).

[2] P. F. Baker, A. L. Hodgkin, and H. Meves, The Effect of Diluting the Internal Solution on the Electrical Properties of a Perfused Giant Axon, *J. Physiol.*, **170**:541 (1964).

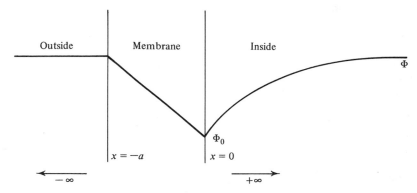

**Fig. 4.6** The effect of a charge layer on potential distribution. The profile is to be superimposed on the normal constant-field potential distribution. [*W. K. Chandler et al., J. Physiol.*, **180**:821 (1965).]

where $\epsilon$ is the dielectric constant of the solution (essentially water) and $\epsilon_m$ that of the membrane (related to 1 $\mu f/cm^2$). Assuming a resting potential of zero in the absence of the charge layer, then in its presence we require that $\Phi \to 0$ as $x \to \infty$. Thus for small $\Phi$, (3.38) can be linearized, and a solution

$$\Phi = \Phi_0 e^{-x/R_m} \qquad x > 0 \tag{4.6}$$

is obtained. The effect of the charge layer in concentrating cations or diluting anions is given by (3.36) and is described by the factor

$$\rho = \exp\left(\frac{-F\Phi_o}{RT}\right) \tag{4.7}$$

This relationship can be applied, by superposition, to a membrane whose resting potential is nonzero, provided $\Phi_0$ is understood to be the inner membrane potential due to the charge layer alone. Application of the constant-field equation gives

$$V = \frac{RT}{F} \ln \frac{P_K(C_K)_o + P_{Na}(C_{Na})_o + P_{Cl}(C_{Cl})_i/\rho}{\rho P_K(C_K)_i + \rho P_{Na}(C_{Na})_i + P_{Cl}(C_{Cl})_o} + \frac{RT}{F} \ln \rho \tag{4.8}$$

$$V = \frac{RT}{F} \ln \frac{P_K(C_K)_o + P_{Na}(C_{Na})_o + P_{Cl}(C_{Cl})_i/\rho}{P_K(C_K)_i + P_{Na}(C_{Na})_i + P_{Cl}(C_{Cl})_o/\rho} \tag{4.9}$$

where the first term in (4.8) accounts for the voltage across the membrane, and the second is the voltage between the bulk and inner edge of the membrane. The quantities $(C_K)_i$, $(C_{Na})_i$, and $(C_{Cl})_i$ are concentrations in the bulk solution and the factor $\rho$ takes into account the aforementioned concentrating effect. The value of $\Phi_0$ can be obtained from (4.6) subject to (4.5). While the charge layer has no direct effect on the potential difference between the internal and external solution, it is seen in (4.9) how it can exert an indirect influence through the permeability parameters. That the latter depend on the potential difference across the membrane rather than the total potential difference seems quite reasonable.

For the case that the internal (KCl) concentration of 300 mM is reduced to 6 mM and with $\sigma = -2.23$ coul/cm$^2$, Chandler et al.[1] [using a more exact solution than (4.6) and noting that the potential variation in the external solution is essentially negligible] show that $\Phi_0$ goes from $-17$ to $-80$ mv. This change of 63 mv corresponds to an observed shift of 65 mv. The factor $\rho$ increases as the concentration is reduced, and this is equivalent to a decrease in $P_{Cl}$ relative to $P_K$ and $P_{Na}$, accord-

[1] W. K. Chandler, A. L. Hodgkin, and H. Meves, The Effect of Changing the Internal Solution on Sodium Inactivation and Related Phenomena in Giant Axons, *J. Physiol.*, **180**:821 (1965).

**Table 4.4  A comparison of calculated and measured resting potential as a function of internal potassium concentration**

| Resting potential, corrected measurement, mv | $(C_K)_i$, $mM$ | $(a_K)_i$, $mM$ | Resting potential, calculated, mv |
|---|---|---|---|
| $-63$ | 456.6 | 232.9 | $-63.3$ |
| $-57$ | 256.6 | 146.3 | $-55.7$ |
| $-46$ | 93.3 | 64.4 | $-45.8$ |
| $-37$ | 56.6 | 41.3 | $-42.0$ |
| $-34$ | 56.6 | 41.3 | $-42.0$ |
| $-33$ | 10.0 | 8.0 | $-35.2$ |

$(a_{Cl})_o = 381$ mM; $(a_{Na})_o = 292$ mM; $(a_K)_o = 6.8$ mM.
Internal medium is sodium- and chloride-free.
Calculated $E = 56 \log [(P_K(a_K)_o + P_{Na}(a_{Na})_o)/(P_K(a_K)_i + P_{Cl}(a_{Cl})_o)]$.

ing to (4.9). Specifically, for the value given, $\rho$ increases fourfold, and this is roughly consistent with the permeability ratio change inferred by Baker et al.[1]

The effect of internal potassium concentration was also investigated by Adelman and Fok.[2] A portion of these results is listed in Table 4.4, where the ratio $P_K:P_{Na}:P_{Cl} = 1:0.06:0.25$ gave a best fit. The potassium-ion activity was obtained from data published by Baker, Hodgkin, and Shaw,[3] which, in turn, were determined from measurements made with a glass electrode. Good agreement is to be noted between calculated and measured values. The importance of including the chloride ion for very high or very low concentrations is demonstrated by these data. For these experiments, it was not necessary to introduce changes in the relative permeabilities when the internal ionic strength was reduced.

Investigation of the effect of various intracellular anions was conducted by Tasaki and Takenaka.[4] Included in these experiments were halogens, aspartate, glutamate, citrate, tartrate, and other anions. They determined that the fluoride, aspartate, and glutamate ions produced the largest action potentials (other ion species in the extra- and intracellular media remaining the same). Replacement of chloride by

[1] Baker, Hodgkin, and Meves, *loc. cit.*

[2] W. J. Adelman and Y. B. Fok, The Effects of Internal Potassium and Sodium on Membrane Electrical Characteristics, *J. Cellular Comp. Physiol.*, **64**:429 (1964).

[3] P. F. Baker, A. L. Hodgkin, and T. I. Shaw, Replacement of the Axoplasm of Giant Nerve Fibers with Artificial Solutions, *J. Physiol.*, **164**:330 (1962).

[4] I. Tasaki and T. Takenaka, Effects of Various Potassium Salts and Proteases upon Excitability of Intracellularly Perfused Squid Giant Axons, *Proc. Natl. Acad. Sci.*, **52**:804 (1954).

the above anions in the external medium produces little effect. From this it is concluded that the protein surface layers are different at the inner and outer surfaces and that the latter contains negative fixed charges. The inner surface also provides fixed charges, and the sign depends on the pH, the chemical species, and the concentration of the salt in the perfusion fluid. The "anion effect" is presumed to arise from interaction between the fixed charges in the inner membrane and the ions in the perfusion fluid. It is noted that the magnitude of the intracellular effect of the anions follows the lyotropic series and that the electric charge at the phase boundary of a complex colloid is similarly dependent.[1]

In addition to the above, a study of sodium-rich perfusate was made utilizing glutamate, aspartate, or fluoride anions. The results are at variance with those obtained by Chandler and Hodgkin (see Table 4.3). Thus for an external medium of $(C_{Na}) = 350$ mM, $(C_{Mg}) = 100$ mM, and $(C_{Ca}) = 20$ mM (all with chloride anion) and for an internal composition of $(C_{Na}) = 350$ mM and $(C_K) = 150$ mM (glutamate anion) a resting potential of $-45$ mv and an action potential of 75 mv were measured.[2] This compares with an action potential of 80 mv reported by Tasaki and Takenaka[3] in experiments utilizing 0.3 $M$ sodium on both sides of the membrane. Under similar conditions Tasaki and Luxoro[4] obtained action potentials of 70 mv using *Dosidicus* axons. The resting potential for the aforementioned experiments was around $-45$ mv. Consequently, for these measurements, the peak of the action potential exceeds the sodium Nernst potential (which is zero for the case of equal internal and external sodium concentration) by around 30 mv. A suggested explanation of these effects is that the sign and magnitude of the fixed charges of the protein layers, as well as the variation of ion mobility as a function of coordinate normal to the membrane, enter into a determination of the properties of the membrane.

Measurements by Chandler and Hodgkin, reproduced in Table 4.3, show an action potential of close to 65 mv when internal and external sodium are approximately equal, and this differs only slightly from the 70 to 80 mv obtained by Tasaki et al. Thus the difference in the conclusions of these two groups lies in the difference in their measurement of the dc resting potential; Chandler and Hodgkin's value is 30 mv lower than that obtained by Tasaki et al.,[5] so that the sodium Nernst potential

[1] H. R. Kruyt, "Colloid Science," Elsevier Publishing Company, New York, 1949.

[2] I. Tasaki, M. Luxoro, and A. Ruarte, Electrophysiological Studies of Chilean Squid Axons under Internal Perfusion with Sodium-rich Media, *Science*, **150**:899 (1965).

[3] *Loc. cit.*

[4] I. Tasaki and M. Luxoro, Intracellular Perfusion of Chilean Giant Squid Axons, *Science*, **145**:1313 (1964).

[5] Tasaki, Luxoro, and Ruarte, *loc. cit.*

is not exceeded.    Tasaki et al. criticize the use by Hodgkin and coworkers of agar bridges and electrodes with unplatinized platinum to measure the resting potential and point out the sensitivity of the platinum to changes in electric current, $O_2$ tension, and concentration of trace polyvalent cations.    In contrast, they use a single KCl–calomel electrode connected alternately to the internal and the external medium by means of a glass capillary filled with 0.6 $M$ KCl solution.    This matter, which remains unresolved, illustrates the difficulty in interpreting measurements by practical electrode systems (the salt-bridge in particular), which was noted in Chap. 2.

In the next several sections, we shall present the formulation of the Hodgkin-Huxley approach to the dynamic analysis of the squid axon.    This work is based on the sodium theory, which, despite limitations that have been noted, nevertheless permits satisfactory results in many cases.    We have also seen that there is much evidence supporting the sodium theory, particularly under typical physiological conditions and for the squid axon.    The Hodgkin-Huxley theory has survived surprisingly well and has been able to predict a wide variety of electrophysiological observations.    It furthermore suggests development of further models of a class in which it is a member.

## 4.4  VOLTAGE–CLAMP EXPERIMENTS

The direct measurement of ($Na^+$) and ($K^+$) currents is of great importance.    While the tracer experiments yield excellent results, they are based on the cumulative effect of a few thousand impulses.    The aforementioned need was met by the development of the *voltage-clamp* technique by Cole and Marmont.[1]    The circuit used is illustrated in Fig. 4.7, as employed in a classical series of experiments by Hodgkin, Huxley, and Katz.[2]

In the voltage clamp, a simple proportional controller is used to keep the membrane potential at a preset value.    This is accomplished by controlling the current flow between axial electrode $a$ (inserted into the nerve axoplasm) and electrode $e$, which is a concentric cylindrical electrode in the extracellular fluid.    This causes the transmembrane potential, as developed between electrodes $b$ and $c$, to be locked to a preset value.    In Fig. 4.7 it can be seen how the error signal $V - V_0$ is developed and applied to the current generator.    The resultant change

[1] K. S. Cole, Dynamic Electrical Characteristics of the Squid Axon Membrane, *Arch. Sci. Physiol.*, **3**:253 (1949).   G. Marmont, Studies on the Axon Membrane.   I. A New Method, *J. Cellular Comp. Physiol.*, **34**:351 (1949).

[2] A. L. Hodgkin, A. F. Huxley, and B. Katz, Measurement of Current-voltage Relations in the Membrane of the Giant Axon of *Loligo*, *J. Physiol.*, **116**:424 (1952).

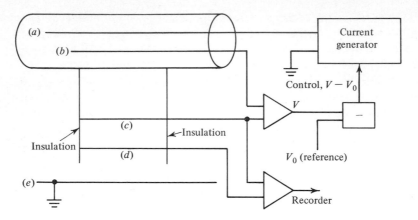

**Fig. 4.7** Schematic diagram showing voltage-clamp apparatus. Current electrodes are (*a*) and (*e*); potential sensing electrodes are (*b*) and (*c*). Transmembrane current is determined from the potential between (*c*) and *d*. Transmembrane voltage $V$ is compared with the desired clamp $V_0$, and the difference causes the transmembrane current to change so that $(V - V_0) \to 0$.

in applied current is such as to reduce $V - V_0$ toward zero. The voltage across the external electrodes *c* and *d* can be used to determine the total net radial current.

Electrodes *a* and *b* are actually interleaved helices which are exposed over an axial extent, as shown. Electrodes *c* and *d* are silver wires, while electrode *e* is a cylindrical silver sheet. Exposed portions of the electrodes were coated electrolytically with chloride. The electrodes *b*, *c*, and *d* are located within a compartment in order to eliminate *end effects* and achieve axial uniformity. This is important since, as we shall soon discuss, an impulse at one point on an axon is ordinarily propagated down the fiber. In this experiment, we wish to eliminate the spatial dependence; this is done in the above arrangement in that the axon is caused to operate synchronously over the extent of the recording electrodes.

A typical record resulting from the application of a step change in membrane voltage is shown in Fig. 4.8. One notes an initial inward current followed by the expected rise to an asymptotic outward current. An initial capacitive surge is completed in 20 $\mu$sec, corresponding to the presence of a capacitor with $C = 0.9$ $\mu$f/cm$^2$, a value in agreement with measurements by other methods noted in Chap. 3. Because of the very short time constant this current drops to zero before the ionic current becomes significant, and hence is normally ignored in studies of the latter.

The initial influx of ionic current is due to the sodium ion.   That is, the application of a voltage step of +70 volts, as illustrated in Fig. 4.8 (a step as small as +15 volts would produce the same result), causes the sodium permeability to increase greatly and rapidly and result in the movement of sodium ions from the highly concentrated exterior to the interior of the cell.   To verify this, if an external solution with choline[1] substituted for sodium is used, then the inward current pulse is eliminated, as shown in Fig. 4.9.   (The outward "hump" is the movement of sodium ions from the axoplasm to the sodium-free solution outside.)

Since the inward current is caused by sodium diffusion (due to the high external and low internal concentration), the equivalent "force" for this flow should be given by the sodium Nernst potential.

[1] Choline is an inert cation which does not affect the resting potential.

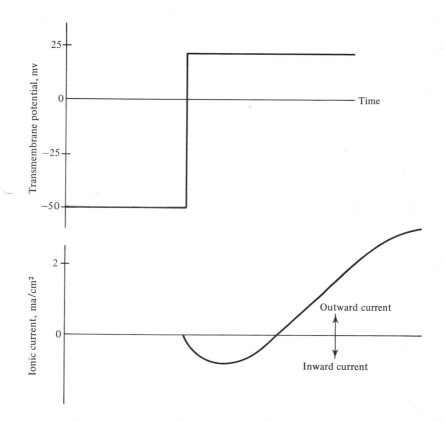

**Fig. 4.8**  Ionic current for the squid axon under application of voltage clamp.

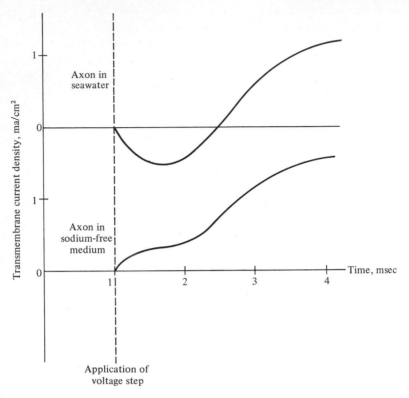

**Fig. 4.9**  Comparison of transmembrane ionic current for the squid axon under application of voltage clamp for an axon in seawater and an axon in a sodium-free medium.

$$V_{Na} = \frac{RT}{F} \ln \frac{(C_{Na})_o}{(C_{Na})_i}$$

Conversely, the greater the positive voltage step applied to the membrane, since this causes sodium outflow, the smaller the inward current flow ought to be.  If a step is applied which clamps the transmembrane potential at the value $V_{Na}$ itself, then the sodium is in electrochemical equilibrium and no inward current should result.  If $V > V_{Na}$ is imposed, then the flow should always be outward.  The curves, Fig. 4.10, of ion flow vs. the voltage step $\Delta V$ show the disappearance of an inward flow at essentially the predicted value.  [The relevant ionic composition and $V_{Na}$ for Fig. 4.10 is $(C_{Na})_o = 460$ mM, $(C_{Na})_i = 55$ mM, $V_{Na} = 56$ mv.]  That is, a step change of $+117$ mv from a resting potential of $-60$ mv corresponds correctly to a sodium Nernst potential of 57 mv for this particular axon.    For a step greater than $+117$ mv, the flow is always outward.

The ionic current is composed mainly of potassium and sodium ions.    The individual components can be found in the following way. First a measurement of ionic current due to a step change of clamped voltage, such as $+56$ mv as illustrated in Fig. 4.11, is taken.    Then the sodium concentration in the external solution is modified so that the sodium Nernst potential is the same as the membrane potential after applying the clamp.    By repeating the measurement under these modified conditions, since sodium is now in equilibrium, only a potassium current is obtained.    (In the illustrated case, the external sodium concentration is reduced from 460 to 46 mM/liter, as a result of which the Nernst potential drops to around 0 volts.    Since the resting potential is $-60$ mv and $\Delta V = 56$ mv, the sodium potential is approximately in equilibrium after application of the voltage clamp.)    The sodium current may be obtained by subtracting the second measurement from the first, a result which assumes that the potassium current is independent of the sodium (or choline) concentrations.[1]    Such a set of measured and computed currents is plotted in Fig. 4.11.    One notes that the sodium current, which

[1] A fundamental assumption of Hodgkin and Huxley is that the "chance that any individual ion will cross the membrane in a specified interval of time is independent of other ions."    This hypothesis was called the "independance principle."    Much experimental work appears to confirm this concept.

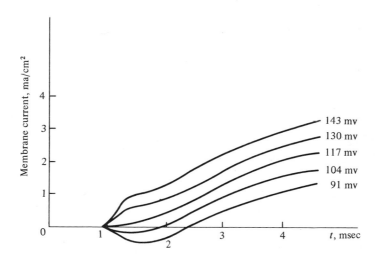

**Fig. 4.10**  Ionic current for the squid axon following the application of a voltage clamp of the value indicated.    The sodium Nernst potential is reached with a step change of 117 mv.    [*After A. L. Hodgkin, A. F. Huxley, and B. Katz, J. Physiol.,* **116**:424 (1952).]

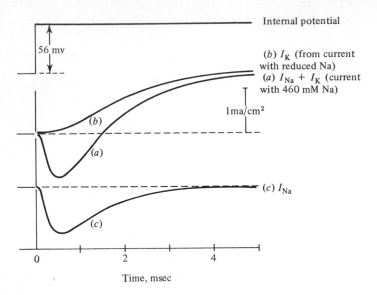

**Fig. 4.11** Separation of ionic transmembrane current into components carried by sodium and potassium. [*A. L. Hodgkin and A. F.
Huxley, J. Physiol.*, **116**:449 (1952).]

is always inward, declines after about 0.5 msec and the main current
becomes an outward potassium current. This current reaches an asymptotic limit, which, other experiments would show, depends on the magnitude of the voltage clamp. In view of the polarity of the voltage clamp
(positive with respect to resting potential of potassium) the direction of
the asymptotic current flow is appropriately an outward current.

So long as $V < V_{Na}$, the sodium current can only be inward, while
the potassium current must be outward. A measure of potassium current,
uncontaminated by sodium current for normal concentrations, can be
achieved by applying a step $\Delta V = 117$ mv, which makes $V = V_{Na}$. The
result is an S-shaped curve with the current changing very little at
first, then increasing to 3 ma/cm$^2$, where it levels off. This same result
is also obtained if sodium is replaced by choline.

Further evidence of the outward potassium current is furnished by
radioactive-tracer studies. Measurements made by Hodgkin and
Huxley[1] showed that the outward flow of current as determined electrically equals the outward flow of potassium ions as found from radioactive tracers. A linear relationship was obtained between these two
techniques for outward current densities up to 200 $\mu$a/cm$^2$.

[1] A. L. Hodgkin and A. F. Huxley, Movement of Radioactive Potassium and Membrane Current in a Giant Axon, *J. Physiol.*, **121**:403 (1953).

## 4.5 HODGKIN–HUXLEY FORMULATION; DEFINITION
## OF MEMBRANE CONDUCTANCE

A convenient way of representing the sodium or potassium permeability of a membrane is in terms of conductance parameters $g_{Na}$ and $g_K$ defined as

$$g_{Na} = \frac{I_{Na}}{V - V_{Na}} \tag{4.10}$$

$$g_K = \frac{I_K}{V - V_K} \tag{4.11}$$

In these equations, $V$ is the transmembrane potential with the outside chosen as the (zero) reference,[1] $V_{Na}$ and $V_K$ are the Nernst potentials

[1] In other words, $V$ equals the potential of the inside minus the potential of the outside.

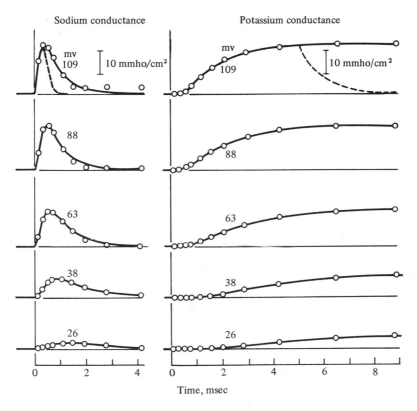

**Fig. 4.12** Changes in sodium and potassium conductance associated with different values of depolarization, $\Delta V$, as noted. Circles are experimentally determined points, while smooth curves are calculated from Eqs. (4.13) to (4.23). [*A. L. Hodgkin and A. F. Huxley, J. Physiol.*, **117**:500 (1952).]

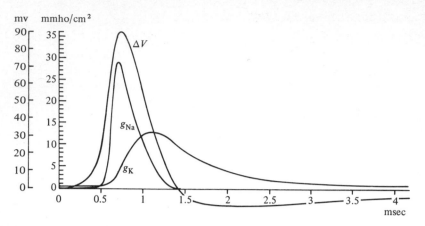

**Fig. 4.13.** Solution of the Hodgkin-Huxley equations showing components of membrane conductance during propagated action potential ($\Delta V$). [*A. L. Hodgkin and A. F. Huxley, J. Physiol.*, **117**:500 (1952).]

of sodium and potassium, respectively, while $I_{Na}$ and $I_K$ are the currents per unit area carried by the sodium and potassium ions and taken as positive for *outward* flow.[1]   The conductances, as defined by (4.10) and (4.11), are in units of mhos per unit area.   These definitions take account of the dependence of $I_{Na}$ on the quantity $(V - V_{Na})$ and $I_K$ on $(V - V_K)$.   That is, $(V - V_{Na})$ and $(V - V_K)$ can be thought of as the "driving force" for the respective ionic currents.   If (3.136) is examined it is clear that neither $g_{Na}$ nor $g_K$ can be expected to be constant since this would not be true even if the permeability were independent of $V$ (which it is not).   Changes in $g_{Na}$ and $g_K$ as functions of time for several voltage steps of magnitude $\Delta V$† are given in Fig. 4.12.   Actually, $g_{Na}$ and $g_K$ depend both on the transmembrane potential and on past history (i.e., on time as a parameter).

A consideration of Fig. 4.12 shows that the sodium conductance, upon application of the indicated step changes, rises to a peak and then declines to the resting value.   This occurs despite the fact that the fiber is clamped at a depolarized level.   The potassium conductance, on the other hand, rises slowly at first, then follows an S curve to a final equilibrium value.   This is maintained as long as the voltage clamp is applied.   The dotted curves in Fig. 4.12 show how the conductances

---

[1] The polarity convention for membrane voltage and current given here corresponds to that adopted in the mid 1950s.   The reader should be alert to a reversed polarity definition in earlier papers.

† The change in potential from that under resting conditions is denoted by $\Delta V$, and $\Delta V$ is positive if the axoplasm potential increases algebraically with respect to the outside.

change when the clamp is removed, the reversible nature of the changes being apparent. The repolarization curves are exponential, with a time constant of 0.1 to 0.2 msec for sodium and 5 to 10 msec for potassium (at 6°C). The rate constants depend on temperature, membrane potential, and calcium concentration. An indication of the changes in sodium and potassium conductances during a normal action potential is furnished by the plot in Fig. 4.13.

## 4.6 QUANTITATIVE DESCRIPTION OF CONDUCTANCES (HODGKIN–HUXLEY EQUATIONS)

A dynamic description of an action potential can be formulated through the use of a membrane model that includes as parameters the sodium and potassium conductances, their respective Nernst potentials, and the membrane capacitance. We consider first an application to a specialized but simple nerve preparation that is defined as being "space clamped." This is created by placing a wire along the axis of a nerve. A stimulus applied between this wire and the external medium will activate the entire axon synchronously; hence, no axial variation in potential can occur. As a consequence there will be no longitudinal current, and only radial variations need be considered (i.e., the problem is one-dimensional). Since there are no external sources[1] and the total current is solenoidal,

---

[1] During the brief stimulating pulse, an external current does exist; it must equal the net membrane current.

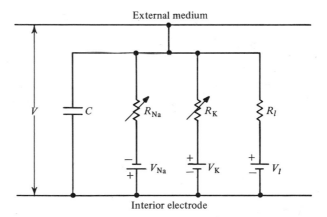

**Fig. 4.14** Equivalent circuit for the spatially clamped squid axon according to the Hodgkin-Huxley analysis. The membrane current consists of a capacitive component as well as sodium, potassium, and miscellaneous ionic contributions.

the total radial current must be zero.    The above remarks are summarized
by the model shown in Fig. 4.14, for which the appropriate equations are

$$I = C\frac{\partial V}{\partial t} + (V - V_K)g_K + (V - V_{Na})g_{Na}$$
$$+ (V - V_l)\bar{g}_l = 0 \quad (4.12)$$

(As noted earlier, $I$ is a current density, $C$ is the capacitance per unit
area, and $g_K$, $g_{Na}$, $\bar{g}_l$ are conductances per unit area.)    An additional
current term has been included in (4.12) which corresponds to leakage
current (ionic current from sources other than sodium or potassium,
e.g., chloride); this term is ordinarily neglected.

From an examination of data acquired from voltage-clamp experiments (such as given in Fig. 4.12), Hodgkin and Huxley[1] devised empirical
formulas which evaluate $g_K$ and $g_{Na}$ as functions of voltage and time.
As a guide to the selection of such formulas, Hodgkin and Huxley
assumed that it required four particles to occupy specific sites to bring
about a change in potassium conductance.    Thus if $n$ is the probability
for a single particle to be in the correct location, then

$$g_K = \bar{g}_K n^4 \tag{4.13}$$

where $\bar{g}_K$ is the maximum potassium conductance.    The value of $n$ is
the solution of

$$\frac{dn}{dt} = \alpha_n(1 - n) - \beta_n n \tag{4.14}$$

where the rate constants $\alpha_n$ and $\beta_n$ depend only on $V$ (for fixed temperature and calcium concentration).

For the sodium conductance, three simultaneous events each of
probability $m$ were required for its increase, while a single event of
probability $(1 - h)$ caused inactivation.    Thus the probability of facilita-
tion and no blocking is $m^3 h$.    Accordingly, the sodium relationship is

$$g_{Na} = \bar{g}_{Na} m^3 h \tag{4.15}$$

where $\bar{g}_{Na}$ is the maximum sodium conductance, while the values of $m$
and $h$ are given by solutions to

$$\frac{dm}{dt} = \alpha_m(1 - m) - \beta_m m \tag{4.16}$$

$$\frac{dh}{dt} = \alpha_h(1 - h) - \beta_h h \tag{4.17}$$

[1] A. L. Hodgkin and A. F. Huxley, A Quantitative Description of Membrane Current
and Its Application to Conductance and Excitation in Nerve, *J. Physiol.*, **117**:500
(1952).

The rate constants $\alpha_m$, $\beta_m$, $\alpha_h$, and $\beta_h$ similarly depend on $V$, temperature, and calcium concentration. Values of these constants, at 6°C, are as follows:

$$\alpha_n = \frac{0.01(10 - \Delta V)}{\exp\left[(10 - \Delta V)/10\right] - 1} \tag{4.18}$$

$$\beta_n = 0.125 \exp \frac{-\Delta V}{80} \tag{4.19}$$

$$\alpha_m = \frac{0.1(25 - \Delta V)}{\exp\left[(25 - \Delta V)/10\right] - 1} \tag{4.20}$$

$$\beta_m = 4 \exp \frac{-\Delta V}{18} \tag{4.21}$$

$$\alpha_h = 0.07 \exp \frac{-\Delta V}{20} \tag{4.22}$$

$$\beta_h = \left[ \exp\left( \frac{30 - \Delta V}{10} \right) + 1 \right]^{-1} \tag{4.23}$$

In the above equations, $\Delta V = V - V_r$ and $V_r$ is the value of resting potential (all in millivolts). That is, $\Delta V$ is the change in the membrane potential from its value at rest; $\Delta V$ is positive when the cell interior becomes more positive with respect to the outside. An explanation of the way in which (4.18) to (4.23) were obtained will now be given.

We note first that with the application of a voltage clamp, Eqs. (4.14), (4.16), and (4.17) become readily integrable, because the rate constants are themselves constant (since $V$ is constant). Upon integration one obtains

$$n(t) = n_\infty - (n_\infty - n_0)e^{-t/\tau_n} \tag{4.24}$$

$$m(t) = m_\infty - (m_\infty - m_0)e^{-t/\tau_m} \tag{4.25}$$

$$h(t) = h_\infty - (h_\infty - h_0)e^{-t/\tau_h} \tag{4.26}$$

where

$$n(0) = n_0 \qquad m(0) = m_0 \qquad h(0) = h_0 \tag{4.27}$$

$$n_\infty = \frac{\alpha_n}{\alpha_n + \beta_n} \qquad m_\infty = \frac{\alpha_m}{\alpha_m + \beta_m} \qquad h_\infty = \frac{\alpha_h}{\alpha_h + \beta_h} \tag{4.28}$$

$$\tau_n = \frac{1}{\alpha_n + \beta_n} \qquad \tau_m = \frac{1}{\alpha_m + \beta_m} \qquad \tau_h = \frac{1}{\alpha_h + \beta_h} \tag{4.29}$$

In the resting state $g_{Na}$ is very small compared with its value for a depolarization of, say, $\Delta V > 30$ mv, so that as a good approximation

$m_0$ can be neglected.  Also for $\Delta V > 30$ mv, inactivation becomes complete at $t \to \infty$; hence $h_\infty = 0$.  Thus we get

$$g_{\mathrm{Na}} = \bar{g}_{\mathrm{Na}} m_\infty{}^3 h_0 \left(1 - e^{-t/\tau_m}\right)^3 e^{-t/\tau_h} \tag{4.30}$$

$$g_{\mathrm{Na}} = g'_{\mathrm{Na}}(1 - e^{-t/\tau_m})^3 e^{-t/\tau_h} \tag{4.31}$$

where

$$g'_{\mathrm{Na}} = \bar{g}_{\mathrm{Na}} m_\infty{}^3 h_0 \tag{4.32}$$

Note that $g'_{\mathrm{Na}}$ is constant.  By plotting curves of $g_{\mathrm{Na}}$ versus $t$ for different values of clamped voltage and by fitting the experimental data with curves of the form expressed by (4.31), it is possible to obtain $g'_{\mathrm{Na}}$, $\tau_m$, and $\tau_h$ corresponding to each value of $V$.

If there were no inactivation (if $h$ remained at the resting value of $h_0$), then the steady-state value of sodium conductance under clamped conditions would approach $g'_{\mathrm{Na}}$, as is clear from (4.31).  If we normalize the sodium conductance (by choosing a suitable $\bar{g}_{\mathrm{Na}}$) so that $m_\infty = 1$ for the largest depolarizations, then we obtain

$$m_\infty = \left[\frac{g'_{\mathrm{Na}}}{(g'_{\mathrm{Na}})_{\max}}\right]^{\frac{1}{3}} \tag{4.33}$$

In (4.33), $(g'_{\mathrm{Na}})_{\max}$ is the asymptotic value of $g'_{\mathrm{Na}}$ for $\Delta V \to \infty$.

With values of $m_\infty$, $\tau_m$, and $\tau_h$ obtained from voltage-clamp experiments as functions of $\Delta V$, it is possible to compute $\alpha_m$, $\beta_m$, $\alpha_h$, and $\beta_h$ as functions of $\Delta V$.  For example, from the relationships in (4.28) and (4.29), namely, that

$$m_\infty = \frac{\alpha_m}{\alpha_m + \beta_m}$$

$$\tau_m = \frac{1}{\alpha_m + \beta_m}$$

one can then obtain the quantities $\alpha_m$ and $\beta_m$ since

$$\alpha_m = \frac{m_\infty}{\tau_m}$$

$$\beta_m = \frac{1 - m_\infty}{\tau_m}$$

Expressions for $\alpha_h$ and $\beta_h$ are similar, with $m_\infty$ and $\tau_m$ replaced by $h_\infty$ and $\tau_h$; determination of $h_\infty$ (for $\Delta V < 30$ mv) follows a procedure which will be described shortly.  A smooth analytic curve for $\alpha_m(\Delta V)$, $\beta_m(\Delta V)$, $\alpha_h(\Delta V)$, and $\beta_h(\Delta V)$ is the basis for expressions (4.20) to (4.23).  A completely analogous procedure is followed to give (4.18) and (4.19).  In this case the temporal behavior of $n$ gives $\tau_n$, while its asymptotic value leads to $n_\infty$ by procedure analogous to that followed for $m_\infty$. (That is, $n_\infty \to 1$ for $\Delta V \to \infty$.)

The above method for the determination of sodium conductance parameters depends on setting $h_\infty = 0$ for $t \to \infty$, and this is valid for $\Delta V > 30$ mv. For $\Delta V < 30$ mv the following procedure is required for a determination of $h_\infty$. In this experiment, the membrane potential is changed in two successive steps. The first step involves application of a conditioning voltage ($\Delta V_1$), while the second step refers to application of a test voltage ($\Delta V_2$). The conditioning voltage is subthreshold, while $\Delta V_2$ elicits an action potential. The membrane current response depends on $\Delta V_1$, $\Delta V_2$, and the duration of $\Delta V_1$. The effect of increasing the duration of the conditioning step on the magnitude of the peak inward current that results from the suprathreshold test step approaches saturation exponentially. In the procedure to be discussed, we shall consider the relationship of the peak inward current versus conditioning voltage $\Delta V_1$ under the condition that the duration of the latter exceeds the *inactivation time constant,* and for a fixed value of $\Delta V_2$. The result is a measure of the *steady-state* inactivation $h_\infty$; a characteristic sequence is plotted in Fig. 4.15. A detailed analysis follows.

If the conditioning experiment is initiated at $t = 0$, then it is terminated at a time $t = T_c$ which is large compared with $\tau_h$, so that, according to (4.26),

$$h(T_c) = (h_\infty)_c$$

At $t = T_c$ the suprathreshold test pulse $\Delta V_2$ is applied, and if we shift the time origin to this event then (4.30) decribes the ensuing temporal behavior of the sodium conductance. Now the initial value of $h$ during the test-pulse experiment, $(h_0)_t$, must be equal to the final value of $h$ reached during conditioning, $(h_\infty)_c$. Furthermore, since all test pulses reach the same transmembrane potential $\Delta V_2$, each response given by (4.31) will have the same time constants but will differ in $g'_{Na}$. Consequently, since the sodium current depends on $g'_{Na}$,

$$(h_0)_t = (h_\infty)_c \propto g'_{Na} \propto (I_{Na})_{peak}$$

In view of the proportionality of $g'_{Na}$ to the peak inward sodium current, and, with such data as illustrated in Fig. 4.15, it is possible to plot this current vs. the conditioning voltage. If the peak inward current is normalized to the maximum value attained, then the resultant curve corresponds to the parameter $h_\infty$. The normalized curve resulting from data in Fig. 4.15 is plotted in Fig. 4.16. Here it is seen that inactivation is almost complete for $\Delta V > 20$ mv, confirming the assumption that $h_\infty = 0$ for $\Delta V > 30$ mv while it is essentially absent for $\Delta V < -30$ mv. At the resting potential, inactivation is approximately 40 percent com-

Membrane potential, $\Delta V$                    Membrane current

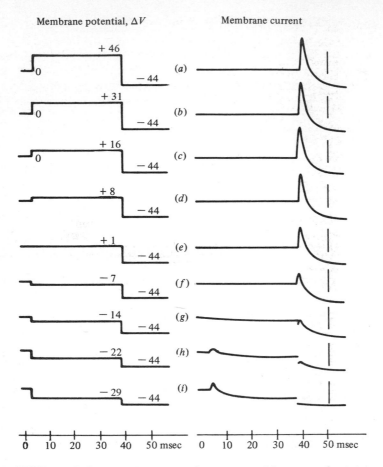

**Fig. 4.15** Influence of the membrane potential on steady-state inactivation. Vertical lines are peak inward current (approximately 0.70 ma/cm²) in the absence of conditioning voltage. [*A. L. Hodgkin and A. F. Huxley, J. Physiol.*, **116**:497 (1952).]

plete ($h_\infty = 0.6$). This result is particularly important since it furnishes the initial value of $h = h_0$ for voltage-clamp experiments which start from the resting state. The curve shown in Fig. 4.16 was obtained from the following equation:

$$h_\infty = \frac{1}{1 + \exp\left[(\Delta V - \Delta V_h)/7\right]} \tag{4.34}$$

which is an excellent fit to the experimental data. In this equation $\Delta V_h$ is the value of $\Delta V$ for $h = 0.5$. Determination of $m_\infty$ and $\tau_m$ for $\Delta V < 30$ mv is now possible since $\tau_h$ and $h_\infty$ are known for this condition.

The computation of ionic potassium and sodium currents from (4.10) and (4.11) using (4.13), (4.15), and (4.24) to (4.29) with (4.18) to (4.23) is straightforward. This result is an excellent fit of the original voltage-clamp data such as in Fig. 4.12.

As noted by Hodgkin,[1] the measurements on changes in sodium and potassium conductances reveal considerable sensitivity to changes in membrane potential. For example, an $e$-fold increase in sodium conductance results from only a 5-mv change in membrane potential. By contrast, a physical device requires a potential change of $kT/q_e$ to produce a conductance change by the factor of $e$ (where $q_e$ is the electronic charge). At room temperature, the latter calculates to 25 mv. Consequently, this suggests that alterations in permeability result from simultaneous activity of several particles or of single particles with several charges. The formulation of $g_K$ as proportional to $n^4$ follows the first possibility, while expressions (4.18) to (4.23) follow the suggestion that each particle is divalent. Thus the quantitative expressions might be thought of as a compromise in the two viewpoints noted above.

[1] A. L. Hodgkin, Ionic Movements and Electrical Activity in Giant Nerve Fibers, *Proc. Roy. Soc.*, **B148**:1 (1958).

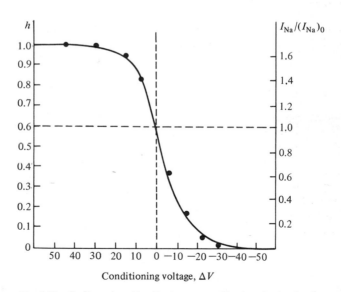

**Fig. 4.16** Sodium inactivation curve. Abscissa is the deviation from the resting potential, $\Delta V$. Dots are experimental points, and the smooth curve satisfies (4.34) for $\Delta V_h = -2.5$ mv. [*A. L. Hodgkin, and A. F. Huxley, J. Physiol.*, **116**:497 (1952).]

The concentration of calcium in the external medium has an important effect on membrane permeability. As a rough indication, the sodium conductance vs. membrane potential curve is shifted 6 to 9 mv along the voltage axis for an *e*-fold increase in calcium concentration. The direction is such as to increase excitability for reduction in calcium. A suggested explanation[1] is that calcium ions are adsorbed on the membrane and thereby constitute a surface charge density. If we follow the analysis in Sec. 4.3, calcium would thus affect permeability and excitability without actually taking part in the conduction of impulses. A similar effect occurs when, for the perfused axon, the perfusate is diluted with an isotonic sugar solution. In this case action potentials can be obtained even though the resting potential is reduced to zero.[2] From Fig. 4.16, it is clear that for an intact axon such a shift in resting potential would completely inactivate the sodium-carrying system. One expects, consequently, that the sodium inactivation curve must shift in the direction of positive transmembrane potential. This is precisely what does take place, as seen in Fig. 4.17, which is the sodium

---

[1] *Ibid.*

[2] P. F. Baker, A. L. Hodgkin, and T. I. Shaw, The Effects of Changes in Internal Ionic Concentrations on the Electrical Properties of Perfused Giant Axons, *J. Physiol.* (*London*), **164**:355 (1962). I. Tasaki, A. Watanabe, and T. Takenaka, Resting and Action Potential of Intracellularly Perfused Squid Giant Axon, *Proc. Natl. Acad. Sci.*, **48**:1177 (1962).

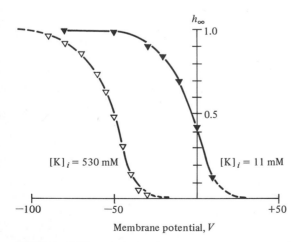

**Fig. 4.17** Sodium inactivation curves for normal- and low-concentration potassium perfusate. [*J. W. Moore, T. Narahashi, and W. Ulbricht, J. Physiol.*, **172**:163 (1964).]

inactivation curve for both normal and low potassium concentration as measured by Moore, Narahashi, and Ulbricht.[1]

When the internal potassium is replaced by sodium, calcium, or choline, the action potential fails when the resting potential rises from $-60$ to about $-40$ mv.† This suggests that the shift of the inactivation curve, as depicted in Fig. 4.17, occurs only when sucrose or glucose is used for the dilution of the internal potassium. Such a conclusion is confirmed by experiments performed by Chandler, Hodgkin, and Meves,[2] thus establishing the shift as a consequence of reduction in ionic strength rather than a decrease in potassium concentration. A tentative explanation of this phenomenon, based on the hypothesis advanced by Baker et al.,[3] has been noted in Sec. 4.3, and is based on the formation of a negative-fixed-charge layer on the inner membrane surface. According to this hypothesis, both the fixed-charge layer and the ionic strength should be considered as determining the potential difference across the *membrane* (as opposed to potential difference between bulk solutions). The shift in threshold and inactivation curve can then be explained by the reasonable supposition that they are related to the *membrane* difference of potential (rather than the total difference in potential).

## 4.7  ACTION POTENTIAL (HODGKIN–HUXLEY EQUATIONS)

For the case where axial variations are suppressed by the use of axial electrodes [i.e., the condition described by (4.12) and Fig. 4.14], a solution for the action potential can be found by the numerical integration of the equation

$$I_s = C\frac{dV}{dt} + \bar{g}_{\mathrm{K}}n^4(V - V_{\mathrm{K}}) + \bar{g}_{\mathrm{Na}}m^3h(V - V_{\mathrm{Na}}) \\ + \bar{g}_l(V - V_l) \quad (4.35)$$

[which is (4.12) combined with (4.13) and (4.15)] along with (4.14), (4.16), and (4.17). The initial conditions are the resting-state values

[1] J. W. Moore, T. Narahashi, and W. Ulbricht, Sodium Conductance Shift in an Axon Internally Perfused with a Sucrose and Low-potassium Solution, *J. Physiol.*, **172**:163 (1964).

† P. T. Baker, A. L. Hodgkin, and H. Meves, Internal Potassium Concentration and the Action Potential of Giant Axons of *Loligo*, *J. Physiol.*, **167**:162 (1963).

[2] W. K. Chandler, A. L. Hodgkin, and H. Meves, The Effect of Changing the Internal Solution on Sodium Inactivation and Related Phenomena in Giant Axons, *J. Physiol.*, **180**:821 (1965).

[3] P. T. Baker, A. L. Hodgkin, and H. Meves, The Effect of Diluting the Internal Solution on the Electrical Properties of a Perfused Giant Axon, *J. Physiol.*, **170**:541 (1964).

of $V$, $m$, $n$, and $h$ at $t = 0$.   The stimulating current density $I_s$ is normally assumed at a constant value for a brief interval and thereafter set equal to zero.   The action potential results from a sufficiently strong initial current stimulus of adequate duration.   An outline of the procedure follows.

We define the following parameters:

$$v = V - V_r = \Delta V \tag{4.36}$$

$$v_{\mathrm{Na}} = V_{\mathrm{Na}} - V_r \tag{4.37}$$

$$v_{\mathrm{K}} = V_{\mathrm{K}} - V_r \tag{4.38}$$

$$v_l = V_l - V_r \tag{4.39}$$

where $V_r$ is the resting-membrane potential, $V$ is the instantaneous membrane potential, and $v$ is the displacement from the resting value.     If $|V| < |V_r|$ (for depolarization), then $v$ is positive.   (This is a reversal of sign from the Hodgkin-Huxley work of 1952 but consistent with current nomenclature.)   Then the problem is to solve, simultaneously, the following:

$$C \frac{dv}{dt} + \bar{g}_{\mathrm{K}} n^4 (v - v_{\mathrm{K}}) + \bar{g}_{\mathrm{Na}} m^3 h (v - v_{\mathrm{Na}}) + \bar{g}_l (v - v_l) = I_s \tag{4.40}$$

$$\frac{dn}{dt} = \alpha_n (1 - n) - \beta_n n \tag{4.41}$$

$$\frac{dm}{dt} = \alpha_m (1 - m) - \beta_m m \tag{4.42}$$

$$\frac{dh}{dt} = \alpha_h (1 - h) - \beta_h h \tag{4.43}$$

subject to initial values $v = v_0$, $n = n_0$, $m = m_0$, and $h = h_0$.   [The last three are simply the values $n_\infty$, $m_\infty$, and $h_\infty$ following a long period at rest, and consequently obtainable from (4.28), using (4.18) to (4.23) with $\Delta V = 0$.]   We shall outline the step-by-step procedure which accomplishes this.   By denoting the beginning and end of a step by $t_0$ and $t_1$ ($= t_0 + \delta t$), the following sequence is carried out.

1. Estimate $v_1$ from $v_0$.   (Except at the start, one can extrapolate from earlier data.)
2. Estimate $n_1$ from $n_0$.   (Extrapolate when possible.)
3. Calculate $(dn/dt)_1$ from (4.41), using $\alpha_n$ and $\beta_n$ for the estimated $v_1$.
4. Calculate a value of $n_1$ from

$$n_1 - n_0 = \frac{\delta t}{2} \left\{ \left( \frac{dn}{dt} \right)_0 + \left( \frac{dn}{dt} \right)_1 - \frac{1}{12} \left[ \Delta^2 \left( \frac{dn}{dt} \right)_0 \right. \right.$$
$$\left. \left. + \Delta^2 \left( \frac{dn}{dt} \right)_1 \right] \right\} \tag{4.44}$$

where $\Delta^2(dn/dt)$ is the second difference of $dn/dt$;[†] its value at $t_1$ is estimated.

5. If this value of $n_1$ differs from that estimated in (2), repeat (3) and (4), using this new value. The process is repeated until $n_1$ from (4.44) corresponds with the value of $n_1$ used in step 2.
6. Find $m_1$ and $h_1$ by procedures analogous to steps 2 to 5.
7. Calculate $\bar{g}_K n_1{}^4$ and $\bar{g}_{Na} m_1{}^3 h_1$.
8. Calculate $(dv/dt)_1$ from (4.40), using the values obtained in step 7 with the original estimation of $v_1$.
9. Calculate a corrected $v_1$ by procedures similar to those given in steps 4 and 5. (According to Hodgkin and Huxley, the discrepancy was never large enough to necessitate repetition of the entire procedure.)

The values of $\delta t$ suggested are $\delta t = 0.01$ msec at the outset, $\delta t = 0.02$ msec during the rising phase of the spike, and $\delta t = 1$ msec during the small oscillations following the spike.

Results obtained with the above procedure by Hodgkin and Huxley[1] are shown in Fig. 4.18. As can be seen, the agreement is fairly good. In addition to this result, the satisfactory nature of the Hodgkin-Huxley formulation will be demonstrated for the propagated action potential.

## 4.8 NERVE CONDUCTION

Consider a nerve fiber, as illustrated in Fig. 4.19. If a stimulus is applied at point 1, a local depolarization results. That is, the sodium membrane permeability is greatly increased in the vicinity of 1, and an inflow of sodium occurs. The sodium inflow establishes a solenoidal current field which extends to the surrounding region, all of which contain electrolytes.

A graphical representation of the lines of current (conduction and ionic plus displacement) at an instant of time for an action potential that is propagating along a cylindrical axon in a volume conductor is shown in Fig. 4.19. The central region, which corresponds to sodium influx, is the site of activation at this moment. Note that in the sur-

[†] $\Delta^2 \left(\dfrac{dn}{dt}\right)_0 = \left(\dfrac{dn}{dt}\right)_2 - 2\left(\dfrac{dn}{dt}\right)_1 + \left(\dfrac{dn}{dt}\right)_0$

[1] A. L. Hodgkin and A. F. Huxley, A Quantitative Description of Membrane Current and Its Application to Conduction and Excitation in Nerve, *J. Physiol.*, **117**:500 (1952).

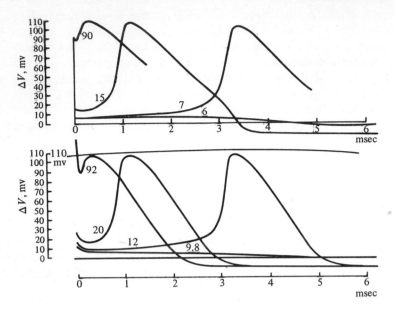

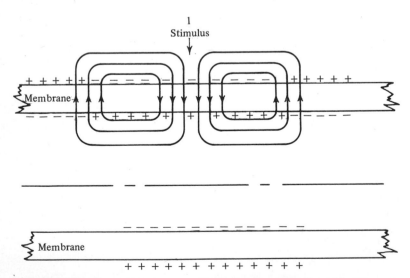

**Fig. 4.18** Upper curves: calculated action potentials from the solution of Eq. (4.40). Lower curves: tracings of measured membrane action potentials; the numbers are the shock strength in millimicrocoulombs per square centimeter. [*A. L. Hodgkin and A. F. Huxley, J. Physiol.,* **117:** 500 (1952).]

**Fig. 4.19** Local currents due to activation at the site indicated by the arrow. Total current is shown; note its solenoidal character.

rounding axonal region the direction of the current is outward. As a consequence a "depolarization" (an algebraic increase in transmembrane potential) is communicated to the neighboring region. Since the current density diminishes with distance from the active region, the strongest depolarization is produced in the immediate vicinity. When this effect becomes sufficiently great that the threshold is exceeded, then a lateral movement of the active zone results (i.e., propagation of the action potential). Note that during this transition the transmembrane current changes from an outward flow to an inward flow.

The above phenomenological description explains how a disturbance, once initiated, can spread to all points of an excitable membrane. Movement is always toward resting membrane since the region from which the activity has come must be in a refractory state.

It has been found experimentally that for an active nerve fiber the action potential propagates with constant velocity and is neither attenuated nor distorted. The transmembrane potential thus satisfies the wave equation and can be described mathematically by the expression

$$V = f(x - \theta t) \tag{4.45}$$

and

$$\frac{\partial^2 V}{\partial t^2} = \theta^2 \frac{\partial^2 V}{\partial x^2} \tag{4.46}$$

where $\theta$ is the velocity of propagation.

The region just ahead of the zone of activation behaves as a passive linear system and, as such, satisfies the electrotonic equations of Sec. 3.13 subject to (4.45) as a constraining condition. For a cylindrical fiber (3.173) provides a mathematical relation that must be satisfied by $V(x,l)$; we require $i_p = 0$ since no external currents are applied under the physiological conditions being considered. Thus

$$\lambda^2 \frac{\partial^2 V}{\partial x^2} - \tau \frac{\partial V}{\partial t} - V = 0 \tag{4.47}$$

Substituting (4.46) into (4.47) yields the following differential equation:

$$\frac{\lambda^2}{\theta^2} \frac{d^2 V}{dt^2} - \tau \frac{dV}{dt} - V = 0 \tag{4.48}$$

The coefficients in (4.48) are assumed constant under the assumed subthreshold condition, so that we get as a solution

$$V = A \exp\left(\frac{\tau\theta^2}{2\lambda^2} + \frac{\theta}{2\lambda}\sqrt{\frac{\tau^2\theta^2}{\lambda^2} + 4}\right) t$$

$$+ B \exp\left(\frac{\tau\theta^2}{2\lambda^2} - \frac{\theta}{2\lambda}\sqrt{\frac{\tau^2\theta^2}{\lambda^2} + 4}\right) t \tag{4.49}$$

For points in advance of the approaching activation "wave" (4.49) applies only for $t \leq T_0$, where $T_0$ is the time at which passive linear relations no longer apply. Since $V$ is bounded, then $B = 0$ is required and

$$V = A \exp \left[ \frac{\theta t}{2\lambda^2} (\theta\tau + \sqrt{\theta^2\tau^2 + 4\lambda^2}) \right] \qquad t \leq T_0 \qquad (4.50)$$

If at a later time $T_1$ the activity has passed and the region under consideration has returned to a passive state, then

$$V = B \exp \left[ \frac{\theta t}{2\lambda^2} (\theta\tau + \sqrt{\theta^2\tau^2 + 4\lambda^2}) \right] \qquad t \geq T_1 \qquad (4.51)$$

We note from (4.50) and (4.51) that the transmembrane potential variation preceding and following membrane activation is exponential. This fact is independent of the temporal form of the action potential. The point of transition from exponential behavior has been suggested as defining the threshold. Investigations based on this definition were carried out by Van der Kloot,[1] Jenerick,[2] Tasaki,[3] and others. The exponential portion of the action potential waveform satisfying (4.50), which occurs just ahead of the nonlinear activity, is known as the *foot of the action potential.*

Equation (4.50) depends on the following several approximations whose appropriateness should be checked in a given situation. Among the assumptions are, first, that the nerve fiber or trunk is uniform and cylindrical, so that the linear core-conductor model may be a good representation; second, that the passive membrane characteristics be suitably defined by (3.172), a relationship that seems satisfactory for nerve but (as noted in Sec. 3.12) not for muscle. In the latter case, the additional shunt element, consisting of a series $R$ and $C$ that is necessary, clearly affects the form of the resultant expression. Finally, use of (4.49) implies no influence from the ends of the fiber (i.e., the fiber is assumed as essentially infinite). This may be satisfactory if the point in question is at least one space constant from either end; significant deviations in the results can be expected for short fibers.

For typical values of the parameters in (4.50) obtained from nerve and muscle fibers,[4] $\tau^2\theta^2/\lambda^2 \gg 4$, so that as a good approximation

$$V = Ae^{Kt} \qquad (4.52)$$

[1] W. G. Van der Kloot and B. Dane, Conduction of the Action Potential in the Frog Ventricle, *Science*, **147**:74 (1964).

[2] H. Jenerick, Phase Plane Trajectories of the Muscle Spike Potential, *Biophys. J.*, **3**:363 (1963).

[3] I. Tasaki and S. Hagiwara, Capacity of Muscle Fiber Membrane, *Am. J. Physiol.*, **188**:423 (1957).

[4] Typical parameter values are $\tau = 15$ msec, $\lambda = 2$ mm, $\theta = 0.4$ m/sec.

where

$$K = \frac{\tau\theta^2}{\lambda^2} = \theta^2 C_m(r_1 + r_2) \tag{4.53}$$

From measurements of the time constant $K$ at the foot of an action potential propagating at the velocity $\theta$, by considering a single fiber in a volume conductor (where $r_2 \approx 0$), the membrane capacitance per unit area can be found from the expression

$$C_m = \frac{K}{\theta^2 r_1} \tag{4.54}$$

which is obtained from (4.53). Measurements utilizing this technique are described by Tasaki and Hagiwara,[1] Fozzard,[2] and others.

The technique given above represents the fourth method described for the measurement of membrane capacitance. The others, discussed in Secs. 3.12 and 3.13, utilize external ac impedance, internal ac impedance, and square-wave-response techniques. For the squid axon all methods give a capacitance around 1 $\mu$f/cm². For muscle fibers the results are not completely comparable because the different techniques tend to emphasize different aspects of the electrical characteristics of the fiber. The step response, for example, tends to reflect the long time constant (as if $r_s = 0$ in Fig. 3.14) and, hence, to be relatively insensitive to the need for a two-time-constant membrane circuit. On the other hand, because of the short electrotonic interval the foot of the action potential response tends to reflect only the short-time-constant component and hence yields small capacitance values. In general, it appears possible to account for the differences obtained and to substantiate the more elaborate two-time-constant model. Total parallel capacitance values of 2 to 12 $\mu$f/cm² are typical for frog muscle fibers, with larger values for crayfish and crab.

An electric-network representation of an axon which is suitable for the study of the propagated action potential can be formed by extension of Fig. 4.14. Such a circuit is given in Fig. 4.20. This figure represents a unit length of membrane, where $r_2$ is the effective resistance per unit length in the external medium, $r_1$ is the resistance per unit length of the axoplasm, and the box labeled H-H represents the Hodgkin-Huxley shunt circuit described in Fig. 4.14. The entire axon may be thought of as made up of an iterative structure of the form of Fig. 4.20. The longitudinal current is $i_1$, and $i_m$ is the shunt (transmembrane) current.

[1] *Loc. cit.*

[2] H. A. Fozzard, Membrane Capacity of the Cardiac Purkinje Fiber, *J. Physiol.*, **182**:255 (1966).

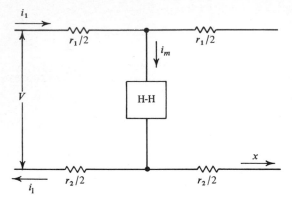

**Fig. 4.20**  Equivalent network for a unit-length axon. Resistances $r_1$ and $r_2$ represent axial resistance per unit length of axoplasm and external medium; H-H is the Hodgkin-Huxley transverse network, as illustrated in Fig. 4.14.

If the model of Fig. 4.20 is used as a reference, but the limiting case of distributed elements (all impedance and admittance parameters are on a per-unit-length basis) is utilized, then we see that[1]

$$\frac{\partial V}{\partial x} = -i_1(r_1 + r_2) \tag{4.55}$$

$$\frac{\partial i_1}{\partial x} = -i_m \tag{4.56}$$

Now the transmembrane current, evaluated as the sum of its components, is [see (4.12)]

$$I_m = C \frac{\partial V}{\partial t} + (V - V_K)g_K + (V - V_{Na})g_{Na} + (V - V_l)\bar{g}_l \tag{4.57}$$

(Note that $i_m$ is the membrane current per unit length, while $I_m$ is the current density in amperes per square centimeter.)  If we take the derivative of (4.55) with respect to $x$ and substitute (4.56), we get

$$i_m = \frac{1}{r_1 + r_2} \frac{\partial^2 V}{\partial x^2} \tag{4.58}$$

For an axon of radius $a$ and axoplasmic resistivity $\rho$, $i_m = 2\pi a I_m$, and $r_1 = \rho/\pi a^2$.  For an extensive extracellular medium, the effective external

[1] This analysis bears a superficial resemblance to the analysis of transmission lines (hence the result is called the *cable equation*).  While the two phenomena are totally different, some assistance in understanding the mathematical formalism might result from consideration of the transmission line.

resistance can be neglected.   Consequently, setting $r_2 = 0$, we have

$$I_m = \frac{\pi a^2}{2\pi a\rho}\frac{\partial^2 V}{\partial x^2} = \frac{a}{2\rho}\frac{\partial^2 V}{\partial x^2} \tag{4.59}$$

Thus, by putting (4.59) into (4.57), we finally get the Hodgkin-Huxley modified cable equation

$$\frac{a}{2\rho}\frac{\partial^2 V}{\partial x^2} = C\frac{\partial V}{\partial t} + g_K(V - V_K) + g_{Na}(V - V_{Na})$$

$$+ \bar{g}_l(V - V_l) \tag{4.60}$$

As noted, $V$ satisfies the wave equation (4.46); thus (4.60) can be expressed as

$$\frac{a}{2\rho\theta^2}\frac{d^2 V}{dt^2} = C\frac{dV}{dt} + g_K(V - V_K) + g_{Na}(V - V_{Na})$$

$$+ \bar{g}_l(V - V_l) \tag{4.61}$$

In order to solve (4.61), one guesses at $\theta$ and proceeds in a way that is very similar to that outlined previously for the action potential under space-clamp conditions.   It was found by Hodgkin and Huxley that incorrect guesses of $\theta$ caused $V$ to approach $\pm \infty$.   The correct value of $\theta$ brings $V$ back to the resting condition when the action potential is completed.   For the parameter $K$ defined as

$$K = \frac{2\rho\theta^2 C}{a} \tag{4.62}$$

(4.53) becomes [also using (4.13) and (4.15)]

$$\frac{d^2 V}{dt^2} = K\left\{\frac{dV}{dt} + \frac{1}{C}\left[\bar{g}_K n^4(V - V_K) + \bar{g}_{Na}m^3 h(V - V_{Na})\right.\right.$$

$$\left.\left. + \bar{g}_l(V - V_l)\right]\right\} \tag{4.63}$$

Figure 4.21 shows a plot of $V(t)$ for $K = 10.47$ msec$^{-1}$,[1] as found experimentally and as computed by Hodgkin and Huxley.[2]   The calculated velocity of 18.8 m/sec compares well with the measured velocity of 21.2 m/sec (for this calculation $a$ and $\rho$ were 238 microns and 35.4 ohm-cm, respectively).

[1] This value of $K$ was the one which was found to make the computed value of $V$ return to the resting value.

[2] A. L. Hodgkin and A. F. Huxley, A Quantitative Description of Membrane Current and Its Application to Conduction and Excitation in Nerve, *J. Physiol.*, **117**:500 (1952).

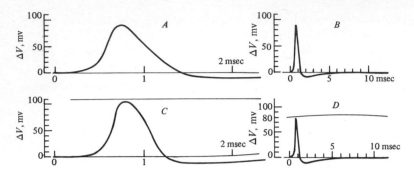

**Fig. 4.21** Curve $A$ is the computed (propagated) action potential from Eq. (4.63) with $K$ of 10.47 $msec^{-1}$. Curve $B$ is the same result to a slower time scale. Curves $C$ and $D$ are measured from different axons. [*A. L. Hodgkin and A. F. Huxley, J. Physiol.*, **117**:500 (1952).]

## 4.9  ACTIVE MEMBRANE PROCESSES

The Hodgkin-Huxley membrane model, as well as models discussed earlier, is capable of describing some but not all membrane processes. This section considers several phenomena not specifically considered thus far and for which the aforementioned membrane models have not been particularly satisfactory.

### SELECTIVITY

The difference in permeability between sodium and potassium ions for a membrane at rest has not been satisfactorily explained. Even less understood are the reasons for the change in relative permeability of these ions during activity (for example, see Fig. 4.12). Although the Hodgkin-Huxley model describes the phenomena in a quantitative way, it fails to provide any physical basis for the changes.

One hypothesis, due to Nachmansohn,[1] is based on the behavior of acetylcholine. According to this view, acetylcholine is present at rest, presumably bound to a protein molecule. During excitation it is released from its bound form and recombines with another protein element of the membrane called the receptor. In some way, this results in a configurational change which is reflected in a change in permeability. The acetylcholine–receptor complex readily dissociates, and the acetylcholine reacts with the enzyme acetylcholinesterase. The cycle is completed when acetylcholine is resynthesized by the enzyme choline acetylase, whereupon it is stored for future activity. Support for this theory is

[1] D. Nachmansohn and I. B. Wilson, The Enzymic Hydrolysis and Synthesis of Acetyl Choline, *Advan. Enzymol.*, **12**:429 (1951).

provided by the work of Schoeffeniels,[1] who measured a $Q_{10}$ for the action-potential duration of eel electroplaque of 3.33. Since $Q_{10}$ of physical processes is around 1.4 but of chemical processes is in excess of 2, the chemical nature of these processes is indicated.

## TRIGGER ACTION

The very rapid increase in sodium permeability following the application of an excitatory stimulus is referred to as the *trigger action* of the membrane. The underlying basis for this important phenomenon is not fully understood, but several hypotheses have been advanced.

As indicated by (4.25), the variation of $m(t)$ with time following application of a voltage step is exponential, with time constant $\tau_m$. Data for the squid axon[2] show that $\tau_m$ does not vary greatly over a wide range of voltages from sub- to suprathreshold. Thus the sodium activation curves for both sub- and suprathreshold conditions are similar in shape (although much different in amplitude). In view of this it is strongly suggested that the processes are similar at the molecular level and that changes in system structure account for the two types of response. According to this view the reduction in membrane potential causes an increase in sodium permeability, which facilitates a further reduction in membrane potential, etc. The system could thus be described as nonlinear with positive feedback. For an excitatory stimulus of sufficient magnitude the regenerative process is understood to continue without limit until the maximum sodium permeability is reached; this is the suprathreshold condition.

The membrane would remain in the "depolarized" condition were it not for a second sodium process, that is, inactivation. The inactivation process is known to be more vigorous for greater depolarizations. But the molecular mechanism that is responsible, as with the activation process, is not known.

## ACTIVE TRANSPORT

It has been pointed out that under resting conditions, there is a leakage of potassium from the inside to the outside of a cell and a movement of sodium ions in the reverse direction. The potassium efflux is around 8 pM/cm² sec, while the sodium influx is about 3.5 pM/cm² sec for frog muscle fibers. The net transport of these ions is greatly increased as a result of activation. For frog muscle fibers, this consists of an efflux of 11.4 pM/cm² of $K^+$ and an influx of 19.4 pM/cm² of $Na^+$ per stimulus.

[1] M. Schoeffeniels, Les Bases physiques, *Arch. Intern. Physiol. Biochim.*, **68**:1 (1960).
[2] A. L. Hodgkin and A. F. Huxley, A Quantitative Description of Membrane Current and Its Application to Conduction and Excitation in Nerve, *J. Physiol.*, **117**:500 (1952).

It is clear that the ability of a fiber to sustain activity depends on a process capable of restoring the resting composition, i.e., an active process. Such a process must result in an influx of potassium and an efflux of sodium against the net electrochemical gradient and hence requires an expenditure of energy which must ultimately come from metabolic processes in the cell.

The nature of the "pump mechanism" is not known, but some of its characteristics can be described. Thus if a cell is repeatedly stimulated, a net loss of $K^+$ from the cell and a gain of $Na^+$ can be readily detected, and the resting potential decreases (in magnitude) in accordance with the changed concentration. If the cell is permitted to rest, the initial concentration is slowly restored. The "sodium pump" thus operates at a high enough rate to maintain normal composition under physiological conditions, but this rate is slow compared with the quantities transported when the cell is repeatedly activated.

If a metabolic inhibitor, such as DNP, is added to the cell, then there is no restoration of the $Na^+$–$K^+$ balance, and after repeated stimulations, the cell "runs down." However, the disabling of the sodium pump does not affect the ability of the cell to sustain an action potential. The latter mechanism appears to be solely dependent on the presence of the appropriate sodium and potassium concentrations. The sodium pump may be thought of as providing a source of energy which performs at a low rate for a long time.

It is possible to give a rough estimate of the quantities involved in ion exchange for the squid axon based on the information already available. The result will serve to illustrate the previous comment that a large number of action potentials are possible before any significant change in ion concentrations results.

We shall assume that the net flow of ionic charge in an action potential goes, essentially, into the membrane capacitance. This would be exactly true under space-clamped conditions and is, otherwise, a rough approximation. Thus if $V_K = -50$ mv and $V_{Na} = +65$ mv and $C = 1$ $\mu$f/cm², then

$$Q = 1 \times 10^{-6} \times 115 \times 10^{-3} = 115 \times 10^{-9} \text{ coul/cm}^2 \qquad (4.64)$$

The number of micromoles equals $115 \times 10^{-9}/0.0965$. Thus the net positive charge that is required to enter equals 1.2 $\mu\mu$M (that is, $1.2 \times 10^{-12}$ moles) per square centimeter of membrane. As noted earlier, sodium influx as measured in the squid axon by radioactive techniques is in the neighborhood of 3.5 $\mu\mu$M. Our result is clearly in the correct range since the potassium efflux offsets the sodium, so that it is reasonable to expect a net positive exchange of 1.2 $\mu\mu$M necessary to charge the transmembrane capacitance to produce a 65-mv (sodium) potential.

The change in concentration caused by the sodium-potassium movements depends on the size of the axon. For an axon whose diameter is $d$ (cm) the entry of 3.0 $\mu\mu\text{M}/\text{cm}^2$ of $\text{Na}^+$ raises the internal concentration by

$$\Delta C_{\text{Na}} = \frac{3 \times 10^{-12} \times \pi d}{\pi d^2/4} = \frac{12 \times 10^{-12}}{d} \qquad (d \text{ in cm}) \qquad (4.65)$$

Thus for $d = 500$ microns $= 500 \times 10^{-4}$ cm, the internal concentration increases by $1.4 \times 10^{-10}$ moles/cm$^3$. But the normal internal concentration of $\text{Na}^+$ is 50 $\mu\text{M}/\text{cm}^3$, which is $3 \times 10^5$ greater. For a 50-micron-diameter fiber, the change in concentration would be only 1 part in 30,000, but this would still be negligible.

When the *Sepia* axon is placed in potassium-free seawater, the rate of sodium efflux is markedly reduced.[1] On the other hand, an increase in external potassium reversibly increases the sodium outflow. Such experiments have been carried out, with similar results, on the frog skin, red cells, and muscle. They suggest that a coupling exists between sodium efflux and potassium influx. One idea is that the pump consists of a carrier molecule which picks up a sodium ion at the inner surface and then moves to the outside, where a chemical change causes the sodium to be released and a potassium ion to be picked up. The carrier is then forced back to the inside, where the form changes again with the consequent release of potassium and acquisition of a sodium ion. Hokin and Hokin[2] propose that the carrier may be phosphatidic acid. In their theory, the phosphatide is formed at the inner membrane surface from diglyceride and ATP and is hydrolyzed to diglyceride and orthophosphate at the external surface of the membrane.

In addition to the Na–K coupled pump, a second parallel pump has also been described. The second pump is insensitive to cardiac glycerides but is dependent on the presence of Na in the external medium. A discussion of the properties and their distinctive features is given in a review paper by Hoffman and Kregnow.[3]

The control mechanism for the active transport has been subject to study. Earlier suggestions were that pump activity depended on internal sodium concentration linearly or to some power. However,

[1] A. L. Hodgkin and R. D. Keynes, Active Transport of Cations in Giant Axons from *Sepia* and *Loligo*, *J. Physiol.*, **128**:28 (1955).

[2] L. E. Hokin and M. R. Hokin, Studies on the Carrier Function of Phosphatidic Acid in Sodium Transport, *J. Gen. Physiol.*, **14**:61 (1960).

[3] J. F. Hoffman and F. M. Kregnow, The Characterizations of a New Energy Dependent Cation Transport Processes in Red Blood Cells, *Ann. N.Y. Acad. Sci.*, **137**:566 (1966).

the work of Woodbury[1] suggests that $(C_{Na})_i$ is not sufficiently sensitive as a control variable. A possible alternative is the dependence of pump rate on transmembrane potential or membrane permeability.

## 4.10 SALTATORY CONDUCTION

The medullated fibers, found only in vertebrates, have a thick covering of a fatty material called *myelin*. This myelin sheath is a good insulator but is broken periodically by nodes at which ionic and electric interchange can take place. These nodes are known as nodes of Ranvier. The mechanism of local excitation, as described in the previous section, must be modified since electrical activity can spread only by jumping to adjacent nodes. The impulse in "hopping" from node to node is termed *saltatory* (from the Latin *saltare*, meaning to dance). Conduction speed is about twenty times greater in myelinated than in unmyelinated fibers of the same diameter and temperature.

An approach to the analytical discussion of saltatory conduction is given by Fitzhugh.[2] We shall consider this work in some detail, both for its intrinsic interest and also because it illustrates how the Hodgkin-Huxley formulation for the squid axon can be generalized to other types of nervous behavior. The equivalent circuit used is shown in Fig. 4.22. At each node, a lumped Hodgkin-Huxley membrane is

[1] J. W. Woodbury, Interrelationships between Ion Transport Mechanisms and Excitatory Events, *Federation Proc.*, **22**:31 (1963).

[2] R. Fitzhugh, Computation of Impulse Initiation and Saltatory Conduction in a Myelinated Nerve Fiber, *Biophys. J.*, **2**:11 (1962).

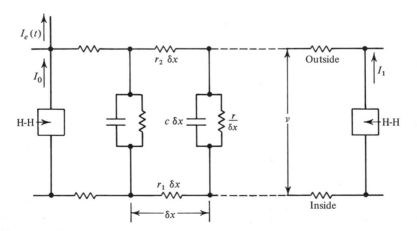

**Fig. 4.22** Equivalent circuit for a myelinated fiber. This differs from Fig. 4.20 in that the active membrane, represented by a transverse H-H element, lies only at the nodes. The internodal membrane is a distributed resistance and capacitance.

assumed. The conditions at this node are expressed by the H-H equations, which are modified to incorporate the actual lumped capacitance and conductance as measured. The internodal region is described by distributed-parameter equations which represent a "leaking cable" model. This involves series resistance and shunt capacitance and conductance, which behave passively and linearly as in electrotonus. The result is the set of differential equations given below.

$$\frac{c\partial v}{\partial t} = \frac{\partial^2 v/\partial x^2}{r_1 + r_2} - \frac{v}{r} \tag{4.66}$$

$$C_{\mathrm{N}}\frac{dv_i}{dt} = I_i - A[\bar{g}_{\mathrm{Na}}m_i{}^3 h_i(v_i - v_{\mathrm{Na}}) + \bar{g}_{\mathrm{K}}n_i{}^4(v_i - v_{\mathrm{K}})$$
$$+ \bar{g}_l(v_i - v_l)] \tag{4.67}$$

$$\frac{dm_i}{dt} = (1 - m_i)\alpha_m - m_i\beta_m \tag{4.68}$$

$$\frac{dh_i}{dt} = (1 - h_i)\alpha_h - h_i\beta_h \tag{4.69}$$

$$\frac{dn_i}{dt} = (1 - n_i)\alpha_n - n_i\beta_n \tag{4.70}$$

Equation (4.66), which is essentially (3.173) but with the nomenclature of Fig. 4.22, describes the time-varying transmembrane potential $v$ in the internodal region. Equations (4.67) to (4.70) are the H-H differential equations for the transmembrane potential in terms of the conductance parameters. Note that these equations apply only at the nodes (are valid only for $i = 0, 1, 2, 3, \ldots$). The node spacing $L$ is 2 mm, so that the $i$th node is at $x = 2i$ mm. $C_{\mathrm{N}}$ is the nodal capacitance, while $A$ is the area of the H-H membrane at each node.

The continuity of potential at the nodes means that

$$\lim_{x \to iL} v(x,t) = v_i(t) \qquad i = 0, 1, 2, 3, \ldots \tag{4.71}$$

Boundary conditions that must be satisfied by (4.66) at each node are

$$I_0 = \frac{r_2 I_e + 2 \lim_{x \to 0} \partial v/\partial x}{r_1 + r_2} \tag{4.72}$$

$$I_i = \frac{\lim_{x \to iL+0} \partial v/\partial x - \lim_{x \to iL-0} \partial v/\partial x}{r_1 + r_2} \tag{4.73}$$

Equation (4.72) is based on the symmetry of $v$ with respect to $x$ and the application of Kirchhoff's current law along with (4.55). The term $r_2 I_e/(r_1 + r_2)$ is an *effective stimulating current*. Other relationships, which completely specify the problem, are

$$v_i(0) = v(x,0) = 0 \qquad \text{all } x$$

$$m_i(0) = m_\infty(0) = \frac{\alpha_m(0)}{\alpha_m(0) + \beta_m(0)}$$

$$h_i(0) = h_\infty(0) = \frac{\alpha_h(0)}{\alpha_h(0) + \beta_h(0)} \qquad\qquad (4.74)$$

$$n_i(0) = n_\infty(0) = \frac{\alpha_n(0)}{\alpha_n(0) + \beta_n(0)}$$

In the above set of equations, the nomenclature for potentials corresponds to that developed earlier; that is, *all potentials* are referred to the resting potential (which may be thought of as arbitrarily assigned the value zero). The argument of zero in (4.74) corresponds to the initial uniform resting potential.

The value used for the myelin-sheath capacitance was 1.6 $\mu\mu$f/mm, the longitudinal resistance was 15 M$\Omega$/mm, and the myelin resistance was 290 M$\Omega$ mm, as measured by Tasaki[1] on frog. The area of the H-H membrane was chosen so that multiplication by the H-H resting conductance of 6.77 $\mu$mhos/mm$^2$ would give 0.02 $\mu$mho, which is the value measured by Tasaki[2] and Tasaki and Freygang.[3] This value of area, $A = 0.003$ mm$^2$, should be thought of as a scale factor rather than representing an actual area. The nodal membrane capacitance used was 1.5 $\mu\mu$f from Tasaki. This is one-twentieth the value obtained from using the H-H capacitance of 1 $\mu$f/cm$^2$ times an $A$ equal to 0.003 mm$^2$.

Computations based on the differential equations, for several initial conditions of $I_e(t)$, were performed by Fitzhugh on an IBM 704. The results are given in Fig. 4.23. These are spatial action potentials ($v$ versus $x$). Each curve shows a cusp at the node, because the finite nodal current demands a discontinuity in $\partial v/\partial x$ [see (4.73)]. The vertical lines correspond to the location of the nodes. This curve is computed for a stimulus which is 0.01 msec in duration, during which time 0.030 amp flows.

Figure 4.23 can be replotted to yield temporal data, as shown in

[1] I. Tasaki, New Measurements of the Capacity and the Resistance of the Myelin Sheath and the Nodal Membrane of the Isolated Frog Nerve Fiber, *Am. J. Physiol.*, **181**:63 (1955).

[2] *Ibid.*

[3] I. Tasaki and W. H. Freygang, The Parallelism between the Action Potential, Action Current, and Membrane Resistance at a Node of Ranvier, *J. Gen. Physiol.*, **39**:211 (1955).

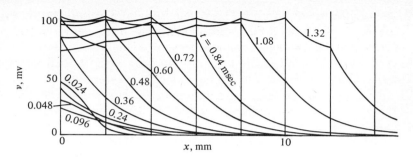

**Fig. 4.23** Computed spatial action potential curves following a 30,000-μa pulse stimulus of 0.01 msec duration at $t = 0$. Curves are the potential distribution at the time indicated. Since the stimulus is applied at $x = 0$, the curves are symmetric about the origin. Vertical lines represent nodes. [*R. Fitzhugh, Biophys. J.*, **2**:11 (1962).]

Fig. 4.24. The left-hand curves are for the node positions. Note how the waveform becomes nearly uniform three or four nodes away from the site of stimulation. The right-hand curves show the internodal response which does not exhibit a propagating phenomena. It appears, rather, to be due to the superposition of two action potentials from adjacent nodes arising from electrotonic conduction.

If the times of peak values in $v_i$ are plotted against node position ($t$ versus $x$), a straight line is obtained, starting at node 1. The slope is the conduction velocity, which was evaluated as 11.9 m/sec and compares fairly well with the figure of 10 m/sec of Tasaki and Fujita.[1]

[1] I. Tasaki and M. Fujita, Action Currents of Single Nerve Fibers as Modified by Temperature Changes, *J. Neurophysiol.*, **11**:311 (1948).

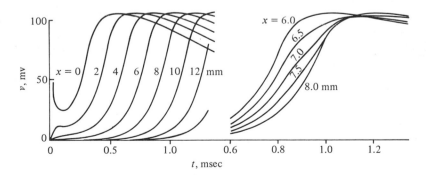

**Fig. 4.24** Temporal action potential curves. The left figure is at the nodes (only) as labeled. The right figure is for several positions between nodes. [*R. Fitzhugh, Biophys. J.*, **2**:11 (1962).]

Experimental work by Huxley and Stampfli[1] of longitudinal current in frog axons shows a propagation along the internode in a time around 0.02 msec, but a jump to the beginning of the next node requires 0.1 msec (node-to-node conduction time). The peak longitudinal current falls nearly linearly from 2.5 to 1.5 m$\mu$a along one internode.

The computed temporal action potential curves most nearly resemble those of Hodler, Stampfli, and Tasaki.[2] Their peak potentials did not vary appreciably as a function of distance along the fiber. The computed variation is from 106.6 to 102.9 mv, which is quite slight.

## 4.11  MEMBRANE PROPERTIES FROM CURRENT–VOLTAGE RELATIONS

A voltage-clamp experiment which has proved useful in characterizing membrane properties is that which yields current-voltage curves. Figure 4.25 illustrates the basis for this relationship. Shown here is the transmembrane current response to application of a voltage clamp; the curve illustrates the definition of current amplitudes $I_{Na}$ and $I_K$, which are the *peak inward* and the *steady-state outward current*, respectively. Actually, $I_{Na}$ is the net algebraic inward current and only an approximation to the sodium current, as is clear from Fig. 4.11. On the other hand, $I_K$ is the saturation potassium current since $I_{Na}$ is essentially

[1] A. F. Huxley and R. Stampfli, Evidence for Saltatory Conduction in Peripheral Myelinated Nerve Fibers, *J. Physiol.*, **180**:315 (1949).

[2] J. Hodler, R. Stampfli, and I. Tasaki, Role of the Potential Curve Spreading along Myelinated Nerve in Excitation and Conduction, *Am. J. Physiol.*, **170**:375 (1952).

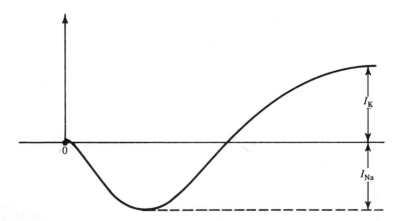

**Fig. 4.25**  Definition of the terms $I_{Na}$ and $I_K$. The curve is of the membrane current following application of a voltage clamp. $I_{Na}$ is referred to as the *peak inward current* and $I_K$ the *steady-state outward current*.

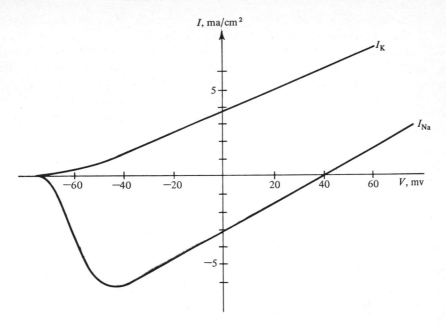

**Fig. 4.26**  Current-voltage relation.   For $I_{Na}$: peak inward current vs. clamped transmembrane voltage $V$ after change from resting conditions.   For $I_K$: steady-state outward current following change from resting conditions to the abscissa value of $V$.

zero.   It is the plot of $I_{Na}$ and $I_K$ versus the (clamped) membrane potential which is designated the *current-voltage relation*.

A typical current-voltage relation for a squid axon under normal conditions is given in Fig. 4.26.   The effect of drugs or modification of ionic constituents (external or internal) on membrane function is often particularly well characterized by the resulting alterations in the current-voltage relation.   As an illustration, Fig. 4.27 shows the effect

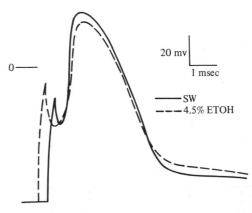

**Fig. 4.27**  Effect of ethanol, ETOH, on the action potential. Solid curve is seawater; broken curve is 4.5 vol percent ethanol. Resting potential is −90 mv; zero potential is indicated. [*J. W. Moore, W. Ulbricht, and M. Takata, J. Gen. Physiol.,* **48**:279 (1964).]

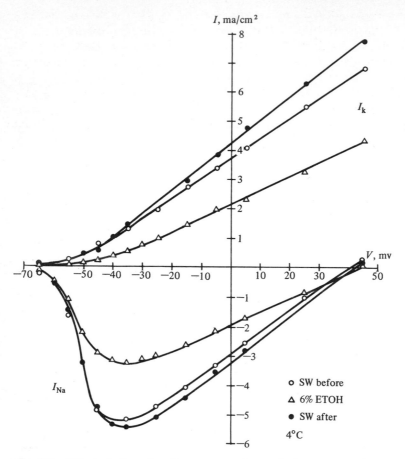

**Fig. 4.28** Effect of ethanol on the current-voltage relation for a squid axon. Curves with open circles are before, triangles during, and filled circles after application of 6% ETOH. [*J. W. Moore, W. Ulbricht, and M. Takata, J. Gen. Physiol.*, **48**:279 (1964).]

of ethanol on the action potential, while Fig. 4.28 shows its effect on the current-voltage relation.   One notes the much greater sensitivity of the latter representation to the changed environment.

The behavior of the sodium and potassium conductances is clearly revealed by the current-voltage curves.   Thus, from Fig. 4.28, one notes a linear relationship between the *peak transient current* ($I_{Na}$) and the steady-state outward current ($I_K$) with the membrane potential for $V > -20$ mv.   Since the equilibrium potential $V_{Na}$ is the value for which $I_{Na} = 0$ (the intercept with the horizontal axis), *chord conductances* $[g_{Na} = I_{Na}/(V - V_{Na})]$ can be determined graphically.   Since in this procedure $V_{Na}$ and $I_{Na}$ are approximations to the quantities defined in (4.10), the graphically determined $g_{Na}$ is also approximate. The chord conductances for $V < -20$ mv will differ from that for $V > -20$ mv. In general, $g_{Na}$ is not a constant, as has been pointed out earlier.

The effect of 0.1% procaine on the current-voltage curve of the squid axon is shown in Fig. 4.29. In this experiment, the voltage step was from a hyperpolarized value ($V_H$) to a particular value of $V_p$. The effect of procaine is to depress nerve activity with little or no effect on the resting-membrane potential. The mechanism by which procaine acts is clearly delineated in the current-voltage curves of Fig. 4.29. For

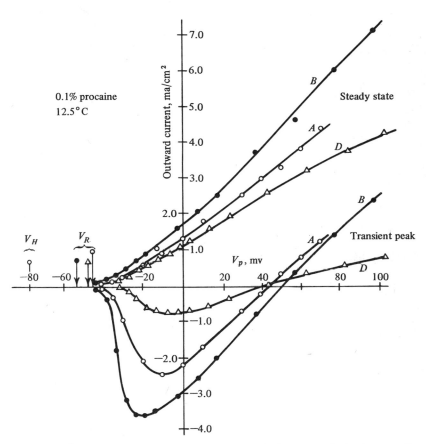

**Fig. 4.29** Current-voltage curves following step from hyperpolarized value $V_H$ to $V_p$. Resting potential is $V_R$. Curves $B$ before, $D$ during, $A$ after presence of 0.1% procaine at pH 7.9. [*R. Taylor, Am. J. Physiol.,* **195**:1071 (1959).]

we note here that its introduction, while affecting potassium ion flow only slightly, almost completely blocks sodium transport.

It is convenient to consider the squid-axon membrane as containing separate sodium and potassium channels. The behavior of the steady-state current $I_K$ characterizes the potassium channel, while the sodium channel is functionally described by $I_{Na}$, the transient peak current. Accordingly, the effect of procaine is to seriously reduce the transient inward current (a reduction of the carrying capacity of the sodium channel) while affecting the potassium channel relatively slightly.

The notion of separate functional channels, which arises naturally from Hodgkin-Huxley theory, is the basis for a theory that the membrane has structural channels of this sort.[1] However, it should be noted that no direct evidence for anatomic channels exists. Furthermore, were such channels to exist, then present information could not rule out the possibility of there being a single path with discrete temporal characteristics.[2]

In contrast to the effect of procaine, the effect of placing the squid axon in a solution containing TEA (tetraethylammonium) is to block the potassium channels while producing little or no effect on sodium. The action potential that results under these conditions has a prolonged duration as a result of the slow recovery due to impaired potassium conduction.[3]

A further example is the effect on nerve of the toxin TTX from the puffer fish. This works, essentially, on the sodium channel, which is almost completely blocked. At the same time, the potassium system remains almost completely intact.[4]

## 4.12 ANALYSIS OF ERRORS IN VOLTAGE–CLAMP MEASUREMENTS

The Hodgkin-Huxley equations are based on voltage-clamp experiments, as discussed in earlier sections. This technique is characterized by the elimination of spatial variations of potential (and, thereby, propagation) and the maintenance of constant temporal membrane potential so that (except for an initial capacitive surge) only an ionic current flows across the membrane. It is the purpose of this section to examine the degree of success with which several voltage-clamp techniques achieve the aforementioned desired properties.

[1] J. F. Lettvin, W. F. Pickard, W. S. McCulloch, and W. Pitts, A Theory of Passive Ion Flux through Axon Membrane, *Nature*, **202**:1338 (1964).

[2] Tetrodotoxin: Comments on Effects on Squid Axons, *Science*, **157**:220 (1967).

[3] C. M. Armstrong and L. Binstock, Anomalous Rectification in the Squid Giant Axon Injected with Tetraethylammonium Chloride, *J. Gen. Physiol.*, **48**:859 (1965).

[4] T. Narahashi, J. W. Moore, and W. R. Scott, Tetrodotoxin Blockage of Sodium Conductance Increase in Lobster Giant Axons, *J. Gen. Physiol.*, **47**:966 (1964).

## TEMPORAL STABILITY

The maintenance of a constant potential, in time, is achieved through the use of a feedback control circuit. The proper functioning of this device rests on an accurate sensing of the transmembrane potential. This in turn depends on the proper location of the potential control electrodes.

Figure 4.30 is a lumped-parameter representation of the radial current path between the current control electrodes for an axon where longitudinal spatial variations are satisfactorily eliminated. The resistances depend on the surface area of the axon and are on a unit-area basis. The condition of spatial uniformity is achieved through the use of an axial current electrode of 75 microns or less in diameter. The aforementioned electrode introduces a resistance into the current flow path, mainly due to interface (surface) effects. In addition, the net resistance of the axoplasm, the membrane, the external seawater, and the external electrode are involved as shown. The current source, depicted in Fig. 4.30, is under the control of potential electrodes whose placement is quite critical; the location used by several investigators is shown. The symbol C refers to early work of Cole,[1] HHK to that of Hodgkin, Huxley, and Katz,[2] and MC to that of Moore and Cole.[3]

[1] K. S. Cole, Dynamic Electrical Characteristics of Squid Axon Membrane, *Arch. Sci. Physiol.*, **3**:253 (1949).

[2] A. L. Hodgkin, A. F. Huxley, and B. Katz, Measurement of Current-voltage Relations in the Membrane of the Giant Axon of *Loligo*, *J. Physiol.*, **116**:424 (1952).

[3] K. S. Cole and J. W. Moore, Ionic Current Measurements in the Squid Giant Axon Membrane, *J. Gen. Physiol.*, **44**:123 (1960).

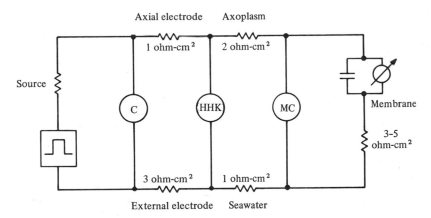

**Fig. 4.30**   Resistance paths between membrane and current source. (*From J. W. Moore and K. S. Cole, in "Physical Techniques in Biological Research," vol.* VI, *Academic Press Inc., New York, 1963.*)

Cole[1] has shown that the peak inward membrane current under carefully controlled conditions may reach 5 ma/cm² or more (rather than 1 ma/cm² or so, as measured by Hodgkin and Huxley).  Consequently, control of such current by means of HHK electrodes could result in a voltage error of as much as 35 mv in view of the (approximately) 7 ohm-cm² effective series resistance beyond the membrane.  In this case the assumption of constant membrane potential is very seriously in error and, indeed, would be of considerable consequence even for 1 ma/cm² peak current.

An improvement in voltage control is achieved by Cole and Moore in that the potential electrodes are micropipettes placed on either side of the membrane itself.  From Fig. 4.30, it is seen that the series resistance, as present in the HHK measurement, is greatly reduced.  The residual membrane resistance of 3 to 5 ohms cannot be readily compensated for by the feedback circuit.  However, compensation can be introduced, particularly if this resistance is accurately known and is fairly constant.  On the other hand, the effect of the remaining 6 to 8 ohm-cm² series resistance can be readily controlled by the utilization of a high-gain feedback arrangement.  This can be deduced from the control circuit shown in Fig. 4.31.

In Fig. 4.31, $R_i$ is the sum of the internal resistances (axoplasm plus axial electrode), while $R_o$ is the net external resistance (seawater and external electrode).  The potential electrodes connect across the membrane and series resistance $R_s$, and the nominal transmembrane potential $V_m$ is developed at $b$ (with respect to ground).  Assuming ideal amplifiers that draw no current, then

$$\frac{-E - S}{R_1} + \frac{V_m - S}{R_1} = 0 \tag{4.75}$$

where $E$ is the desired value of clamped potential.  We also have

$$V_0 - V_m = RI \qquad (R = R_i + R_o) \tag{4.76}$$

and

$$V_0 = -\mu S \tag{4.77}$$

If $V_0$ and $S$ are eliminated, we get

$$V_m = \frac{E}{1 + 2/\mu} - \frac{RI}{1 + \mu/2} \tag{4.78}$$

which can be interpreted as an equivalent emf of $E/(1 + 2/\mu)$ and internal resistance $R/(1 + \mu/2)$ driving a "load" consisting of the membrane and $R_S$.  If $\mu > 200$, say, then $V_m$ must lie within 1 percent of the

[1] K. S. Cole, An Analysis of the Membrane Potential along a Clamped Squid Axon, *Biophys. J.*, 1:401 (1961).

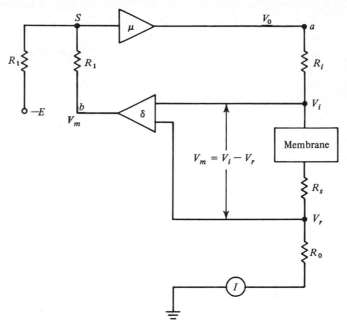

**Fig. 4.31.**   Control circuit for voltage clamp.

value $E$, and the effective resistance can be neglected (it will be less than 0.1 ohm-cm$^2$).   Thus full compensation can be effected by feedback control for the internal plus the external resistance.

An indication of the need for potential control accuracy is given by the current-voltage curve of Fig. 4.32.   When the membrane potential is in the range of $-40$ to $-20$ mv, small voltage perturbations yield significant changes in current.   For example, a change of 1 mv produces a peak current change of 0.5 ma/cm$^2$ at the reference point of 2.5 ma/cm$^2$.   Consequently, a tolerance in membrane potential of less than 1 mv is highly desirable.   As noted above, the series membrane resistance $R_S$ is more difficult to compensate for, yet it may cause an error of perhaps 15 mv (this could arise if $R_S$ had a value of, say, 3 ohm-cm$^2$ with 5 ma/cm$^2$ peak current).   Compensation for $R_S$ requires its accurate measurement, but this, in turn, depends on good control.

In connection with the curve for $I_{Na}$ in Fig. 4.32, an incremental resistance $\partial E/\partial I$ can be computed, and we note that it will be negative over the critical region where the clamp voltage lies between $-40$ and $-20$ mv.   The exact meaning of this resistance is not clear since the curve is a quasi-steady-state relationship, but its slope appears to be related to the limiting values of the equivalent external resistance

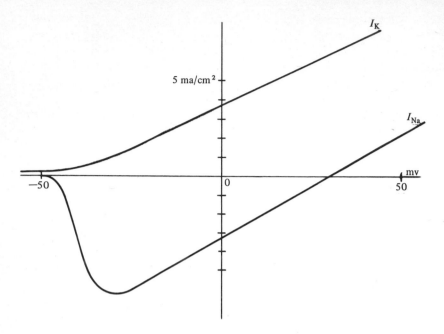

**Fig. 4.32** Current-voltage relationship for the squid axon under normal conditions.

required for stability. The value of the negative resistance is around $-2$ ohm-cm², so that the load line, (4.78), will give a unique (single) intersection if $R/(1 + \mu/2) < 2$ ohm-cm².† Such a condition is considered stable since small changes in the effective emf cause small changes in the steady-state peak inward current. For an effective resistance greater than 2 ohm-cm² three intersection points can result, one of which would be unstable. For the latter point, a disturbance in $E$ causes the system to undergo a substantial shift to one of the two stable points. With $\mu > 200$ an effective resistance of 0.1 ohm-cm² results, as noted earlier, and this clearly satisfies the stability requirements. Experiments conducted by Cole and Moore on the squid axon with point potential control show temporal variations of around 5 mv.

### SPATIAL UNIFORMITY

For a study of the spatial distribution of potential in the axon with an axial electrode,[1] the modified equivalent circuit of Fig. 4.33 is appro-

† Since the data are taken under conditions where $R_S$ is not included in feedback control, this result applies directly to the point-control condition.
[1] K. S. Cole, An Analysis of the Membrane Potential along a Clamped Squid Axon, *Biophys. J.,* **1**:401 (1961).

priate.   We consider subthreshold (electrotonic) conditions and, in comparison with earlier representations, note the inclusion of radial (surface)
resistances for both the internal and the external electrode and the
requirement that the electrodes themselves be equipotential.   Application of Kirchhoff's equations gives

$$\frac{dV_2}{dx} = -i_1 r_2 \tag{4.79}$$

$$\frac{di_1}{dx} = \frac{V_0 - V_2}{r_1} - \frac{V_2 - V_4}{r_3} \tag{4.80}$$

$$\frac{dV_4}{dx} = -i_2 r_4 \tag{4.81}$$

$$\frac{di_2}{dx} = \frac{V_2 - V_4}{r_3} - \frac{V_4}{r_5} \tag{4.82}$$

where series resistances and shunt conductances are on a per-unit-length
basis.   Let

$$\alpha = \frac{r_2}{r_1} \qquad \beta_1 = \frac{r_2}{r_3} \qquad \beta_2 = \frac{r_4}{r_3} \qquad \gamma = \frac{r_4}{r_5} \tag{4.83}$$

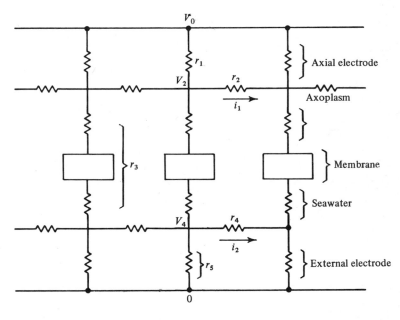

**Fig. 4.33**   Equivalent circuit for an axon, which includes axial and radial
resistance elements.   Pictured is the region for which space-clamped conditions are desired.

Then, from (4.79) to (4.82), we obtain

$$\frac{d^2V_2}{dx^2} = -\alpha(V_0 - V_2) + \beta_1(V_2 - V_4) \tag{4.84a}$$

$$\frac{d^2V_4}{dx^2} = -\beta_2(V_2 - V_4) + \gamma V_4 \tag{4.84b}$$

The solution of (4.84) for $V_m(x) = V_2(x) - V_4(x)$ is

$$V_m(x) = Ae^{-x/\lambda_1} + Be^{-x/\lambda_2} + V_2(\infty) - V_4(\infty) \tag{4.85}$$

where

$$\frac{1}{\lambda_1{}^2} = \frac{\alpha + \beta_1 + \beta_2 + \gamma + D}{2} \qquad \frac{1}{\lambda_2{}^2} = \frac{\alpha + \beta_1 + \beta_2 + \gamma - D}{2}$$

and

$$D^2 = (\alpha + \beta_1 + \beta_2 + \gamma)^2 - 4(\alpha + \beta_1)(\beta_1 + \gamma) + 4\beta_1\beta_2$$

The values of $\lambda_1$ and $\lambda_2$ are the constants for the space-clamped axon and are a measure of the extent over which passive spatial variations could occur. This might be thought of as establishing the most rapid change in potential that could occur if the activation were nonuniform. Typical values for the space constants are in the neighborhood of 0.5 mm. Consequently, it may be assumed that regions which are, say, 1 mm or more from the control point are not necessarily well controlled. In the latter case, potential and current uniformity must rely entirely on physical uniformity of axon and electrodes. As a rough analysis of this factor, we set down the following relationship between the control potential $V_0$ and the membrane potential $V_m$ which is in the form

$$V_0 = V_m + RI_m \tag{4.86}$$

where $R$ is the resistance between both electrodes and the membrane in ohm-centimeters, and $I_m$ is the membrane current in milliamperes per centimeter. The above equation can be written as

$$V_0 = V_m + 2\pi a J_m R \tag{4.87}$$

where $a$ is the axon radius and $J_m$ is the membrane current density. We see that for small variations

$$\delta V_0 = \delta V_m + 2\pi Ra J_m \left( \frac{\delta R}{R} + \frac{\delta a}{a} + \frac{\delta J_m}{J_m} \right) = 0 \tag{4.88}$$

since $V_0$ is a constant. Thus

$$\delta V_m = 35 \left( \frac{\delta R}{R} + \frac{\delta a}{a} + \frac{\delta J_m}{J_m} \right) \tag{4.89}$$

for typical values of $R = 45$ ohm-cm, $a = 0.025$ cm, and $J_m = 5$ ma/cm$^2$. Variations in radius were found to be as much as 20 percent while in some instances $J_m$ had fluctuations (particularly for operation in the negative-resistance region) of $2:1$. For this region, earlier measurements of current using 5-mm-length axons were discarded and replaced by experiments utilizing axons of 1.5 mm length.

The conclusions from analysis along the lines indicated in this section seem to be that[1]

1. The membrane potential can be well controlled at the control point.
2. Considerable variation of membrane potential away from this point is expected.
3. Reasonably accurate current measurements can be made within a few tenths of a millimeter of the control point.

Several of the disadvantages in the voltage-clamp technique discussed in this section can be minimized by using sucrose-gap technique.[2] This method can be described as essentially creating an artificial node. Since it does not require an axial wire, it is much simpler to implement. Furthermore, by confining operation to a short segment of membrane it minimizes axial variations.

### 4.13  MODELS FOR SQUID AXON

An analysis of the Hodgkin-Huxley equations by Fitzhugh[3] has served to provide a broader point of view in which to characterize nerve membrane. His methods utilize phase-plane techniques which permit summarization of the physiological states of the nerve fiber (namely, resting, active, refractory, enhanced, depressed, etc.). In particular, a mathematical model is developed, based on a modification of Van der Pol's equation, which consists of a pair of nonlinear differential equations with either a stable singular or a limit cycle. In this scheme, the Hodgkin-Huxley model is viewed as a member of a class of nonlinear systems which show excitable and oscillatory behavior.

An alternative mathematical model to that developed by Hodgkin and Huxley has been proposed by Hoyt.[4] In this work, emphasis is

---

[1] *Ibid.*

[2] F. J. Julian, J. W. Moore, and D. E. Goldman, Membrane Potentials of the Lobster Giant Axon Obtained by Use of the Sucrose Gap Technique, *J. Gen. Physiol.*, **45**:1195 (1962).

[3] R. Fitzhugh, Mathematical Models of Threshold Phenomena in the Nerve Membrane, *Bull. Math. Biophys.*, **17**:257 (1955); Impulses and Physiological States in Theoretical Models of Nerve Membrane, *Biophys. J.*, **1**:445 (1961).

[4] R. Hoyt, The Squid Axon: Mathematical Models, *Biophys. J.* **3**:399 (1963).

given to the physical interpretation of the mathematical formulation. Only variables $m$ and $n$ are utilized, but the parameter $m$ (related to sodium conductance) satisfies a second-order differential equation, while $n$ (as in H-H) satisfies a first-order equation. In this model, the potassium and sodium conductances are specified as functions of $m$ and $n$, rather than by the power relationship of Hodgkin and Huxley. Good fit to clamp data and prediction of action potentials are possible, while at the same time a physical model which differs significantly from H-H is involved.

Another alternative model was developed by Agin.[1] Here the rate constants $\alpha$ and $\beta$ are related to averaging functions, so that by making certain probabilistic assumptions about an abstract physical model the form of the Hodgkin-Huxley equations may be derived.

In addition, one should note the general critique provided by Wei.[2] A recent review by Noble[3] with an extensive biblography may also be useful.

## 4.14 OTHER APPLICATIONS

Frankenhaeuser[4] and others have succeeded in applying the voltage-clamp technique to frog myelinated nerve. This has permitted setting up descriptions of the behavior of the excitable membrane at a single node of Ranvier, just as was done for the space-clamped squid axon. Since the current-voltage curves are much more nonlinear than for the squid axon, it was decided not to base the mathematical formulation on $g_{Na}$ but rather on a parameter $P_{Na}$ of the constant-field theory. Furthermore, $P_{Na}$ was found to best fit the following relation:

$$P_{Na} = P_{Na}m^2h$$

where $m$ and $h$ have the same meaning as in the Hodgkin-Huxley theory. The potassium permeability was similarly determined from

$$P_K = P_K n^2 k$$

[1] D. Agin, Some Comments on the Hodgkin-Huxley Equations, *J. Theoret. Biol.*, **5**:161 (1963); Hodgkin-Huxley Equations: Logarithmic Relation between Membrane Current and Frequency of Repetitive Activity, *Nature*, **201**:625 (1964).

[2] L. Y. Wei, A New Theory of Nerve Conduction, *IEEE Spectrum*, **3**:123 (1966).

[3] D. Noble, Application of Hodgkin-Huxley Equations to Excitable Tissues, *Phys. Rev.*, **46**:1 (1966).

[4] B. Frankenhaeuser and A. Persson, Voltage Clamp Experiments on the Myelinated Nerve Fiber, *Acta Physiol. Scand. Suppl.* 42, **145**:45 (1957); B. Frankenhaeuser, Computed Action Potentials in Nerve from *Xenopus laevis*, *J. Physiol. (London)*, **180**:780 (1965).

where $k$ is an inactivation parameter for potassium which is normally negligible (i.e., constant) over periods of time of interest (i.e., for $\Delta t <$ 200 msec).

Two modified Hodgkin-Huxley theories have also been developed for cardiac muscle, one to apply to mammalian Purkinje fibers[1] and the other to frog ventricular fibers.[2]

## BIBLIOGRAPHY

Bures, J., M. Petran, and J. Zachar: "Electrophysiological Methods in Biological Research," Academic Press Inc., New York, 1967.

Hodgkin, A. L.: "The Conduction of the Nervous Impulse," Charles C Thomas, Publisher, Springfield, Ill., 1964.

Katz, B.: "Nerve, Muscle, and Synapse," McGraw-Hill Book Company, New York, 1966.

Newer Properties of Perfused Squid Axons, *Conf. J. Gen. Physiol.*, **48**(5) (May, 1965).

Ochs, S.: "Elements of Neurophysiology," John Wiley & Sons, Inc., New York, 1965.

Schoffeniels, E.: "Cellular Aspects of Membrane Permeability," Pergamon Press, New York, 1967.

Stevens, C. F.: "Neurophysiology: A Primer," John Wiley & Sons, Inc., New York, 1966.

Symposium on Physical and Mathematical Approaches to the Study of the Electrical Behavior of Excitable Membranes, *J. Cellular Comp. Physiol.*, **66**(3) (December, 1965).

[1] D. Noble, A Modification of the Hodgkin-Huxley Equations Applicable to Purkinje Fiber Action and Pacemaker Potentials, *J. Physiol.*, **160**:317 (1962).

[2] A. J. Brody and J. W. Woodbury, Effects of Sodium and Potassium on Repolarization in Frog Ventricular Fibers, *Ann. N.Y. Acad. Sci.*, **65**:687 (1957).

# 5
# Volume-conductor Fields

## 5.1 INTRODUCTION

In Sec. 4.8, the mechanism of nerve propagation was described as resulting from the depolarizing effect of action currents on the surrounding resting membrane.  Through this mechanism the site of depolarization is capable of spreading contiguously until all parts of an excitable fiber, and in some cases adjacent fibers, have been activated.

While the current density arising from membrane activity is strongest in the immediate vicinity of the active region, there will be current everywhere in the surrounding medium.  We call the latter region the *volume conductor*, and designate thereby the aggregate of passive tissue which supports current from active sources.  Electrodes placed within such a volume conductor measure the net electric potential field (related to the current flow field) which is produced by the summation of all action sources.  Electrical activity in the brain, for example, can be detected with macroscopic electrodes implanted subcortically, or can even be measured with surface electrodes.

It is highly desirable to be able to relate the measured potential differences that characterize a current flow field to the sources of that field.   This is important for the elucidation of the bioelectric phenomena under study and also for clinical applications, as in electrocardiography, electroencephalography, electromyography, and others.   The object of this chapter is to develop mathematical-analytical models of the current flow field in a volume conductor that results from the activity of excitable tissue within.

## 5.2  QUASI-STATIC FORMULATION[1]

In order to describe, mathematically, the potential field or the current flow field in a volume conductor, due to distributed time-varying bioelectric sources, we shall have to utilize pertinent electromagnetic theory. We shall formulate the problem rigorously at first and then go on to show that because of the electrical properties of physiological systems, a considerably simplified representation can be obtained.   The final conclusion is that although the sources are time-varying, one may proceed as if steady-state conditions existed at any instant.   The problem is, consequently, a quasi-static one, and Laplace's equation applies.

We start with a general formulation for the electric field in an infinite volume conductor due to an applied (impressed) current density $\mathbf{J}_i$ whose temporal behavior is harmonic at an angular frequency $\omega$. The medium is assumed to be linear, homogeneous, isotropic, and characterized by physical parameters $\mu$, $\sigma$, and $\epsilon$.   To the degree that the medium possesses linearity (which we assume) the analysis can be applied equally well to a linear combination of harmonic frequencies, hence to a periodic or to an aperiodic source through the use of a Fourier series or integral, respectively.

The electric and magnetic fields are found by solution of the inhomogeneous Helmholtz equations.   However, it is more advantageous to first find expressions for the scalar and vector potentials and then construct the $\mathbf{E}$ and $\mathbf{H}$ fields from them.   The scalar potential $\Phi$ and the vector potential $\mathbf{A}$ are given by[2]

$$\mathbf{A}(x',y',z') = \frac{\mu}{4\pi} \int_V \frac{\mathbf{J}_i(x,y,z)e^{-jkR}}{R}\, dV \tag{5.1}$$

$$\Phi(x',y',z') = \frac{1}{4\pi(\sigma + j\omega\epsilon)} \int_V \frac{I_v(x,y,z)e^{-jkR}}{R}\, dV \tag{5.2}$$

---

[1] The material of this section is taken from R. Plonsey and D. Heppner, Considerations of Quasi-stationarity in Electrophysiological Systems, *Bull. Math. Biophys.*, **29**:657 (1967).

[2] R. Plonsey and R. Collin, "Principles and Applications of Electromagnetic Fields," McGraw-Hill Book Company, New York, 1961.

where

$$R^2 = (x - x')^2 + (y - y')^2 + (z - z')^2 \tag{5.3}$$

the unprimed variables referring to the source point, the primed variables to the field point, and $\nabla \cdot \mathbf{J}_i = -I_v$. (This last expression constitutes a definition of $I_v$ as a volume current source density.) We also have

$$k^2 = \omega^2 \mu \left( \epsilon + \frac{\sigma}{j\omega} \right) = -j\omega\mu\sigma \left( 1 + \frac{j\omega\epsilon}{\sigma} \right) \tag{5.4}$$

We can write (5.4) in the following form:

$$k^2 = -j\omega\mu\sigma_c \tag{5.5}$$

where $\sigma_c$ is a complex conductivity that includes pure conductive effects, dielectric losses, and dielectric displacement; mathematically,

$$\sigma_c = \sigma \left( 1 + \frac{j\omega\epsilon}{\sigma} \right)$$

Finally, the electric field is found from the vector and scalar potentials $\mathbf{A}$ and $\Phi$ by

$$\mathbf{E} = j\omega\mathbf{A} - \nabla\Phi \tag{5.6}$$

It should be noted that complex phasor notation is used. That is, the quantities $\mathbf{A}$, $\mathbf{E}$, $\Phi$, $I_v$, $\mathbf{J}_i$, and $\sigma_c$ are complex and are functions of spatial coordinates and $\omega$. To obtain physical quantities, one multiplies by $e^{j\omega t}$ and considers only the real (or imaginary) parts.

The usual implications of quasi-stationarity involve assumptions which simplify the above equations. These assumptions will be enumerated in the following four main headings: Capacitive Effects, Propagation Effects, Inductive Effects, and Boundary Conditions. To evaluate the error involved in making these simplifications, it will be necessary to utilize representative data for the parameters appearing in (5.1) to (5.6), and this will be considered first.

### ELECTRICAL PROPERTIES OF BIOLOGICAL MATERIALS

Typical values for the conductivity of biological materials as reported by Rush[1] et al. are given in Table 5.1. In computations involving conductivity, we shall choose $\sigma = 0.2$ mho/m as representing a mean value. Conductivities of other biological tissues may be found in the compendium of Geddes and Baker.[2]

[1] S. Rush, J. A. Abildskov, and R. McFee, Resistivity of Body Tissues at Low Frequencies, *Circulation Res.*, **12**:40 (1963).

[2] L. A. Geddes and L. E. Baker, The Specific Resistance of Biological Material— A Compendium of Data for the Biomedical Engineer and Physiologist, *Med. Biol. Eng.*, **5**:271 (1967).

**Table 5.1  Conductivity of biological tissues**

| Tissue | Mean conductivity, mhos/m |
|---|---|
| Blood | 0.67 |
| Lung | 0.05 |
| Liver | 0.14 |
| Fat | 0.04 |
| Human trunk | 0.21 |

It is also required that the complex conductivity $(1 + j\omega\epsilon/\sigma)$ be considered. This, in turn, involves utilization of typical ratios of displacement to conduction current, that is, $j\omega\epsilon/\sigma$. Table 5.2 lists representative values as reported by Schwan and Kay.[1] Further discussion is given in the following sections.

The highest component frequency of significance in bioelectric systems is of the order of 1 kHz. This probably relates to a rise time for action potentials in the order of 1 msec. We shall choose 1 kHz in the following numerical computations requiring a maximum value of frequency.

Finally, we note the absence of magnetic materials in biological systems, so that the permeability is the free-space value of $4\pi \times 10^{-7}$ henry/m. The maximum value of $R$ corresponds to an overall dimension of typical physiological systems; for simplicity $R_{max} = 1$ m, and this should be a conservative estimate for most conditions.

## CAPACITIVE EFFECTS

The nature of the conductive medium is described by its conductivity and dielectric permittivity. The mathematical expression is seen in

[1] H. P. Schwan and C. F. Kay, The Conductivity of Living Tissues, *Ann. N.Y. Acad. Sci.*, **65**:1007 (1957).

**Table 5.2  Averages of ratio of capacitive to resistive currents for various frequencies and body tissues**

| | 10 Hz | 100 Hz | 1,000 Hz | 10,000 Hz |
|---|---|---|---|---|
| Lung | 0.15 | 0.025 | 0.05 | 0.14 |
| Fatty tissue | | 0.01 | 0.03 | 0.15 |
| Liver | 0.20 | 0.035 | 0.06 | 0.20 |
| Heart muscle | 0.10 | 0.04 | 0.15 | 0.32 |

Eq. (5.2), where the coefficient $(\sigma + j\omega\epsilon)$ can be written as $\sigma(1 + j\omega\epsilon/\sigma)$ and the conductivity viewed as a complex phasor quantity. In the quasi-static approximation, a purely resistive medium is required. This is necessary in order to justify the assumption that with arbitrary time variations the field quantities at all points are in synchrony. The quantity $(1 + j\omega\epsilon/\sigma)$ will be essentially real so long as $|j\omega\epsilon/\sigma| \ll 1$. Consultation of Table 5.2 reveals that the inequality is satisfied fairly well. The values of this ratio at 10 Hz, which do not appear to be completely negligible, are reported to be conservative, and the conclusion of Schwan and Kay is that the medium can be considered to be resistive.

## PROPAGATION EFFECTS

The time required for changes at the source to "propagate"[1] to a field point is represented by the phase delay $e^{-jkR}$ in (5.1) and (5.2). Since

$$e^{-jkR} = 1 - jkR - \frac{(kR)^2}{2!} - j\frac{(kR)^3}{3!} + \cdots \tag{5.7}$$

this effect can be ignored if $|kR| \ll 1$, since $e^{-jkR}$ is then approximately unity. In this case the integral in (5.2) is precisely that for static fields. Utilizing the data noted earlier and setting the magnitude of $(1 + j\omega\epsilon/\sigma)$ equal to the conservative value 2 yields

$$kR_{\max} = (1 - j)\sqrt{2{,}000\pi \times 4\pi \times 10^{-7} \times 0.2} = 0.0397(1 - j)$$

Thus the magnitude of $e^{-jkR}$ is unity to within a 4 percent error, while the phase-angle error of 0.0397 rad (2.3°) is clearly negligible. The numerical result is roughly the same as that reported by Geselowitz[2] in a similar analysis.

## INDUCTIVE EFFECTS

The inductive effect corresponds to the component of electric field that arises from magnetic induction. This is given, mathematically, in (5.6) by the term $j\omega\mathbf{A}$. We wish to compare the importance of this term relative to that expressed by $\nabla\Phi$. We shall do this by considering the specific case of a differential current source element and assume that if $|\omega\mathbf{A}| \ll |\nabla\Phi|$, then distributed sources such as might arise in electro-

---

[1] In view of the resistive nature of the medium a true wave propagation does not, in fact, take place. Instead, one has an electromagnetic diffusion. This is revealed in that $k$ is complex with equal real and imaginary parts. Thus both a phase change and attenuation result.

[2] D. B. Geselowitz, The Concept of an Equivalent Cardiac Generator, in "Biomedical Sciences Instrumentation," vol. 1, p. 325, Plenum Press, New York, 1963.

physiology and which are the superposition of such elements would also satisfy the inequality.[1]

Thus, for a source element $\mathbf{J}_i\, dV$ we have, from (5.1),

$$\mathbf{A} = \frac{\mu}{4\pi}\frac{\mathbf{J}_i\, dV}{R} \tag{5.8}$$

Now from (5.1) and (5.2), since $\Phi$ and $\mathbf{A}$ must satisfy the Lorentz condition, namely,

$$\nabla' \cdot \mathbf{A} = -j\omega\epsilon\left(1 + \frac{\sigma}{j\omega\epsilon}\right)\mu\Phi \tag{5.9}$$

where $\nabla' \equiv \mathbf{a}_x\, \partial/\partial x' + \mathbf{a}_y\, \partial/\partial y' + \mathbf{a}_z\, \partial/\partial z'$, we have

$$\Phi = \frac{\mu}{j4\pi\omega\epsilon_c\mu}\nabla' \cdot \left(\frac{\mathbf{J}_i\, dV}{R}\right)$$

$$= -\frac{dV}{j4\pi\omega\epsilon_c}\mathbf{J}_i \cdot \nabla\left(\frac{1}{R}\right) = \frac{dV}{j4\pi\omega\epsilon_c}\frac{\mathbf{J}_i \cdot \mathbf{a}_R}{R^2} \tag{5.10}$$

and

$$\epsilon_c = \epsilon\left(1 + \frac{\sigma}{j\omega\epsilon}\right)$$

Note that the unprimed del operator, defined by

$$\nabla = \mathbf{a}_x\frac{\partial}{\partial x} + \mathbf{a}_y\frac{\partial}{\partial y} + \mathbf{a}_z\frac{\partial}{\partial z}$$

operates on *source* coordinates only. This is the basis for the derivation of (5.10) as well as the identity

$$\nabla\left(\frac{1}{R}\right) = -\nabla'\left(\frac{1}{R}\right)$$

If we let the $z$ axis coincide with the direction of $\mathbf{J}_i$, then

$$\Phi = \frac{J_i\, dV\, \cos\theta}{j4\pi\omega\epsilon_c R^2} \tag{5.11}$$

and

$$|\nabla\Phi| = \frac{J_i\, dV}{4\pi\omega\epsilon_c R^3} \tag{5.12}$$

---

[1] Special electric devices, such as inductors, are capable of setting up quasi-static electric fields that arise almost entirely from time-varying magnetic fields. Bioelectric sources may be characterized by current double layers located at cellular membranes; special geometry which would enhance magnetic effects, such as present in a solenoid, do not arise. Consequently, the ratio $|\omega\mathbf{A}/\nabla\Phi|$ for a current element should be a satisfactory measure for typical electrophysiological distributed sources. That is, the contribution from each current element to the vector resultant of $\mathbf{A}$ or $\nabla\Phi$ should be roughly proportional to the sum.

The ratio of interest is then

$$\left|\frac{\omega \mathbf{A}}{\nabla \Phi}\right| = |\omega^2 \mu \epsilon_c R^2| = |kR|^2 \tag{5.13}$$

Thus, for the inductive effect to be negligible $|kR|^2 \ll 1$ must be satisfied. So long as propagation effects can be ignored, ($|kR| \ll 1$), this condition will automatically be met. For the numerical values used previously, we obtain

$$|kR_{\mathrm{max}}|^2 = 0.0032 \tag{5.14}$$

**BOUNDARY CONDITIONS**

Since the total current (conduction plus displacement) is solenoidal,[1] the normal component at the interface between two media must be continuous. At the boundary between different tissues, since the displacement current can be ignored (see Table 5.2), the rigorous condition that

$$\sigma_1 \left(1 + \frac{j\omega\epsilon_1}{\sigma_1}\right) E_{1n} = \sigma_2 \left(1 + \frac{j\omega\epsilon_2}{\sigma_2}\right) E_{2n} \tag{5.15}$$

reduces to

$$\sigma_1 E_{1n} = \sigma_2 E_{2n} \tag{5.16}$$

where $\sigma_1$ and $\sigma_2$ are the conductivities of regions 1 and 2, respectively, and $E_{1n}$ and $E_{2n}$ the respective normal electric fields. The boundary condition expressed by (5.16) is the same as that for stationary (dc) conditions.

In many problems, however, one of the regions has zero conductivity, as, for example, the space which surrounds the human body in the consideration of the electrocardiographic system. Thus, $\sigma_2 = 0$ results in $E_{1n} = 0$ according to (5.16); however, the rigorous formulation of (5.15) shows that

$$\sigma_1 \left(1 + \frac{j\omega\epsilon_1}{\sigma_1}\right) E_{1n} = j\omega\epsilon_2 E_{2n} \qquad (\sigma_2 = 0) \tag{5.17}$$

It seems reasonable to suppose that $E_{1n} \approx 0$ provided that $(\omega\epsilon_2/\sigma_1) \ll 1$. Since $\epsilon_2$ is actually a free-space dielectric constant, we utilize $\epsilon_0 = 9 \times 10^{-12}$ farad/m and, with $\omega = 2{,}000\pi$ and $\sigma_1 = 0.2$ mho/m, $\omega\epsilon_0/\sigma_1 = 3 \times 10^{-7}$, which is clearly negligible.

The meaning of the above criterion may be clarified by noting that boundary conditions serve to take into account secondary surface sources

---

[1] A fundamental property of the magnetic field $\mathbf{H}$, according to Maxwell's equation, is that $\nabla \times \mathbf{H} = \mathbf{J}$, where $\mathbf{J}$ is the total current (conduction plus displacement plus applied). Consequently, $\nabla \cdot (\nabla \times \mathbf{H}) = 0 = \nabla \cdot \mathbf{J}$, so that $\mathbf{J}$ is solenoidal.

(at the boundary). The field in the conductive medium can be thought of as containing a component which arises from the primary sources $E_0$ and a portion due to the aforementioned secondary sources $E_s$; the total field in the external medium is designated $E_e$. Application of (5.15) then gives

$$\sigma_1\left(1 + \frac{j\omega\epsilon_1}{\sigma_1}\right)E_{0n} + \sigma_1\left(1 + \frac{j\omega\epsilon_1}{\sigma_1}\right)E_{sn} = j\omega\epsilon_2 E_{en} \tag{5.18}$$

Now $E_{0n}$, $E_{sn}$, and $E_{en}$ are in the same order of magnitude, so that if $|j\omega\epsilon_2/\sigma_1| \ll 1$,

$$\sigma_1\left(1 + \frac{j\omega\epsilon_1}{\sigma_1}\right)E_{0n} = -\sigma_1\left(1 + \frac{j\omega\epsilon_1}{\sigma_1}\right)E_{sn} \tag{5.19}$$

i.e., the total normal current due to the applied field is equal and opposite that due to the secondary field. Depending on the method used, (5.19) leads either to an integral-equation formulation for the secondary field or to a condition for determination of series coefficients in a separable coordinate system. Equation (5.19) is equivalent to the requirement

$$(E_0 + E_s)_n = 0 \tag{5.20}$$

which corresponds to the stationary condition ($E_{1n} = 0$).

### SUMMARY

We summarize below the criteria for making the associated simplifications.

| Condition | Criteria | |
|---|---|---|
| Neglect propagation effects | $kR_{\max} \ll 1$ | (5.21) |
| Neglect capacitance effects | $\omega\epsilon/\sigma \ll 1$ | (5.22) |
| Neglect inductive effects | $(kR)^2 \ll 1$ | (5.23) |
| Set $E_{1n} = 0$ ($\sigma_2 = 0$) | $\omega\epsilon_0/\sigma_1 \ll 1$ | (5.24) |

A consequence of (5.21) to (5.23) is that

$$\Phi = \frac{1}{4\pi\sigma}\int_V \frac{I_v(x,y,z)}{R}\,dV \tag{5.25}$$

$$E = -\nabla\Phi \tag{5.26}$$

The total current $J$ is the sum of the source current $J_i$ and (since the displacement current is negligible) the conduction current $\sigma E$; consequently,

$$J = J_i + \sigma E \tag{5.27}$$

Since $\mathbf{J}$ is solenoidal,

$$\nabla \cdot \mathbf{J} = 0 = \nabla \cdot \mathbf{J}_i + \nabla \cdot \sigma \mathbf{E} \tag{5.28}$$

and for homogeneous media we have, utilizing (5.26) and (5.28),

$$\nabla \cdot \mathbf{J}_i = \sigma \nabla^2 \Phi \tag{5.29}$$

Taking the laplacian of (5.25), we get

$$\nabla^2 \Phi = -\frac{I_v}{\sigma} \tag{5.30}$$

which, as noted earlier, identifies $I_v$ $(I_v = -\nabla \cdot \mathbf{J}_i)$.† Equation (5.30) represents the conventional, basic, quasi-static mathematical formulation of the volume-conductor problem, subject to (5.19) and/or (5.24) as boundary conditions.

If conditions (5.21) to (5.24) are satisfied, as is expected under normal conditions, then all field quantities arising from a given source element will have the same temporal behavior as that element, i.e., will be in synchrony. This can be shown formally by taking the Fourier transform of any field quantity (at any point) to obtain its temporal behavior. In all cases, it will be observed that the frequency-dependent term is the same, namely, $I_v(x,y,z,\omega)\,dV$, and consequently the temporal dependence is everywhere the same. This means that one can view the problem at any instant of time as if steady-state conditions were in effect corresponding to a stationary source $I_v(x,y,z,t_0)\,dV$ or $\mathbf{J}_i(x,y,z,t_0)\,dV$. That is, from (5.25) we have

$$\Phi(x',y',z',\omega) = \frac{1}{4\pi\sigma} \int_V \frac{I_v(x,y,z,\omega)}{R}\,dV$$

so that if the Fourier transform of both sides is taken, one obtains

$$\Phi(x',y',z',t_0) = \frac{1}{4\pi\sigma} \int_V \frac{1}{R}\left[\int_{-\infty}^{\infty} I_v(x,y,z,\omega)e^{-j\omega t_0}\,d\omega\right]dV$$

$$= \frac{1}{4\pi\sigma} \int_V \frac{I_v(x,y,z,t_0)}{R}\,dV$$

This result demonstrates that (5.25) to (5.30) can be viewed as denoting instantaneous quantities, rather than complex phasors.

The weakest condition in the set of equations (5.21) to (5.24) is that $\omega\epsilon/\sigma \ll 1$. If the remaining criteria are satisfied, one can still

---

† In deriving (5.29) the only assumption made is that the vector-potential contribution to $\mathbf{E}$ can be neglected. Since (5.25) is a solution to (5.29), this demonstrates that a sufficient condition for propagation effects to be negligible is that inductive contributions are negligible. The converse is also true, as noted earlier, except for special source geometry.

proceed by formally solving Laplace's equation but replacing $\sigma$ everywhere by the complex phasor $\sigma_c$, where, as noted,

$$\sigma_c = \sigma \left( 1 + \frac{j\omega\epsilon}{\sigma} \right)$$

Under these conditions (5.15) continues to be satisfied at an arbitrary interface and (5.20) at an outer boundary. In this case, the temporal behavior of the field does not coincide, necessarily, with that of a source element. The actual Fourier transforms must be taken if this approach is followed, since both $I_v$ and $\sigma_c$ are frequency-dependent, and dispersive effects must be anticipated.

Although homogeneous conditions have been assumed in this section, the results are readily generalized to a region consisting of phase inhomogeneities. The details may be found in the paper by Plonsey and Heppner.[1]

In the material that follows, we shall utilize Poisson's equation as stated in.(5.30) and shall derive the electric field from (5.26). Furthermore, from (5.29) and (5.30), the volume source density may also be expressed in terms of the impressed current by

$$\nabla \cdot \mathbf{J}_i = -I_v \tag{5.31}$$

And finally, the current flow field will be related to the potential field by virtue of (5.27).

## 5.3 BIOELECTRIC SOURCES

In Sec. 5.2 the sources for the fields in a volume conductor were specified either by a current density, as in (5.1), or by the divergence of a current density, as in (5.2) [by taking note of (5.31)]. We wish to examine the nature of these and other measures of bioelectric sources in this section. These field theoretical descriptions are very useful, and we shall wish to relate them to actual electrophysiological phenomena and their measurement.

Using the quasi-static field formulation developed in Sec. 5.2, we consider the instantaneous current density as consisting of ohmic and nonohmic components, as stated in (5.27). With this decomposition, the total current density $\mathbf{J}$ is given by

$$\mathbf{J} = \sigma\mathbf{E} + \mathbf{J}_i \tag{5.32}$$

where $\mathbf{J}_i$ is the nonohmic part. The term $\sigma\mathbf{E}$ is a differential expression of Ohm's law and evaluates the current density that arises in a passive

[1] *Loc. cit.*

conducting medium of conductivity $\sigma$ because of the presence of an electric field **E**. The electric field, and the current $\sigma\mathbf{E}$ associated with it, are a response to the *primary sources* represented by $\mathbf{J}_i$. This latter component designates the "impressed" current density and is a consequence of the conversion of energy (chemical, mechanical, thermal, etc.) to an electric form.

As an illustration of the interpretation of the terms in (5.32), consider an idealized infinite axon immersed in a uniform conducting medium of infinite extent. In the external medium and the axoplasm $\mathbf{J}_i = 0$, while the current density is given by $\sigma\mathbf{E}$, where $\sigma$ represents the conductivity of either the internal or the external medium. On the other hand, within the membrane the current density is described by an interaction of electrical and diffusion effects as well as a capacitive-displacement current. The Nernst-Planck equation quantitatively describes the first two aspects, for we have, from (3.28), the noncapacitive current $\mathbf{J}'$ as

$$\mathbf{J}' = \sigma\mathbf{E} - \sum_j D_j F z_j \nabla C_j \tag{5.33}$$

where

$$\sigma = \frac{F^2}{RT} \sum_j D_j z_j{}^2 C_j = \sum_j F|z_j|u_j C_j \tag{5.34}$$

The diffusion current in (5.33) (the second term) is clearly a contribution to the impressed current and results from the conversion of chemical potential energy into electric energy. The conductivity which appears in (5.34) is not uniform within the membrane, as it is elsewhere in the system. The capacitive displacement current, an effect considered in (4.12), represents a storage or release from storage of electric energy.[1] The instantaneous rate of decrease in capacitive energy may be either positive or negative and in either case represents a contribution (positive or negative) to the total impressed current density. Both capacitive currents and diffusion currents constitute the basic impressed current of biological systems, and their location is confined to the membrane.

In (5.32), the current density $\mathbf{J}$, since it includes all possible con-

---

[1] This appears to violate an earlier remark that the impressed current arises from conversion from a nonelectric form. If the harmonic time-varying formulation is retained, then the medium can be described in terms of its conductivity and permittivity so that the displacement current forms part of a complex "ohmic" term denoted by $\sigma_c$ in Sec. 5.2. However, in using a conductive-medium reference and a quasi-static formulation, the displacement current must be considered as an additive quantity where significant. It is not negligible relative to the conduction current *within the membrane* because of the very low conductivity of the membrane.

tributions, must be solenoidal. Consequently, as we obtained earlier in (5.28),

$$\mathbf{\nabla} \cdot \mathbf{J} = 0 = \mathbf{\nabla} \cdot (\sigma\mathbf{E}) + \mathbf{\nabla} \cdot \mathbf{J}_i \qquad (5.35)$$

If we consider a homogeneous region of conductivity $\sigma$ in which impressed currents exist, then (5.29) applies, which, slightly rearranged, is

$$\nabla^2\Phi = \frac{\mathbf{\nabla} \cdot \mathbf{J}_i}{\sigma} \qquad (5.36)$$

and is an expression of Poisson's equation. The role of $\mathbf{J}_i$ as a source function is clearly seen in (5.36), which is a partial differential equation for $\Phi$ with $(\mathbf{\nabla} \cdot \mathbf{J}_i)/\sigma$ as a nonhomogeneous (source) term.

In the study of electrostatic fields, the potential due to a charge density $\rho(x,y,z)$ in a uniform dielectric medium of permittivity $\epsilon$ is

$$\Phi(x',y',z') = \frac{1}{4\pi\epsilon} \int_V \frac{\rho \, dV}{r} \qquad (5.37)$$

where $(x',y',z')$ locates the field point and

$$r = \sqrt{(x - x')^2 + (y - y')^2 + (z - z')^2} \qquad (5.38)$$

is the distance from the source point $(x,y,z)$ to the field point. But $\Phi(x,y,z)$ also satisfies Poisson's equation, namely,

$$\nabla^2\Phi = -\frac{\rho}{\epsilon} \qquad (5.39)$$

If (5.36) is compared with (5.39), one notes that the equations are interchangeable if $\epsilon$ is replaced by $\sigma$, and $\rho$ by $-\mathbf{\nabla} \cdot \mathbf{J}_i$. If this duality is applied to (5.37), then one obtains

$$\Phi(x',y',z') = \frac{1}{4\pi\sigma} \int_V \frac{-\mathbf{\nabla} \cdot \mathbf{J}_i}{r} \, dV \qquad (5.40)$$

as a solution to (5.36). We note from (5.40), as well as the duality of $-\mathbf{\nabla} \cdot \mathbf{J}_i$ with $\rho$, that $(-\mathbf{\nabla} \cdot \mathbf{J}_i)$ plays the role of a divergence-type source density. Utilizing the designation of (5.31), we have

$$-\mathbf{\nabla} \cdot \mathbf{J}_i = I_v \qquad (5.41)$$

where $I_v$ has the dimensions of current per unit volume. With this notation

$$\Phi(x',y',z') = \frac{1}{4\pi\sigma} \int \frac{I_v}{r} \, dV \qquad (5.42)$$

That (5.40) is a solution to (5.36) can be demonstrated by considering first a point source $\delta$, where $\delta = 0$ except at the origin, where it is infinite, but $\int_V \delta dV = 1$.   Designate by $\Phi_0$ the solution to the following partial differential equation:

$$\nabla^2 \Phi_0 = \frac{\delta}{\sigma} \tag{5.43}$$

Because of the spherical symmetry $\Phi_0$ depends only on $r$ and writing the laplacian in spherical coordinates, we get

$$\frac{1}{r^2} \frac{d}{dr} \left( r^2 \frac{d\Phi_0}{dr} \right) = \frac{\delta}{\sigma} \tag{5.44}$$

If we integrate with respect to $r$, then

$$r^2 \frac{d\Phi_0}{dr} = \frac{1}{\sigma} \int \delta r^2 \, dr$$

$$= \frac{1}{4\pi\sigma} \int 4\pi r^2 \, \delta dr = \frac{1}{4\pi\sigma} \int \delta dV = \frac{1}{4\pi\sigma} \tag{5.45}$$

A second integration yields

$$\Phi_0 = -\frac{1}{4\pi\sigma r} \tag{5.46}$$

Note that in the notation used here, we have $r = \sqrt{x'^2 + y'^2 + z'^2}$, and $(x',y',z')$ is the point at which $\Phi_0$ is evaluated, the delta source being at $(0,0,0)$.

Now for a source $(\nabla \cdot \mathbf{J}_i/\sigma)$, we can consider it the superposition of delta sources typified by $(\nabla \cdot \mathbf{J}_i/\sigma) \, dV$ at $(x,y,z)$ whose contribution to the potential field $d\Phi_0(x',y',z')$ at $(x',y',z')$ is given by [using (5.46) with origin at $(x,y,z)$ and field point at $(x',y',z')$]

$$d\Phi_0(x',y',z') = -\frac{(\nabla \cdot \mathbf{J}_i) \, dV}{4\pi\sigma r} \tag{5.47}$$

and $r = \sqrt{(x' - x)^2 + (y' - y)^2 + (z' - z)^2}$.   Integration   of   (5.47) gives (5.40) as required.

It often proves convenient to represent the bioelectric sources as a volume-source-density function $I_v$.   On this basis, the potential field is found as a solution to Poisson's equation

$$\nabla^2 \Phi = -\frac{I_v}{\sigma} \tag{5.48}$$

On the other hand, representation of the sources as an impressed current density $\mathbf{J}_i$ permits an identification with physical processes in the membrane, as suggested by (5.33).[1]  Still another interpretation of $\mathbf{J}_i$ is obtained through the following vector identity:

$$\nabla \cdot \left(\frac{\mathbf{J}_i}{r}\right) = \frac{1}{r}\,\nabla \cdot \mathbf{J}_i + \mathbf{J}_i \cdot \nabla\left(\frac{1}{r}\right) \tag{5.49}$$

If the integral of both sides of (5.49) is taken, which extends throughout the total volume in which sources are located, and if the divergence theorem is used, then

$$\int_S \frac{\mathbf{J}_i \cdot d\mathbf{S}}{r} = 0 = \int_V \frac{\nabla \cdot \mathbf{J}_i}{r}\,dV + \int_V \mathbf{J}_i \cdot \nabla\left(\frac{1}{r}\right) dV \tag{5.50}$$

The surface integral evaluates to zero since $\mathbf{J}_i = 0$ over the bounding surface which lies beyond all sources.  Consequently,

$$\int_V -\frac{\nabla \cdot \mathbf{J}_i}{r}\,dV = \int_V \mathbf{J}_i \cdot \nabla\left(\frac{1}{r}\right) dV \tag{5.51}$$

where $\nabla(1/r) = -(1/r^2)\mathbf{a}_r$, and $\mathbf{a}_r$ is a unit vector from the field to the source point.  Substituting (5.51) into (5.40) gives

$$\Phi(x',y',z') = \frac{1}{4\pi\sigma} \int_V \mathbf{J}_i\,dV \cdot \nabla\left(\frac{1}{r}\right) \tag{5.52}$$

Now the potential field of a dipole source of strength $\mathbf{p}$ is[2]

$$\Phi(P) = \frac{1}{4\pi\sigma}\,\mathbf{p} \cdot \nabla\left(\frac{1}{r}\right) \tag{5.53}$$

---

[1] Since the membrane conduction current is relatively small, $\mathbf{J}_i$ is nearly equal to $\sigma\,\partial\Phi/\partial n$ (of the bounding medium).  In the solution of a problem consisting of, say, a cell lying in a conducting medium it is important to remember that the membrane region has a relatively low specific conductivity.  This introduces an additional complication in that the source $\mathbf{J}_i$ lies in an inhomogeneous region.  The final result is contained in (5.190), which is developed in a formal way in Sec. 5.12.  This shows that if the cytoplasmic and external conductivities are the same, the effect of the discontinuity in conductivity introduced by the membrane itself is contained in the additional (secondary) source term given by a double layer of strength $V$ (the transmembrane potential).

[2] A dipole is formed from a positive and a negative point source, each of equal magnitude (say $m$), separated by a vanishing small directed distance l such that $\mathbf{p} = \lim_{l\to 0} m\mathbf{l}$.  The quantity $\mathbf{p}$ is the dipole moment.  The field produced by the dipole can be evaluated by superposing the contribution from the component point sources.  This can be accomplished by first considering both positive *and* negative point sources at the origin; the net field is zero since component fields are equal and opposite.  Displacement of the positive source a distance l, hence forming a dipole, results in noncanceling component fields.  The resultant is precisely the *change* in the field of the positive source due to displacing it.  Utilizing (5.46), we get $\Phi = \partial/\partial l(m/4\pi\sigma r)|\mathbf{l}| = (m/4\pi\sigma)\nabla(1/r) \cdot \mathbf{l}$.

so that by comparison with (5.52) $\mathbf{J}_i \, dV$ may be interpreted as an elemental dipole source. Put another way, $\mathbf{J}_i$ has the dimensions of a dipole moment per unit volume. Since isolated point sources with fields as given in (5.46) cannot exist in electrophysiological systems, the characterization of the bioelectric sources as a dipole distribution is a very natural one. The electrostatic dual to $\mathbf{J}_i$ is, in the sense discussed here, the dielectric polarization (usually denoted $\mathbf{P}$).

## 5.4  FIELD OF SINGLE CELL[1]

Rather than evaluate the impressed current density directly from fundamental electrochemical properties, we can obtain expressions for the potential field which imply certain fictitious or *equivalent* sources. The latter are normally expressed in terms of potential and/or current distributions over specialized or arbitrary surfaces. We shall consider here a single cell immersed in a uniform conducting medium and shall obtain expressions for the potential field in terms of the potential distributions over the outer and inner membrane surfaces. We shall see that the latter constitutes an effective (or equivalent) dipole-layer source.

Figure 5.1 illustrates a single cell lying in a volume conductor. The external membrane surface is $S_o$ and the internal membrane surface is $S_i$. We let $P(x', y', z')$ be an arbitrary field point in the external medium

[1] R. Plonsey, An Extension of the Solid Angle Potential Formulation for an Active Cell, *Biophys. J.*, **5**:663 (1965).

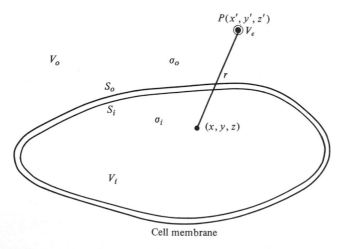

**Fig. 5.1**  Single cell in a volume conductor of infinite extent.

and $r = \sqrt{(x - x')^2 + (y - y')^2 + (z - z')^2}$ be the distance from it to an arbitrary internal point $(x,y,z)$. We now apply Green's theorem[1]

$$\int_{V_i} (\Phi\nabla^2\psi - \psi\nabla^2\Phi)\, dV = \int_{S_i} \left(\frac{\Phi\, \partial\psi}{\partial n} - \frac{\psi\, \partial\Phi}{\partial n}\right) dS \tag{5.56}$$

and choose the scalar function $\psi = 1/r$ and $\Phi$ equal to the potential field, $\Phi_i$, in $V_i$. The field $\Phi_i$ arises from sources within the membrane under either resting or active conditions. Since $r$ cannot go to zero by virute of $P$ being outside the region of integration, we have $\nabla^2(1/r) = 0$. Also since no sources exist within $V_i$ (the axoplasm is assumed to be a passive conductor), $\nabla^2\Phi_i = 0$. Consequently,

$$\int_{S_i} \Phi_i \frac{\partial(1/r)}{\partial n}\, dS = \int_{S_i} \frac{1}{r}\frac{\partial\Phi_i}{\partial n}\, dS \tag{5.57}$$

We can replace

$$-\frac{\partial\Phi_i}{\partial n}\bigg|_{S_i} = \frac{J_m}{\sigma_i} \tag{5.58}$$

where $\sigma_i$ is the conductivity in the axoplasm and $J_m$ is the outward membrane current density. From (5.57) and (5.58), we have

$$\int_{S_i} \Phi_i \frac{\partial(1/r)}{\partial n}\, dS = -\frac{1}{\sigma_i}\int_{S_i} \frac{J_m}{r}\, dS$$

If we multiply both sides by $\sigma_i/4\pi\sigma_o$, where $\sigma_o$ is the external conductivity, then we obtain

$$\frac{1}{4\pi\sigma_o}\int_{S_i} \frac{J_m}{r}\, dS = -\frac{1}{4\pi}\frac{\sigma_i}{\sigma_o}\int_{S_i} \Phi_i \frac{\partial(1/r)}{\partial n}\, dS \tag{5.59}$$

Green's theorem is now applied to the external region. In this case we choose $\Phi = \Phi_o$, where $\Phi_o$ is the external potential field due to the bioelectric sources, while $\psi = 1/r$ as before. Since there are no sources

---

[1] This is readily derived in the following way: Let vector field $\mathbf{A} = \psi\nabla\Phi$ and vector field $\mathbf{B} = \Phi\nabla\psi$, where $\Phi$ and $\psi$ are arbitrary scalar potential fields with continuous second derivatives. We use the divergence theorem to get $\int_V \nabla \cdot \mathbf{A}\, dV = \int_S \mathbf{A} \cdot d\mathbf{S}$, and a similar expression for $\mathbf{B}$. Now $\nabla \cdot \mathbf{A} = \nabla \cdot (\psi\nabla\Phi) = \psi\nabla^2\Phi + \nabla\psi \cdot \nabla\Phi$, so that, using the divergence theorem, we have

$$\int_V (\psi\nabla^2\Phi + \nabla\psi \cdot \nabla\Phi)\, dV = \int_S \psi\nabla\Phi \cdot d\mathbf{S} \tag{5.54}$$

Considering expressions for $\mathbf{B}$ gives (interchange $\Phi$ and $\psi$)

$$\int_V (\Phi\nabla^2\psi + \nabla\psi \cdot \nabla\Phi)\, dV = \int_S \Phi\nabla\psi \cdot d\mathbf{S} \tag{5.55}$$

If (5.54) is subtracted from (5.55), the result is Green's theorem (5.56).

in the external region, $\nabla^2 \Phi_o = 0$. The following result is obtained:

$$\Phi_o(P) = -\frac{1}{4\pi} \int_{S_i} \frac{1}{r} \frac{\partial \Phi_o}{\partial n} \, dS + \frac{1}{4\pi} \int_{S_o} \Phi_o \frac{\partial}{\partial n} \left(\frac{1}{r}\right) dS \tag{5.60}$$

where $\Phi_o(P)$ is the potential at the point $P(x',y',z')$ (see Fig. 5.1). The sign given in this equation assumes the positive normal to be taken *outward* from $V_i$ and corresponds to the direction given in Eq. (5.57). The additional term, as compared with Eq. (5.57), is a consequence of the singularity of $\nabla^2(1/r)$ at $P$, which now lies in the region of integration.

In verifying (5.60) the result of substituting $\Phi = \Phi_o$ and $\psi = 1/r$ into (5.56) is straightforward except for the volume integral, which becomes

$$\int \Phi_o \nabla^2 \left(\frac{1}{r}\right) dV$$

The integrand is zero except at $r = 0$, where it becomes infinite. However, the singularity is integrable. If we surround the point $(x',y',z')$ by a differential spherical volume $V_\epsilon$, then since there are no contributions to the integral from the region external to $V_\epsilon$ [$\nabla^2(1/r) = 0$ in this case],

$$\int_{V_o} \Phi_o \nabla^2 \left(\frac{1}{r}\right) dV = \int_{V_\epsilon} \Phi_o \nabla^2 \left(\frac{1}{r}\right) dV = \Phi_o(x',y',z') \int_{V_\epsilon} \nabla^2 \left(\frac{1}{r}\right) dV$$

The last equality follows since $\Phi_o$ is well behaved and is essentially constant [of value $\Phi_o(x',y',z')$] in $V_\epsilon$. Now

$$\int_{V_\epsilon} \nabla^2 \left(\frac{1}{r}\right) dV = \int_{V_\epsilon} \nabla \cdot \nabla \left(\frac{1}{r}\right) dV = \int_{S_\epsilon} \nabla \left(\frac{1}{r}\right) \cdot d\mathbf{S}$$

by the divergence theorem. Using spherical coordinates

$$\lim_{r \to 0} \int_{S_\epsilon} = 4\pi r^2 \left(-\frac{1}{r^2}\right) = -4\pi$$

Hence

$$\int_{V_o} \Phi_o \nabla^2 \left(\frac{1}{r}\right) dV = -4\pi \Phi_o(x',y',z')$$

Substitution into (5.56) results in an expression in which the normal, **n**, is outward to $V_o$ (inward to $V_i$). To obtain (5.60) the sign of the left-hand side is changed while the direction of the positive normal on the right-hand side is reversed.

Since the membrane is very thin, we may assume negligible longitudinal current, in which case the transverse current must be continuous.

That is,

$$-\frac{\partial \Phi_o}{\partial n}\bigg|_{S_o} = \frac{J_m}{\sigma_o} \tag{5.61}$$

where $J_m$ is the same as that in Eq. (5.58) and $\sigma_o$ is the external conductivity. Using Eq. (5.61) and adding Eqs. (5.59) and (5.60) yields the following expression for the potential $\Phi_o(P)$:

$$\Phi_o(P) = \frac{1}{4\pi\sigma_o}\left(\int_{S_o}\frac{J_m}{r}\,dS - \int_{S_i}\frac{J_m}{r}\,dS\right)$$
$$- \frac{1}{4\pi}\left[\frac{\sigma_i}{\sigma_o}\int_{S_i}\Phi_i\frac{\partial(1/r)}{\partial n}\,dS - \int_{S_o}\Phi_o\frac{\partial(1/r)}{\partial n}\,dS\right] \tag{5.62}$$

We shall consider field points whose distance to the membrane is large compared with the membrane thickness $m_d$. Since $m_d$ is actually extremely small, this is not a significant restriction. Under this condition, the integrals in the first group can be combined into a double-layer representation. Also, the integrals in the second group may both be integrated over a mean surface $S$ which lies between $S_o$ and $S_i$. Equation (5.62) then takes on the following form (where the positive surface normal is outward from the cell, i.e., from $S$):

$$\Phi_o(P) = \frac{1}{4\pi\sigma_o}\int_S m_d J_m \nabla\left(\frac{1}{r}\right)\cdot d\mathbf{S}$$
$$- \frac{1}{4\pi}\int_S\left(\frac{\sigma_i}{\sigma_o}\Phi_i - \Phi_o\right)\nabla\left(\frac{1}{r}\right)\cdot d\mathbf{S} \tag{5.63}$$

The quantity

$$-\nabla\left(\frac{1}{r}\right)\cdot d\mathbf{S} = \frac{\mathbf{a}_r\cdot d\mathbf{S}}{r^2} \tag{5.64}$$

where $\mathbf{a}_r$ is a unit vector from the field to source point, corresponds to the solid angle subtended at $P$ by a surface element $dS$. If we designate this quantity by $d\Omega$, then Eq. (5.63) can be expressed as

$$\Phi_o(P) = \frac{1}{4\pi}\int_S\left(\frac{\sigma_i}{\sigma_o}\Phi_i - \Phi_o - \frac{m_d J_m}{\sigma_o}\right)d\Omega \tag{5.65}$$

A sketch of the above parameters is given in Fig. 5.2.

Equation (5.65) can be simplified since the third term is ordinarily negligible. To show this, we first note that the contribution to $\Phi_o(P)$ comes only from variations in the integrand over $S$; constant terms evaluate to zero since the integral of $d\Omega$ over an external closed surface is zero. Assuming the external potential $\Phi_o$ to vary relatively little, we can consider the first two terms to have a range of about 5 to 100 mv. For a membrane of 100 Å thickness, 1 ma/cm² maximum current den-

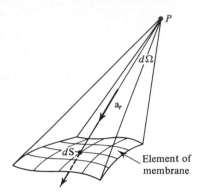

Element of
membrane    **Fig. 5.2**  Solid-angle geometry.

sity, and with $\sigma_o$ equal to 0.05 mho/cm, the third term evaluates to $2 \times 10^{-5}$ mv, a completely negligible quantity. Thus, to an excellent approximation,

$$\Phi(P) = \frac{1}{4\pi} \int_S \left( \frac{\sigma_i}{\sigma_o} \Phi_i - \Phi_o \right) d\Omega \tag{5.66}$$

When $\sigma_i = \sigma_o$, then denoting $(\Phi_i - \Phi_o) = V$, where $V$ is the transmembrane potential, we get

$$\Phi(P) = \frac{1}{4\pi} \int_S V \, d\Omega \qquad (\sigma_i = \sigma_o) \tag{5.67}$$

A further approximation is based on $\Phi_o$ being relatively constant so that its contribution is small. In this case, we can take it to be a zero reference and get

$$\Phi(P) = \frac{1}{4\pi} \frac{\sigma_i}{\sigma_o} \int_S V \, d\Omega \tag{5.68}$$

For an extensive medium we shall show that, at least in a specific example, this approximation is well justified. The coefficient in (5.68) can be thought of as expressing a division of the applied voltage between the internal and external loads.

The general expression of (5.66) can be rewritten based on the definition of the solid angle given in (5.64). The result is

$$\Phi(P) = \frac{1}{4\pi\sigma_o} \int_S (\sigma_o \Phi_o - \sigma_i \Phi_i) \, d\mathbf{S} \cdot \nabla \left( \frac{1}{r} \right) \tag{5.69}$$

Comparison with (5.53) shows that $(\sigma_o \Phi_o - \sigma_i \Phi_i) \, d\mathbf{S}$ behaves like a dipole of magnitude $(\sigma_o \Phi_o - \sigma_i \Phi_i) \, dS$ having the direction of the outward surface normal. The quantity $(\sigma_o \Phi_o - \sigma_i \Phi_i)$ can be interpreted, therefore, as a dipole moment per unit area or *double layer* in a $\sigma_o$ medium. The result fits the general discription given in (5.52) except that the source is no longer directly identified with the underlying phenomena responsible for converting energy into the electrical form. Consequently, the source is an equivalent one and is capable of giving the correct values everywhere except in the membrane or, because of the approximations, very close to the membrane.

From a field theoretic standpoint the membrane is treated as an infinitely thin surface across which the potential function $(\sigma\Phi)$ is discontinuous but whose normal derivative is continuous. Such a boundary condition is characteristic of a double-layer source located at the surface.[1] Moreover, the strength of the double layer must equal the discontinuity in $(\sigma\Phi)$, and this is precisely the condition imposed by (5.69).

For a medium containing many small active cells, such as for heart muscle, (5.69) applies to each such cell. However, each differential cell can be approximated by a single dipole which is the integral of

$$(\sigma_o\Phi_o - \sigma_i\Phi_i)\mathbf{n}\, dS$$

over its surface. If the density is high enough, this permits specifying a dipole moment per unit volume $\mathbf{J}_i$. The potential field may now be evaluated by (5.52). Alternatively, the divergence-type source

$$-\boldsymbol{\nabla} \cdot \mathbf{J}_i = I_v$$

may be used in (5.42). Note that $\mathbf{J}_i$ is an effective dipole moment per unit volume, and the representation here is an equivalent one. The accuracy of the computed field outside the source region depends only on the accuracy with which $\mathbf{J}_i$ represents the source density.

## 5.5 VOLUME-CONDUCTOR FORMULATION UNDER EXCISED CONDITIONS

To actually apply (5.66) to a nerve fiber, let us say, it is necessary to measure $\Phi_i$ and $\Phi_o$. Such data could be obtained for the squid axon. In other cases it might prove easier to excise the nerve and measure the potential distribution over the nerve surface while it is thus isolated. A relation between this measurement and the fields generated by the nerve *in situ* was developed by Lorente de Nó[2] in what has become a classical paper on the volume-conductor formulation, and we proceed now to its study.

We start by considering a region $V$ of finite extent which has a con-

[1] W. K. H. Panofsky and M. Phillips, "Classical Electricity and Magnetism," 2d ed., p. 20, Addison-Wesley Publishing Company, Inc., Reading, Mass., 1962.

[2] R. Lorente de Nó, "A Study of Nerve Physiology," chap. 16, Rockefeller Institute Studies, vol. 132, New York, 1947.

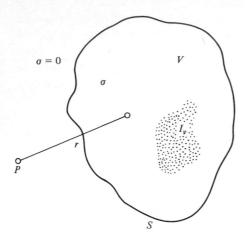

$\sigma = 0$

$V$

$\sigma$

$I_v$

$r$

$P$

$S$

**Fig. 5.3**  Bioelectric sources $I_v$ in a bounded volume conductor.

ductivity $\sigma$ and contains the sources $I_v$, and which is surrounded by a region of zero conductivity, as illustrated in Fig. 5.3.

We apply Green's theorem (5.56) with $\Phi$ chosen as the potential field within $V$ and $\psi = 1/r$, where $r = \sqrt{(x - x')^2 + (y - y')^2 + (z - z')^2}$ is the distance from a fixed point $(x',y',z')$ external to $V$ to an arbitrary point $(x,y,z)$ within $V$. We get

$$\int_V \left[ \Phi \nabla^2 \left( \frac{1}{r} \right) - \frac{1}{r} \nabla^2 \Phi \right] dV = \int_S \left[ \Phi \frac{\partial (1/r)}{\partial n} - \frac{1}{r} \frac{\partial \Phi}{\partial n} \right] dS \qquad (5.70)$$

Now $\nabla^2(1/r) = 0$ since $r = 0$ is excluded. Also $\partial \Phi / \partial n = 0$ on $S$, as specified by (5.24). Consequently,

$$-\int_V \frac{\nabla^2 \Phi}{r} dV = \frac{1}{\sigma} \int_V \frac{I_v}{r} dV = \int_S \Phi \frac{\partial (1/r)}{\partial n} dS \qquad (5.71)$$

Consider, now, the problem of the same source in a medium of infinite extent. Then the potential $\Phi_V$, at the field point $(x',y',z')$, is given by (5.42),

$$\Phi_V = \frac{1}{4\pi\sigma} \int_V \frac{I_v}{r} dV$$

If we combine (5.71) with (5.42), we determine that

$$\Phi_V = \frac{1}{4\pi} \int_S \Phi_S \frac{\partial (1/r)}{\partial n} dS \qquad (5.72)$$

where the notation $\Phi_S$ is used to emphasize that this potential corresponds to excised conditions. In vector form, (5.72) becomes

$$\Phi_V = \frac{1}{4\pi} \int_S \Phi_S \nabla \left( \frac{1}{r} \right) \cdot d\mathbf{S} \qquad (5.73)$$

Thus, the potentials in the infinite volume can be obtained from the surface potentials on the insulated body. The relationship (5.72) or (5.73) has the physical interpretation that the surface potential of the insulated body, $\Phi_S$, acts as a double-layer source for the potentials in the infinite volume conductor. The result in (5.72) was obtained in a somewhat different way by Lorente de Nó;[1] the method given here is adapted from Geselowitz.[2]

Equation (5.73) is valid only for $\sigma_i = \sigma_o = \sigma$, but under this condition both (5.68) and (5.73) should give the same results. A comparison shows that this requires $V = -\Phi_S$. Under *in situ* conditions, the load on the bioelectric generators is essentially the axoplasm alone, while under excised conditions, it is the thin external conducting layer alone; hence this result is quite reasonable.

A specific application considered by Lorente de Nó is where $\Phi_S$ is the surface potential (measured or calculated) for an excised nerve trunk. Since the latter consists of a large number of parallel fibers within an interstitial fluid, it is assumed that the bioelectric sources $I_v$ associated with the fiber membranes are to a first approximation, the same under *in situ* or excised conditions. Thus, the function $I_v$ is the same in (5.71) and (5.42), and (5.72) is a valid result.

In applying (5.72), the measurement of $\Phi_S$ is taken with respect to an arbitrary reference. One should recognize that the potential field is unchanged if a constant $C$ is added to $\Phi_S$. That this is the case is readily proved since we then have, from (5.72),

$$\Phi_V = \frac{1}{4\pi} \int_S (\Phi_S + C) \frac{\partial(1/r)}{\partial n} \, dS$$

$$= \frac{1}{4\pi} \int_S \Phi_S \frac{\partial(1/r)}{\partial n} \, dS + \frac{C}{4\pi} \int_S \nabla \left(\frac{1}{r}\right) \cdot d\mathbf{S}$$

If Gauss' theorem is used to evaluate the second integral, the integrand in the resulting volume integral is $\nabla^2(1/r) = 0$. Consequently, there is no contribution from this term, and we verify the assertion above.

In the application of (5.73) the volume $V$ is ordinarily chosen to correspond to the volume occupied by the cell or group of cells. In order to proceed further, analytically, we shall restrict our attention to circular cylindrical volumes (i.e., axon structures) whose cross section does not vary axially and where the length is much greater than the diameter. A specific example is that of a nerve trunk and is illustrated in Fig. 5.4.

---

[1] *Ibid.*

[2] D. Geselowitz, Multipole Representation for an Equivalent Cardiac Generator, *Proc. IRE*, **48**:75 (1960).

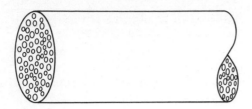

**Fig. 5.4** Cylindrical nerve trunk of essentially infinite length.

Let the axial variable be $z$ and *define* a function $\Phi$ such that

$$\Phi(x,y,z) = \Phi_S(z) \qquad \text{for } (x,y,z) \text{ within } V \tag{5.74}$$

That is, $\Phi$ is uniform over the cross section and takes on the value of $\Phi_S$ at the surface (it is assumed that $\Phi_S$ varies with $z$ only and not circumferentially). Of course, $\Phi$ as defined in (5.74) is simply a mathematical abstraction and represents an actual potential only on $S$. But with $\Phi$ so defined, the divergence theorem can be used to transform (5.73) into the following expression for potential external to $V$:

$$
\begin{aligned}
\Phi_V &= \frac{1}{4\pi} \int_V \boldsymbol{\nabla} \cdot \left[ \Phi \boldsymbol{\nabla} \left( \frac{1}{r} \right) \right] dV \\
&= \frac{1}{4\pi} \int_V \boldsymbol{\nabla}\Phi \cdot \boldsymbol{\nabla} \left( \frac{1}{r} \right) dV + \frac{1}{4\pi} \int_V \Phi \nabla^2 \left( \frac{1}{r} \right) dV
\end{aligned}
\tag{5.75}
$$

The second integral goes to zero since $r \neq 0$, and we have

$$\Phi_V = \frac{1}{4\pi} \int_V \left[ \frac{\partial \Phi_S}{\partial z} \mathbf{a}_z \cdot \boldsymbol{\nabla} \left( \frac{1}{r} \right) \right] dV \tag{5.76}$$

where, by virtue of the definition of $\Phi$ given in (5.74), $\partial/\partial x = \partial/\partial y = 0$, so that $\boldsymbol{\nabla}\Phi = (\partial\Phi/\partial z)\mathbf{a}_z$. In (5.76) the potential $\Phi_V$ is expressed in terms of an equivalent dipole moment $(\partial\Phi_S/\partial z)\mathbf{a}_z$.

We may simplify (5.76) by writing out

$$\mathbf{a}_z \cdot \boldsymbol{\nabla} \left( \frac{1}{r} \right) = \frac{\partial(1/r)}{\partial z} \tag{5.77}$$

which upon substitution gives

$$\Phi_V = \frac{1}{4\pi} \int_V \frac{\partial \Phi_S}{\partial z} \frac{\partial(1/r)}{\partial z} dV \tag{5.78}$$

$$\Phi_V = \frac{1}{4\pi} \int_A dA \int_z \frac{\partial \Phi_S}{\partial z} \frac{\partial(1/r)}{\partial z} dz \tag{5.79}$$

where $dV = dA\,dz$, and $dA$ is a cross-sectional area. If we integrate the above equation by parts (with respect to $z$), then

$$\Phi_V = \frac{1}{4\pi}\int_A dA \left.\frac{\partial \Phi_S}{\partial z}\frac{1}{r}\right|_{z_1}^{z_2} - \frac{1}{4\pi}\int_A dA \int_z \frac{1}{r}\frac{\partial^2 \Phi_S}{\partial z^2}\,dz \qquad (5.80)$$

$$\Phi_V = \frac{1}{4\pi}\int_{A_2}\left(\frac{\partial \Phi_S}{\partial z}\frac{1}{r}\right)dA - \frac{1}{4\pi}\int_{A_1}\left(\frac{\partial \Phi_S}{\partial z}\frac{1}{r}\right)dA$$
$$- \frac{1}{4\pi}\int_V \frac{\partial^2 \Phi}{\partial z^2}\frac{1}{r}\,dV \quad (5.81)$$

The surfaces $A_2$ and $A_1$ are the end (cap) surfaces of the cylinder.

The core-conductor equivalent circuit, which was developed in Chap. 3, may be utilized to effect other formulations. An illustration of this circuit is repeated in Fig. 5.5 with the nomenclature of Fig. 4.18. This circuit represents the electrical behavior of the unmyelinated fiber. For a nerve trunk containing a large number of fibers, this diagram can be considered as an equivalent representation for a group of similar fibers. Then $i_1$ and $i_m$ are total nerve-trunk quantities and the shunt elements do not have a simple "Hodgkin-Huxley" interpretation. More precisely, the parameters $r_2$, $r_1$, $i_1$, $i_m$, and $Z$ are those quantities which, for the homogeneous nerve trunk, yield the observed quantities ($\Phi_S$, $\partial \Phi_S/\partial z$, and $\partial^2 \Phi_S/\partial z^2$). Thus, from Fig. 5.5, we have

$$\frac{\partial \Phi_S}{\partial z} = r_2 i_1 \qquad (5.82)$$

where $r_2$ is the external resistance per unit length, and $i_1$ is the effective longitudinal current for the trunk. The following circuit equations

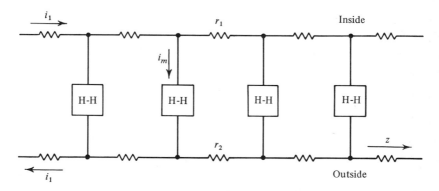

**Fig. 5.5** Electric circuit of linear-core-conductor model of a cylindrical axon or nerve trunk.

must also be satisfied, namely,

$$\frac{\partial i_1}{\partial z} = -i_m \tag{5.83}$$

and consequently

$$\frac{\partial^2 \Phi_S}{\partial z^2} = -i_m r_2 \tag{5.84}$$

For a very long nerve so that the end caps make a negligible contribution to local fields of interest, (5.81) can be written

$$\Phi_V = \frac{r_2}{4\pi} \int_V \frac{i_m}{r} \, dV \tag{5.85}$$

by utilizing (5.84).

The physical interpretation of (5.85) is that the equivalent transverse current $i_m$ acts like a current volume density source. It is tacitly understood that $V$ is circular cylindrical and that in the volume integration $i_m$ is treated as if it were a constant over any cross-sectional plane.

The role played by $i_m$ in (5.85) is that of an equivalent source. That is, $i_m$ is not a true physical observable but is simply a quantity that makes Fig. 5.5 a correct analog. In particular, if $r_2$ is specified, then since $\partial^2 \Phi_S/\partial z^2$ could be measured, $i_m$ is fixed by (5.84). Similar remarks apply to (5.82) since $i_1$ is not physically measurable, as defined in Fig. 5.5, but is simply the solution of (5.82) given $r_2$ and $\partial \Phi_S/\partial z$. These assumptions are linked to the approximations inherent in the core-conductor model, where the problem has been simplified in order to arrive at the one-dimensional circuit analysis of Fig. 5.5. However, to the extent that each fiber has the same cross-sectional area and that both axoplasmic and interstitial currents are axial, then $i_1$ is the net interstitial current (or the negative of the sum of all axoplasmic currents), and $i_m$ is the net transmembrane current taking all fibers into account.

In utilizing the results of this section, the several assumptions and approximations that are made should be kept in mind. First, it is assumed that the electrophysiological sources responsible for $I_v$ are the same whether the nerve or muscle is immersed in an infinite volume conductor or is excised. This requirement must be fulfilled in order that $I_v$ in (5.71) and (5.42) be identical. In addition, we have assumed the conductivity within and outside of $V$ to be the same, and have implied, furthermore, that it is a constant (i.e., that the medium is uniform, isotropic, and linear).[1] Finally, in application to a practical problem,

[1] This assumption of homogeneity is actually not necessary, and inhomogeneities can be included. A discussion of the generalization to such conditions is included in Sec. 5.12.

the external region is assumed extensive enough that it may be considered infinite. A discussion of some of these questions can be found in Lorente de Nó,[1] Benjamin et al.,[2] and Schwan.[3] The effect of finite external dimensions can sometimes be considered by the method of images,[4] or in all cases by inclusion of suitable equivalent sources.[5]

## 5.6 VOLUME-CONDUCTOR POTENTIALS FROM EQUIVALENT SOURCES

We turn attention again to the formulation of the potential field in terms of bioelectric sources under actual (*in situ*) conditions. The expressions that were developed in Sec. 5.3 relate the potential field in a direct way to the bioelectric generators. In Sec. 5.4 the double-layer sources, while of an equivalent nature, are nevertheless closely related to activity within the membrane. In this section we develop expressions for equivalent sources that are capable of replacing extensive distributed (true) sources.

We begin by considering, as in Fig. 5.6, bioelectric sources $I_v$ contained within a volume conductor $V$ (bounded by surfaces $S_0$ and $S_\infty$) of conductivity $\sigma$ and of infinite extent. The surface $S_0$ is any arbitrarily chosen surface which may enclose some or all of the sources, while $S_\infty$ designates the surface at infinity. Consider an arbitrary point $P(x',y',z')$ external to $S_0$, and let an element of integration be defined by the unprimed variables $(x, y, z)$.

Now if we surround $P$ by a very small spherical surface of radius $\epsilon$, whose surface is designated $S_\epsilon$, and let the region bounded by $S_\infty$, $S_0$, and $S_\epsilon$ be designated $(V - \epsilon)$, then in the latter region Green's theorem (5.56) becomes

$$\int_{V-\epsilon} (\Phi \nabla^2 \psi - \psi \nabla^2 \Phi)\, dV = \int_{S_0} \left( \Phi \frac{\partial \Phi}{\partial n} - \psi \frac{\partial \Phi}{\partial n} \right) dS$$

$$+ \int_{S_\epsilon} \left( \Phi \frac{\partial \psi}{\partial n} - \psi \frac{\partial \Phi}{\partial n} \right) dS + \int_{S_\infty} \left( \Phi \frac{\partial \psi}{\partial n} - \psi \frac{\partial \Phi}{\partial n} \right) dS \quad (5.86)$$

We now let $\Phi$ be the desired solution to the potential field in $V$, i.e.,

$$\nabla^2 \Phi = - \frac{I_v}{\sigma} \quad (5.87)$$

[1] *Op. cit.*

[2] J. M. Benjamin et al., The Electrical Conductivity of Living Tissues as It Pertains to Electrocardiography, *Circulation*, **2**:321 (1950).

[3] Herman P. Schwan, Determination of Biological Impedances, in W. L. Nastuk (ed.), "Physical Techniques in Biological Research," vol. VI, Academic Press Inc., New York, 1963.

[4] R. Plonsey, Current Dipole Images and Reference Potentials, *IEEE Trans. Biomed Electron.*, **BME10**:3 (1963).

[5] This will be discussed further in Sec. 5.12.

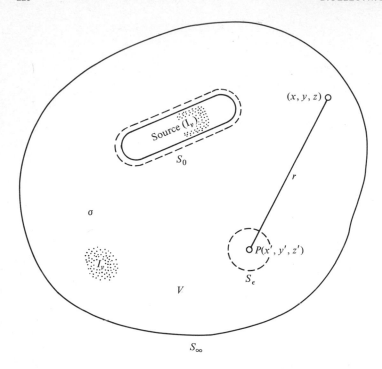

**Fig. 5.6**  Bioelectric sources in a volume conductor.

where $I_v$ is that portion of the volume source in $V$. The function $\psi$ is defined as

$$\psi = \frac{1}{r} = \frac{1}{\sqrt{(x - x')^2 + (y - y')^2(z - z')^2}} \tag{5.88}$$

The behavior of $\Phi$ and $\psi$ at infinity is that they diminish at least as $1/r^2$ and $1/r$, respectively. Consequently, if we think of the surface at infinity as a spherical surface of radius $R \to \infty$, then the integrals over $S_\infty$ go to zero as $0(1/R)^2$.†

Substituting (5.87) and (5.88) into the remaining terms of (5.86) leaves

$$\int_{V-\epsilon} \Phi \nabla^2 \left(\frac{1}{r}\right) dV + \frac{1}{\sigma} \int_{V-\epsilon} \frac{I_v}{r} dV = \int_{S_0} \Phi \nabla \left(\frac{1}{r}\right) \cdot d\mathbf{S}$$

$$- \int_{S_0} \frac{\partial \Phi}{\partial n} \frac{1}{r} dS + \int_{S_\epsilon} \Phi \nabla \left(\frac{1}{r}\right) \cdot d\mathbf{S} - \int_{S_\epsilon} \frac{\partial \Phi}{\partial n} \frac{1}{r} dS \tag{5.89}$$

† J. A. Stratton, "Electromagnetic Theory," pp. 166–169, McGraw-Hill Book Company, New York, 1941.

Since $r \neq 0$ in the region $V - \epsilon$, then $\nabla^2(1/r) = 0$ and the first integral in the expression above vanishes. In the integral over $S_\epsilon$, $\nabla\Phi$ is well behaved on the vanishingly small spherical surface of radius $r = \epsilon \to 0$, so that it may be removed from the integral evaluated at the point. That is,

$$\lim_{r \to 0} \int_{S_\epsilon} \frac{\partial \Phi}{\partial n} \frac{1}{r} \, dS = \lim_{r \to 0} \int_S \frac{\nabla\Phi \cdot d\mathbf{S}}{r} = \lim_{r \to 0} \nabla\Phi \cdot \left( \int_{S_\epsilon} \frac{d\mathbf{S}}{r} \right) \qquad (5.90)$$

Since the area varies as $r^2$, the magnitude of the last integral vanishes at least as $0(r)$. Actually, this integral is identically zero, as is clear from symmetry.

The same procedure when applied to the remaining integral over $S_\epsilon$ yields

$$\lim_{r \to 0} \Phi \Big|_P \int_{S_\epsilon} \nabla \left( \frac{1}{r} \right) \cdot d\mathbf{S} = \lim_{r \to 0} \Phi(P) \frac{4\pi r^2}{r^2} \qquad (5.91)$$

In the limit $\epsilon \to 0$, $(V - \epsilon) \to V$, and we get finally

$$\Phi(P) = \frac{1}{4\pi\sigma} \int_V \frac{I_v}{r} \, dV + \frac{1}{4\pi} \int_{S_0} \frac{\partial \Phi}{\partial n} \frac{1}{r} \, dS - \frac{1}{4\pi} \int_{S_0} \Phi \nabla \left( \frac{1}{r} \right) \cdot d\mathbf{S} \qquad (5.92)$$

The mathematical analysis here is similar to that used in connection with obtaining (5.60). When $S_0$ does not contain all the sources, (5.92) shows that the portion within $V$ contributes by means of the volume integral while sources within $S_0$ contribute through the pair of surface integrals.

For the case where $S_0$ encloses all sources, as, for example, where $S_0$ contains a single axon or nerve trunk, then $I_v = 0$ in $V$. In addition, we then have

$$J_m = \frac{\sigma \partial \Phi}{\partial n} \qquad (5.93)$$

where $J_m$ is the normal component of current density flowing *out* of $S_0$. Equation (5.92) now becomes

$$\Phi(P) = \frac{1}{4\pi\sigma} \int_{S_0} \frac{J_m}{r} \, dS + \frac{1}{4\pi} \int_{S_0} \Phi_{s0} \nabla \left( \frac{1}{r} \right) \cdot d\mathbf{S} \qquad (5.94)$$

where $d\mathbf{S}$ has been reversed so that it is an *outward* normal from $S_0$, and $\Phi_{s0}$ has been so designated to emphasize that it is the surface potential on $S_0$. $J_m$ behaves like a current source surface density, while $\Phi_{s0}$ can be considered as the magnitude of an equivalent double-layer potential source.

If $S_0$ in Fig. 5.6 is the external surface of a single axon, then $J_m$ in (5.94) is identical to the transmembrane current density. Equation

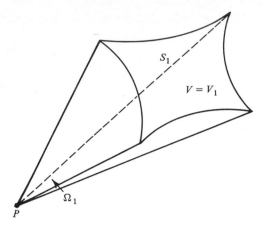

**Fig. 5.7** Computation of poten
tial from the solid-angle formula

(5.94) would be equally valid if $S_0$ were chosen to be a surface some distance beyond the actual axon surface, but in this case $J_m$ can no longer be
identified as the membrane current density. For a cylindrical trunk that
is bounded by $S_0$, $J_m$ is the actual current density crossing into the surrounding volume conductor. Whether or not $J_m$ and $\Phi_{s0}$ have a direct
physical interpretation, they nevertheless behave as (equivalent) sources
for the field external to $S_0$.

### 5.7  SOLID-ANGLE COMPUTATION

The expression for potential given by (5.68) has a simple interpretation,
provided that the magnitude of the transmembrane potential $V$ is a
constant over the entire surface or on each of several sections that constitute the total surface. To see this, let us consider a portion of $S$ over
which we assume $V = V_1$ to be constant. This surface, designated $S_1$, is
illustrated in Fig. 5.7 and is of arbitrary shape. The contribution to the
potential at $P$ due to the transmembrane potential over $S_1$ can then be
written from (5.68) as

$$\Phi_1(P) = \frac{V_1}{4\pi} \frac{\sigma_i}{\sigma_o} \int_S d\Omega \tag{5.95}$$

$$\Phi_1(P) = \frac{V_1 \sigma_1}{4\pi\sigma_o} \int_{S_1} \frac{\mathbf{a}_r \cdot d\mathbf{S}}{r^2} \tag{5.96}$$

We recall that the direction of $d\mathbf{S}$ is from inside to outside of the membrane, while the sign of $V$ is the potential inside minus the potential outside, and $\mathbf{a}_r$ is from the field point to the surface element.

The integral in (5.96) is, of course, the total solid angle subtended at $P$ by the surface $S_1$.   Thus

$$\Phi_1(P) = \frac{V_1 \Omega_1 \sigma_i}{4\pi\sigma_o} \qquad (5.97)$$

The angle $\Omega_1$ is positive if the radius vector $\mathbf{a}_r$ makes an acute angle with the surface normal.   The total potential at $P$ may be found, in this way, by superposing the contributions from each of the remaining surface segments.   If $V$ is a continuous function of position, then, of course, the solid-angle expression in (5.96) can be thought of as a limiting case of superposition of an infinite number of elements; when there are just a few terms, however, a useful conceptualization exists.

While the point of view noted above was used to discuss (5.68), it would hold equally well for the several potential expressions which are based on double-layer sources.   For example, Eqs. (5.69) and (5.73) are in the same form as (5.68).   Where the primary bioelectric sources themselves lie in a surface so that they constitute an impressed current dipole moment per unit area, $J_{is}\mathbf{a}_n$ (with direction assumed normal to the surface), then from an extension of (5.52)

$$\Phi(x',y',z') = \frac{1}{4\pi\sigma} \int_S J_{is}\, d\mathbf{S} \cdot \nabla\left(\frac{1}{r}\right) = -\frac{1}{4\pi\sigma} \int_S J_{is}\, d\Omega \qquad (5.98)$$

We see that (5.98) also is in the same general form as (5.68) and hence subject to the comments above.

For a circular cylindrical axon with an idealized transmembrane potential that is constant over both the depolarized portion $(+V_d)$ and the inactive portion $(-V_r)$,[†] as shown in Fig. 5.8, then the contribution to the total potential is from two terms,

$$\Phi = \Phi_1 + \Phi_2$$

† $V_r$ and $V_d$ are taken as positive magnitudes.

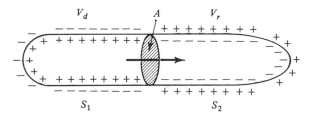

**Fig. 5.8**   Idealized action potential on a cylindrical axon.

where $\Phi_1 = \sigma_i V_d \Omega_1 / 4\pi\sigma_o$ and $\Phi_2 = -\sigma_i V_r \Omega_2 / 4\pi\sigma_o$.  Since $\Omega_1 = -\Omega_2 = \Omega$,

$$\Phi = \frac{\sigma_i \Omega (V_d + V_r)}{4\pi\sigma_o}$$

Furthermore, the solid angle subtended by the crosshatched (cross-sectional) area $A$ is also $\Omega$.  Thus, the potential at an arbitrary point $P$ may be thought of as arising from a double layer on the plane cross-sectional area $A$ across which $\Phi$ is discontinuous by $(V_d + V_r)$.  For long, thin axons, this area is very small and could be replaced for most field points by an idealized linear dipole element, as is described below. For a nerve trunk, the external field then has a simple physical source consisting of linear dipole elements in each fiber, where the dipole element is located at the "boundary" between active and inactive tissue. These dipoles propagate down their respective fibers at the velocity of propagation of the corresponding action potential.

Since the potential at $P$ is related to a double layer of strength $(V_d + V_r)$ on $A$, the following mathematical expression can be written:

$$\Phi(P) = -\frac{\sigma_i}{4\pi\sigma_0} \int_A (V_d + V_r) \frac{\mathbf{a}_r \cdot d\mathbf{S}}{r^2} \tag{5.99}$$

Now if the diameter of $A$ is small compared with the distance to the field point $P$, then in the above integration $r$ may be taken as a constant equal to the distance $R$ from the center of the disk.  If we let $d\mathbf{S} = \mathbf{a}_z \, dS$ so that the axis of the axon is the polar axis, then the above equation can be written

$$\Phi(P) = -\frac{1}{4\pi\sigma_0} \left[ \frac{\sigma_i A (V_d + V_r) \mathbf{a}_z}{R^2} \right] \cdot \mathbf{a}_r = \frac{\sigma_i A (V_d + V_r) \cos\theta}{4\pi\sigma_0 R^2} \tag{5.100}$$

where $\theta$ is the polar angle.  (The center of the disk is the center of a spherical coordinate system.)  Considered with respect to (5.53), $\sigma_i A (V_d + V_r) \mathbf{a}_z$ in (5.100) is to be taken as a current dipole moment. Note that a dimensional check is obtained.  If a cell is at rest, then $V_d$ vanishes and the area $A$ goes to zero, so that the external region is at a constant (zero) potential.

The equations for $\Phi(P)$ expressed in terms of the subtended solid angle have been developed from (5.59) and (5.60).  Since the latter pair are valid *only* for $P$ in the region exterior to $S_0$, a similar restriction applies to the derived forms.  Thus, in particular, (5.96) applies only for $P$ outside $S$.  If the surface $S$ is closed and if $V$ is constant, then $\Omega = 0$ and $\Phi(P)$ is zero (or a constant since we may also add an arbitrary constant to potential solutions).

When $\sigma_i = \sigma_o$ it can be shown that

$$\Phi(P) = \frac{1}{4\pi} \int_S V \, d\Omega$$

(where $V = \Phi_i - \Phi_o$) is valid for $P$ both outside and inside $S$, assuming $S$ to be closed.[1]  In particular, for $P$ outside $\Phi = 0$, while for $P$ inside $(\Omega = 4\pi)\Phi = V$.  The transmembrane potential arrived at, namely, $V$, correctly matches the assumed value.

An extension of (5.69) to field points inside a cell is most easily accomplished by applying the view that the membrane is infinitely thin and by imposing the condition on $\sigma\Phi$ that it be discontinuous but with a continuous normal derivative.  As explained previously, the discontinuity $(\sigma_o\Phi_o - \sigma_i\Phi_i)$ constitutes a double layer source for $\sigma\Phi$ so that in the cell interior, where $\sigma = \sigma_i$

$$\sigma_i\Phi_{\text{int}} = \frac{1}{4\pi} \int_S (\sigma_o\Phi_o - \sigma_i\Phi_i) \, d\mathbf{S} \cdot \boldsymbol{\nabla}\left(\frac{1}{r}\right)$$

If we include the condition that $\Phi_o$ is negligible, and let $V = \Phi_i$, then

$$\Phi_{\text{int}} = \frac{1}{4\pi} \int_S V \, d\Omega$$

showing that (5.67) is true even if $\sigma_i \neq \sigma_o$ for *interior* points.


## 5.8  EXTERNALLY RECORDED ACTION POTENTIALS

As an example in the application of the previous theory, we consider the recordings of an action potential of a nerve, as illustrated in Fig. 5.9. Shown here are three successive instants of time of a stylized propagating action potential.  The shaded area indicates the portion of the membrane that has undergone activation, in the sense of the approximation discussed in connection with Fig. 5.8.  The arrow, in each case, represents the equivalent dipole as discussed in Sec. 5.7 and evaluated in (5.100).  The resultant signal is, consequently, that derived from a dipole which is moving along the axis at a uniform velocity.  For recording electrodes, as illustrated in Fig. 5.9, where one is considered to be at a large distance (remote), the potential variation is easily evaluated since it depends only upon the relatively simple potential field of a dipole.  It could be found, for example, by assuming the dipole and its field stationary while the electrode samples the field with a negative velocity.  The result of

[1] T. C. Ruch and J. F. Fulton, "Medical Physiology and Biophysics," p. 86, W. B. Saunders Company, Philadelphia, 1960.

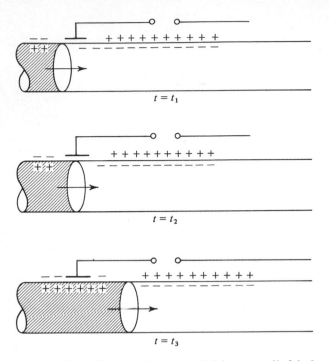

**Fig. 5.9**  Recording of action potential from a cylindrical axon.

such an analysis is illustrated in Fig. 5.10.  This potential variation is called a *diphasic* action potential.

   If the repolarization takes place fairly soon after depolarization, a complex potential waveform involving the interaction of both phenomena will be recorded.  This is illustrated in Fig. 5.11, where the shaded area represents the extent of the active region.  That is, the unshaded area either has not yet been activated or has essentially recovered after having been active.  In this case, the potential variation at the electrode pair arises from movement of the two dipoles shown, where the quantitative description of each is given by (5.100).  The net result is as illustrated in Fig. 5.11.  The broad negative peak occurs when electrode *A* lies between the two dipoles.  This result is called a *triphasic potential*.

   The actual measurement of potential is only approximated by the aforementioned discussion.  This is due to the highly idealized variation in membrane potential.  The true variation in transmembrane potential is, of course, not discontinuous as depicted in Figs. 5.9 and 5.11.

   The two-dipole approximation can be understood by reference to

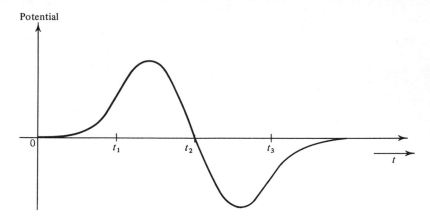

**Fig. 5.10** Diphasic recorded action potential.

(5.76).[1] The source may be thought of as a distributed sequence of dipoles ($\partial \Phi_S / \partial z \; dz$) lying along the axis. In general, $\Phi_S$ is monophasic so that the dipole strength function has but one change in sign. Each of the two dipoles of the model thus constitutes a lumped approximation to the positive (negative) distribution. For field points at distances greater than one-half the extent of the distribution, the "lumped" dipole approximation becomes quite good.

In the above examples one electrode is assumed at a remote location. Consequently, it does not enter in the determination of the resultant difference of potential; in a sense it "sees" essentially zero solid angle throughout the experiment. For two electrodes close together, the position of both in the potential field must be considered in the determination of potential difference.

A more accurate procedure than that described above is given by Lorente de Nó[2] and is based on (5.81). This equation is first simplified by neglecting both the contribution from the ends and the variation of $r$ in the integration over a cross section. The result is

$$\Phi(P) = -\frac{A}{4\pi} \int \left( \frac{\partial^2 \Phi_S}{\partial z^2} \right) \frac{1}{r} \; dz \tag{5.101}$$

[1] By comparing (5.68) with (5.73), it is clear that for *in situ* conditions an expression similar to (5.76) can be derived, namely,

$$\Phi = -\frac{1}{4\pi} \int_V \frac{\partial V}{\partial z} \, \mathbf{a}_z \cdot \nabla \left( \frac{1}{r} \right) dV$$

Thus the above remarks apply to this situation also.

[2] R. Lorente de Nó, "A Study of Nerve Physiology," chap. XVI, pp. 398ff, Rockefeller Institute Studies, vol. 132, New York, 1947.

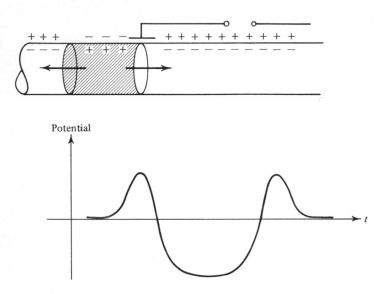

**Fig. 5.11**  Triphasic recorded action potential.

where $r$ is the distance from the field point to the axis.   The above equation was used to compute the potential field of the alpha fibers of the sciatic nerve of the bullfrog based on measured values of $\partial \Phi_S / \partial z$.   A typical record of the latter is shown in Fig. 5.12 along with its derivative and integral.   Since only a small fraction of the fibers were activated in making this measurement, it is probable that they were fairly uniform in cross section; consequently, the recording can be considered typical of each.   The potential field that results is computed from (5.101) and plotted in Fig. 5.13.   The curve below the axis is $|\partial \Phi_S / \partial z|$ and is the absolute value of the measured quantity, obtained by using two closely spaced electrodes on a uniform portion of the excised nerve (where the temporal-spatial relationship of a propagating wave could be assumed).

Since the propagated action potential travels down the nerve without attenuation or dispersion, the potential variation along a line parallel to the axis of Fig. 5.13 can also be interpreted as the potential variation with respect to time of an arbitrary field point at the assumed distance from the axis.   That is, for a specific radius $\rho_0$, one can measure $\Phi(t,\rho_0)$ and the propagation velocity $v$; from this the general solution, $\Phi[(t + z/v),\rho_0]$, with the properties noted, can be constructed.   Potential

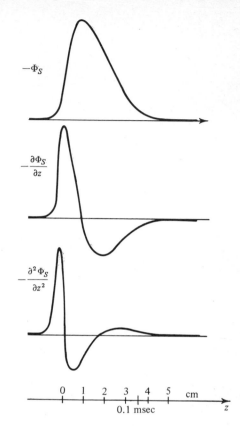

**Fig. 5.12** The external action potential $\Phi_S$ of bullfrog alpha fibers, and its first and second derivatives. (*Reprinted by permission of the Rockefeller University Press from R. Lorente de Nó, "A Study of Nerve Physiology," chap. 16, Rockefeller Institute Studies, vol. 132, New York, 1947.*)

variations at a distance of 1.6 and 4 mm are shown in Fig. 5.14a and b. Note that the abscissa of Fig. 5.13 is calibrated in terms of both distance and time in view of the relationship between these variables.

The results in Fig. 5.14a and b are plotted to the same vertical scale. One notes that the amplitude diminishes with increasing distance from the axis of the nerve, as would be expected. This radial dependence was examined by Lorente de Nó; and although it was found to depend on $z$, its general character was that of exponential decay with increasing distance from the nerve. Some additional conclusions can be obtained by examining the axial dipole source distribution given by $\partial\Phi_S/\partial z$ in Fig. 5.12. This reveals a concentration of positive and negative dipole elements (which, as noted above, can be approximated by the two lumped dipoles). The net positive and negative magnitudes (the integral of $\partial\Phi_S/\partial z$ between zero and the peak of $\Phi_S$) are equal. Consequently the

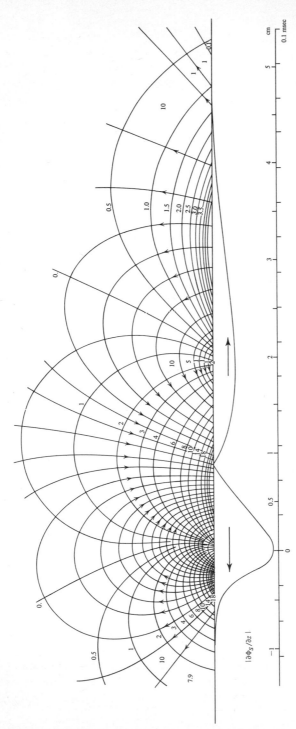

**Fig. 5.13** Potential field of a cylindrical nerve computed [by using (5.101)] from excised surface-potential data (given in Fig. 5.12). *(Reprinted by permission of the Rockefeller University Press from R. Lorente de Nó, Rockefeller Institute Studies, vol. 132, New York, 1947.)*

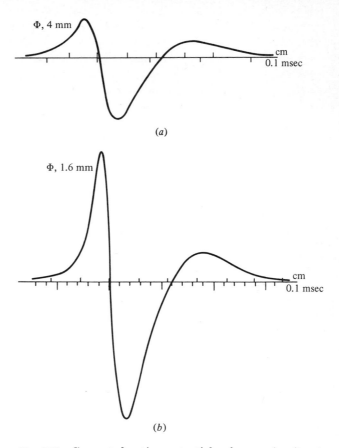

Φ, 4 mm

cm
0.1 msec

(a)

Φ, 1.6 mm

cm
0.1 msec

(b)

**Fig. 5.14** Computed action potentials of nerve *in situ* at points 4 and 1.6 mm from the axis of the nerve. (*Reprinted by permission of the Rockefeller University Press from R. Lorente de Nó, Rockefeller Institute Studies, vol. 132, New York, 1947.*)

radial field behavior must be characteristic of a zero net dipole moment, hence a *quadrupole*, and vary as $1/\rho^3$ for sufficiently large $\rho$.[1]

If Fig. 5.14 is compared with $\partial^2\Phi_S/\partial z^2$ in Fig. 5.12, one notes a striking similarity. This might have been anticipated by examination of (5.101). Here it is clear that the contribution to $\Phi(P)$, when $P$ is not too far from the axis [for (5.101) to be strictly valid $P$ must lie on a radius which exceeds, say, five times the radius of the axon], is essentially the

[1] R. Plonsey, On Volume Conductor Fields of Compound Nerve Action Potentials, *Med. Biol. Eng.* (March, 1969).

value of $\partial^2\Phi_S/\partial z^2$ in its immediate neighborhood. Consequently, the potential variation measured by external electrodes follows roughly the variation in $\partial^2\Phi_S/\partial z^2$. Furthermore, to the extent that (5.84) is valid, it is also proportional to the membrane current. The waveform of Fig. 5.14 is, of course, a triphasic action potential. It differs from Fig. 5.11 in that it does not depend on an idealized transmembrane potential but rather utilizes the actual shape of $\Phi_S$.

In Fig. 5.15 several measured action potentials are shown. The numbered curves were obtained from the bullfrog sciatic nerve *in situ*, and the numbers correspond to the locations on the sketch at which the exploring electrode was placed. We note the relatively good agreement in the general nature of the measured potentials with predictions from Fig. 5.14. (The reference electrode was located at an inactive point on the other leg.)

The action potentials that would be measured in the region external to a squid axon are similar to the triphasic potential of Fig. 5.14. This is based on Eq. (5.85), which, by the same argument that led to (5.101), can be written $\Phi(P) = Ar_2/4\pi \int (i_m/r)\, dz$, so that if $r$ is not too large, the character of $\Phi$ will be similar to that of $i_m$. The latter has the measured temporal variation shown in Fig. 5.16, which is essentially triphasic.

## 5.9 INFINITE CYLINDRICAL AXON

For a single axon (or a single cell) in a volume conductor the potential field is related to the potential and normal current density at the axon

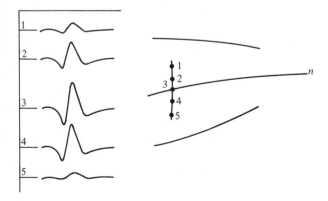

**Fig. 5.15** Action potentials of the bullfrog sciatic nerve *in situ*. Numbered recordings are taken from the corresponding sites. (*Reprinted by permission of the Rockefeller University Press from R. Lorente de Nó, Rockefeller Institute Studies, vol. 132, New York, 1947.*)

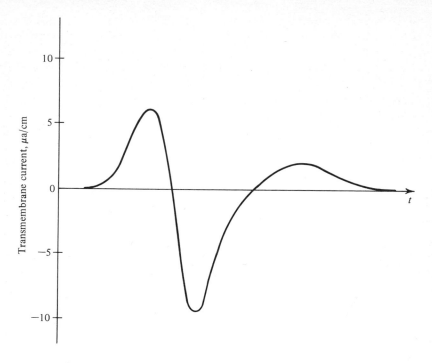

**Fig. 5.16** Transmembrane current per unit length for the crayfish lateral giant axon as computed rigorously from transmembrane potential measurements. [*J. Clark and R. Plonsey, The Extracellular Potential Field of the Single Active Nerve Fiber in a Volume Conductor, Biophys. J.,* **8**:842 (1968).]

surface as described by (5.94). If we assume the external region to be extensive and to have a conductivity similar to that of the axoplasm, it behaves, roughly speaking, as a short circuit. That is, potential variations generated by the bioelectric activity tend to be localized in the membrane and the axoplasm, with relatively small potential changes in the medium. Thus, in the work of Hodgkin and Huxley,[1] $r_2$ of Fig. 5.5 is set equal to zero. As a second example, Taylor, Moore, and Cole[2] find that neglecting potential variations in the external seawater causes no more than a 5 percent error in the computation of characteristic length. Further evidence that potentials in the external medium are of very small magnitude is given in Sec. 5.11. Now if $\Phi_{so}$ can be taken to

[1] A. L. Hodgkin and A. F. Huxley, A Quantitative Description of Membrane Current and Its Application to Conduction and Excitation in Nerve, *J. Physiol.,* **117**:500 (1952).

[2] R. E. Taylor, J. W. Moore, and K. S. Cole, Analysis of Certain Errors in Squid Axon Voltage Clamp Measurements, *Biophys. J.,* **1**:161 (1960).

be a constant, normally zero, the second integral in (5.94) can be transformed by using the divergence theorem, and we obtain the result that

$$\frac{\Phi_{s0}}{4\pi} \int_{S_0} \nabla \left(\frac{1}{r}\right) \cdot d\mathbf{S} = \frac{\Phi_{s0}}{4\pi} \int_V \nabla^2 \left(\frac{1}{r}\right) dV = 0 \qquad (5.102)$$

since $\nabla^2(1/r) = 0$. Choosing $\Phi_{s0}$ equal to zero (reference) yields the same conclusion. The approximation permits, in effect, neglecting the second term of (5.94), so that

$$\Phi(P) \approx \frac{1}{4\pi\sigma} \int_{S_0} \frac{J_m}{r} dS \qquad (5.103)$$

This result suggests that the potential fluctuation with time measured at the surface of a cell should be proportional to the variation in transmembrane current. This conclusion is also reached in an analog model study made by McAllister[1] and also in the approximate analysis of the previous section.

The reasoning leading to (5.103) is fairly satisfactory; however, one should really show that the second term of (5.94) not only is small but is small in comparison with the first term. A quantitative comparison of the first and second terms of (5.94) can be accomplished by applying Green's theorem to the internal region (axoplasm). The result is (5.59), which can be written

$$\frac{1}{4\pi\sigma_i} \int_{S_0} \frac{J_m}{r} dS + \frac{1}{4\pi} \int_{S_0} \Phi_{si} \nabla \left(\frac{1}{r}\right) \cdot d\mathbf{S} = 0 \qquad (5.59)$$

where the notation $\Phi_{si}$ emphasizes that it is a surface-potential distribution on the inner surface and where $S_0 \equiv S_i$ is assumed. For the simplifying condition of $\sigma_i \approx \sigma_o$ we can subtract (5.59) from (5.94) and obtain

$$\Phi(P) = -\frac{1}{4\pi} \int_{S_0} \Phi_{si} \nabla \left(\frac{1}{r}\right) \cdot d\mathbf{S} + \frac{1}{4\pi} \int_{S_0} \Phi_{s0} \nabla \left(\frac{1}{r}\right) \cdot d\mathbf{S} \qquad (5.104)$$

The magnitude of the potential variation of $\Phi_{si}$ is normally much greater than $\Phi_{s0}$; and since the magnitude of the integrals depends on the magnitude of potential fluctuation, the first term will, in general, dominate the second. One can conclude from this that (5.103) should be a good approximation.

Comparison of the two terms in (5.94) can also be facilitated, analytically, for the circular cylindrical axon of essentially infinite length. With this geometry, we may repeat the formal mathematical

[1] A. J. McAllister, Analog Study of a Single Neuron in a Volume Conductor, *Naval Med. Res. Inst., Bethesda, Md.*, **16**:1011 (15 Dec., 1958).

transformation of (5.73) into (5.76) but applied to the second integral in (5.94). If, in addition, the action potential is assumed to lie completely within the nerve, the contribution from the end surfaces is zero, with the result that

$$\Phi(P) = \frac{1}{4\pi\sigma_o} \int_{S_0} \frac{J_m}{r} \, dS - \frac{1}{4\pi} \int_V \frac{\partial^2 \Phi}{\partial z^2} \frac{1}{r} \, dV \tag{5.105}$$

In (5.105) $\Phi$ is defined by

$$\Phi(x,y,z) = \Phi_{s0}(z) \qquad \text{for } (x,y,z) \text{ within } V$$

If the equivalent circuit of Fig. 5.5 can be applied, then we have

$$\frac{\partial^2 \Phi_{s0}}{\partial z^2} = -i_m r_2 \tag{5.106}$$

where $r_2$ is an effective external resistance, and $i_m$ is the normal current outflow from $S_0$ per unit length. Now since $J_m$ is the normal current density, we have

$$2\pi a J_m = i_m \tag{5.107}$$

For field points whose distance from the axis is large compared with the diameter of $S_0$, $r$ in (5.105) is approximately constant in the integration over a cross section. Consequently, both integrals in (5.105) can be approximated by the following line integrals, where $a$ is the radius:

$$\Phi(P) = \frac{1}{4\pi\sigma_o} \int_Z \frac{i_m}{r} \, dz + \frac{1}{4\pi} \int_Z \frac{i_m \pi a^2 r_2}{r} \, dz \tag{5.108}$$

If the internal axial current is assumed uniform, then the internal resistance will be

$$r_1 \approx \frac{1}{\sigma_i \pi a^2}$$

for an internal conductivity $\sigma_i$. Equation (5.108) can now be rewritten as

$$\Phi(P) = \left(1 + \frac{r_2\sigma_o}{r_1\sigma_i}\right) \frac{1}{4\pi\sigma_o} \int_Z \frac{i_m}{r} \, dz \tag{5.109}$$

In this result, it is now possible to actually compare the contribution from the two terms in (5.94) since the magnitude of the second relative to the first is $r_2\sigma_o/r_1\sigma_i$. Now $\sigma_o \approx \sigma_i$, but for an extensive external medium, it seems reasonable for $r_2 \ll r_1$. Consequently, it again appears that one can neglect the second term of (5.94).

A more satisfactory procedure is to obtain a direct comparison of $J_m$ and $\partial^2 \Phi / \partial z^2$, which appear in (5.105). This can be done since specification of $\Phi_{s0}$ determines both $J_m$ and $\partial^2 \Phi / \partial z^2$. Since the relationships are rather complicated, we proceed by means of a numerical example as given by Plonsey.[1]

The procedure employed is to evaluate the potential field, given the membrane surface potential, and is based on the standard technique of the separation of variables of Laplace's equation in cylindrical coordinates. The form of solution for the external region is

$$A(k)K_0(k\rho)e^{-jkz} + B(k)K_0(k\rho)e^{-jkz}$$

where $K_0$ is the modified Bessel function of the second kind, and the cylindrical radial variable is denoted by $\rho$. The function $K_0$ is chosen both because it has an appropriate asymptotic behavior as $\rho \to \infty$ and because it is linked to a desired complex exponential form in $z$. Since the separation constant $k$ may take on any positive real value, the most general solution is

$$\Phi(\rho,z) = \int_0^\infty A(k)K_0(k\rho)e^{jkz}\,dk + \int_0^\infty B(k)K_0(k\rho)e^{-jkz}\,dk \qquad (5.110)$$

This can be rewritten in the more compact form

$$\Phi(\rho,z) = \frac{1}{2\pi}\int_{-\infty}^\infty C(k)K_0(|k|\rho)e^{-jkz}\,dk \qquad (5.111)$$

where

$$\frac{C(k)}{2\pi} = \begin{cases} B(k) & \infty > k > 0 \\ A(-k) & 0 > k > -\infty \end{cases} \qquad (5.112)$$

The function $C(k)$ can now be expressed in terms of the given boundary condition, namely, that

$$\Phi_{s0}(z) = \Phi(a,z)$$

where $a$ is the radius of the cylindrical nerve. We now have

$$\Phi_{s0}(z) = \frac{1}{2\pi}\int_{-\infty}^\infty C(k)K_0(|k|a)e^{-jkz}\,dk \qquad (5.113)$$

By recognizing $\Phi_{s0}(z)$ as a Fourier transform of $C(k)K_0(|k|a)$ and by taking the inverse transform we get

$$C(k)K_0(|k|a) = \int_{-\infty}^\infty \Phi_{s0}(z)e^{jkz}\,dz = F(k) \qquad (5.114)$$

and

$$C(k) = \frac{F(k)}{K_0(|k|a)} \qquad (5.115)$$

[1] R. Plonsey, Volume Conductor Fields of Action Currents, *Biophys. J.*, **4**:317 (1964).

The potential anywhere in the external medium is then

$$\Phi(\rho,z) = \frac{1}{2\pi} \int_{-\infty}^{\infty} \frac{F(k)K_0(|k|\rho)}{K_0(|k|a)} e^{-jkz} \, dk \tag{5.116}$$

We consider field points which are at a distance from the nerve that is large compared with the radius of the latter. Then (5.105) can be written

$$\Phi(\rho,z) \equiv \Phi(P) = -\frac{1}{4\pi} \int_z \left( 2\pi a \frac{\partial \Phi}{\partial \rho} \Big|_{\rho=a} \right) \frac{1}{r} \, dz$$

$$- \frac{1}{4\pi} \int_z \left( \pi a^2 \frac{\partial^2 \Phi}{\partial z^2} \Big|_{\rho=a} \right) \frac{1}{r} \, dz \tag{5.117}$$

Now we wish to compare the relative magnitudes of $\mathcal{J}_1(z)$ and $\mathcal{J}_2(z)$,† defined as

$$\mathcal{J}_1(z) = \frac{a}{2} \frac{\partial \Phi}{\partial \rho} \Big|_{\rho=a} = -\frac{a}{4\pi} \int_{-\infty}^{\infty} |k|F(k) \frac{K_1(|k|a)}{K_0(|k|a)} e^{-jkz} \, dk \tag{5.118}$$

$$\mathcal{J}_2(z) = \frac{a^2}{4} \frac{\partial^2 \Phi}{\partial z^2} \Big|_{\rho=a} = -\frac{a^2}{8\pi} \int_{-\infty}^{\infty} k^2 F(k) e^{-jkz} \, dk = \frac{a^2}{4} \frac{\partial^2 \Phi_{s0}}{\partial z^2} \tag{5.119}$$

The integral formulations in (5.118) and (5.119) follow directly from (5.116) by carrying out the indicated differentiation under the integral sign.

Specific conditions are chosen to correspond to typical squid-axon data. We take $a = 0.02$ cm, velocity of propagation of 20 m/sec, and a monophasic-action-potential duration of 1 msec. The spatial extent of the propagated action potential is then approximately 2 cm. To simplify the computation, the action potential can be approximated by the equation

$$\Phi_S(z) = A e^{-4z^2} \tag{5.120}$$

where $z$ is in centimeters and $\Phi_S(z)$ is the monophasic action potential of the axon in air. Equation (5.120) is plotted in Fig. 5.17. One notes that the spatial extent is nominally the desired 2 cm. The main deviation from a true action potential is in (5.120) being symmetrical whereas the rise time should be much shorter than the decay.[1] However, this approximation yields a significant computational simplification; for the purpose of establishing the relative strength of $\mathcal{J}_1(z)$ and $\mathcal{J}_2(z)$, the error should not be too serious.

---

† Strictly, the comparison is between $\int \mathcal{J}_1/r \, dz$ and $\int \mathcal{J}_2/r \, dz$. However, the functional forms of $\mathcal{J}_1$ and $\mathcal{J}_2$ are fairly similar, so that a comparison of their relative magnitudes is sufficient. In any event, this question can be reexamined once the functions $\mathcal{J}_1(z)$ and $\mathcal{J}_2(z)$ are found.

[1] Compare with Fig. 5.12.

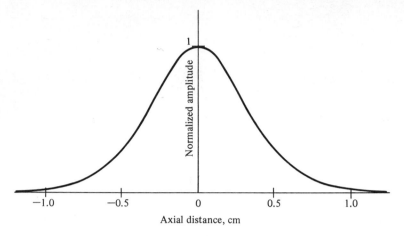

**Fig. 5.17**   Simplified form of a monophasic potential for an excised axon.

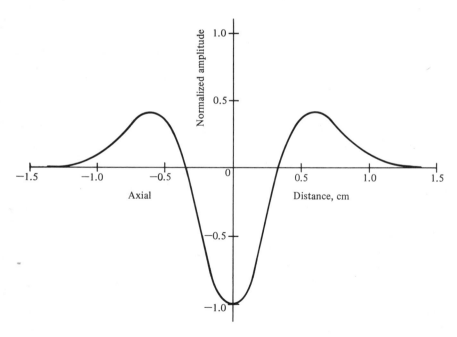

**Fig. 5.18**   *In situ* surface-potential variation as approximated from the monophasic action potential of Fig. 5.17.

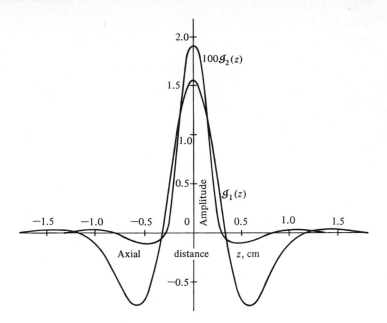

**Fig. 5.19** Integral quantities $\mathcal{I}_1(z)$ and $\mathcal{I}_2(z)$, as defined in the text. [*R. Plonsey, Volume Conductor Fields of Action Currents, Biophys. J.,* **4**:317 (1964).]

With (5.81) as a basis, the surface potential of the axon in an infinite conductor, $\Phi_{s0}(z)$, is clearly proportional to $\partial^2 \Phi_S/\partial z^2$. Thus we have

$$\Phi_{s0}(z) = \frac{Bd^2(e^{-4z^2})}{dz^2} \tag{5.121}$$

where $B$ is a new constant. A plot of (5.121) is shown in Fig. 5.18. The form is the expected triphasic potential. One special advantage of the analytic expression of (5.121) is that $F(k)$ can be readily found from (5.114).

After a change in variable and taking advantage of the symmetry, we finally get

$$\mathcal{I}_1(z) = 17{,}600 \int_0^{0.35} y^2 e^{-156y^2} \left\{ \frac{1 - (y^2/2)\,[\ln(1/y) + \tfrac{1}{2}]}{[\ln(1/y) + 0.116][1 + (y^2/4)]} \right\} \cos(50yz)\,dy \tag{5.122}$$

where the amplitude constant $B$ has been set equal to unity. The upper limit of the integral in (5.122) was chosen to include all essential contributions to $\mathcal{I}_1(z)$. The quantities in the braces are approximations (series form) to the Bessel functions which are accurate to within $2\frac{1}{2}$ percent over the range of integration. The value of $\mathcal{J}_2(z)$ is obtained directly from (5.121), although it was also evaluated, as a check, by the Fourier integral of (5.119). Its evaluation and that of (5.122) are plotted in Fig. 5.19. The magnitude of $\mathcal{J}_1$ is clearly dominant over $\mathcal{J}_2$, thus substantiating the result expressed by (5.103).

## 5.10 EVALUATION OF THE CORE–CONDUCTOR MODEL[1]

In the previous material the electrical properties of the single unmyelinated nerve fiber lying in a homogeneous medium of infinite extent have been described by the core-conductor model of Hermann;[2] the result is embodied in the network of Fig. 4.18, which is replicated here as Fig. 5.20.[3] The electrical properties of the system are denoted, as before, by an external resistance per unit length $r_2$, an axoplasmic resistance per unit length $r_1$, and a shunt (Hodgkin-Huxley) circuit per unit length $Z_m$.

Now the tacit assumption in this model is that only axial variations need be considered, thereby justifying its one-dimensional structure. However, one expects nonuniform skew current lines at least in

[1] J. Clark and R. Plonsey, A Mathematical Evaluation of the Core Conductor Model, *Biophys. J.*, **6**:95 (1966).

[2] L. Hermann, in L. Hermann (ed.), "Handbuch der Physiologie," Leipzig, Vogel. 2, 1879.

[3] This circuit has also been used to represent the average electrical characteristics of a homogeneous group of fibers in a nerve trunk. Because the external axial current tends to be constrained within the relatively narrow interstitial space, this represents a different condition from the single fiber whose external medium is essentially unlimited. The remarks in this section do not apply, therefore, to the nerve trunk where the core-conductor model is probably reasonable when approximations of homogeneity are met.

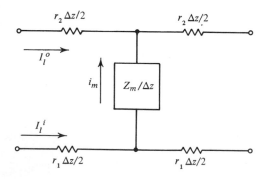

**Fig. 5.20** Electric circuit for the linear-core-conductor model of a cylindrical axon or nerve trunk.

the external medium.[1]   In this section we consider the basic justification of a core-conductor model from field theoretic principles.

Application of Kirchhoff's law to Fig. 5.20 results in the familiar cable equations,

$$\frac{\partial I_l^o}{\partial z} = i_m \tag{5.123}$$

$$I_l^o = -I_l^i \tag{5.124}$$

$$\frac{\partial V^o}{\partial z} = -I_l^o r_2 \tag{5.125}$$

$$\frac{\partial V^i}{\partial z} = -I_l^i r_1 \tag{5.126}$$

where $V^o$ represents the potential along $r_2$ (the external medium) while $V^i$ is the potential along $r_1$ (the axoplasm) relative to an arbitrary reference at infinity.   $I_l^o$ is the total longitudinal external current, $I_l^i$ is the longitudinal internal current, and $i_m$ is the transmembrane current per unit length.   We also adopt the nomenclature that the potential at an arbitrary point in the external medium be designated $\Phi^o(\rho,\phi,z)$, so that

$$\nabla^2\Phi^o(\rho,\phi,z) = 0 \qquad \rho > a \tag{5.127}$$

while in the axoplasm

$$\nabla^2\Phi^i(\rho,\phi,z) = 0 \qquad \rho < a \tag{5.128}$$

and $a$ is the radius.

In order to link the quantities described in Eqs. (5.123) through (5.126) to the three-dimensional field solution, we make the following fairly obvious definitions.   With regard to potential, we let the potential variation $V^o$ of the cable equation correspond to the actual potential of the outer surface of the membrane, while $V^i$ is identified with the potential at the inner-membrane surface.   That is,

$$\Phi^o(a,\phi,z) \equiv \Phi_s^o \equiv V^o \tag{5.129}$$

$$\Phi^i(a,\phi,z) \equiv \Phi_s^i \equiv V^i \tag{5.130}$$

Strictly, we should put $\rho = a^+$ in Eq. (5.129) and $\rho = a^-$ in Eq. (5.130), but we are assuming the membrane to be very thin; no ambiguity results since the superscript indicates the region from which the membrane is approached.

The external axial current $I_l^o$ is associated with the total axial current in the external medium (corresponding to a particular $z$).   That

[1] F. Offner, *Electroencephalog. Clin. Neurophysiol.*, **6**:507 (1954).

is, since the axial electric field is $-\partial\Phi^o/\partial z$ and consequently the axial current density is $-\sigma_o\,\partial\Phi^o/\partial z$, we have

$$I_l^o = -\sigma_o \int_a^\infty 2\pi\rho \left(\frac{\partial\Phi^o}{\partial z}\right) d\rho \qquad (5.131)$$

In a similar way, we define

$$I_l^i = -\sigma_i \int_0^a 2\pi\rho \left(\frac{\partial\Phi^i}{\partial z}\right) d\rho \qquad (5.132)$$

While the above current definitions may be justified on intuitive grounds, they prove also to be such as to ensure satisfaction of Eqs. (5.123) and (5.124). We can show this in the following way. Taking partial derivatives of Eq. (5.131) with respect to $z$ gives

$$\frac{\partial I_l^o}{\partial z} = -\sigma_o \int_a^\infty 2\pi\rho\, \frac{\partial^2\Phi^o}{\partial z^2}\, d\rho \qquad (5.133)$$

From the expansion of $\nabla^2\Phi$ in cylindrical coordinates and using the axial-symmetry condition $(\partial/\partial\phi = 0)$, we find in view of Eq. (5.127)

$$\frac{\partial^2\Phi^o}{\partial z^2} = -\frac{1}{\rho}\frac{\partial}{\partial\rho}\left(\rho\,\frac{\partial\Phi^o}{\partial\rho}\right) \qquad (5.134)$$

Substituting Eq. (5.134) into Eq. (5.133) yields

$$\frac{\partial I_l^o}{\partial z} = 2\pi\sigma_o \int_a^\infty \frac{\partial}{\partial\rho}\left(\rho\,\frac{\partial\Phi^o}{\partial\rho}\right) d\rho = 2\pi\sigma_o \left[\rho\,\frac{\partial\Phi^o}{\partial\rho}\right]_a^\infty \qquad (5.135)$$

The radial behavior of $\Phi^o$ has been shown to vary as $1/\rho^3$ for large $\rho$. Consequently, for $\rho \to \infty$, $\partial\Phi^o/\partial\rho \to 1/\rho^4$. Thus

$$\frac{\partial I_l^o}{\partial z} = -2\pi\sigma_o a\, \frac{\partial\Phi^o(\rho,z)}{\partial\rho}\bigg|_{\rho=a} = i_m^o \qquad (5.136)$$

where $-\sigma_o[\partial\Phi^0(\rho,z)/\partial\rho]\big|_{\rho=a}$ is identified as the outward current density at the outer membrane surface from which the identification of transmembrane current per unit length, $i_m$, is made. This result confirms Eq. (5.123).

Since biological membranes are very thin and of high resistance, they carry negligible longitudinal current. Furthermore, since the current is solenoidal $(\nabla \cdot \mathbf{J} = 0)$, the net current crossing a plane $z =$ const equals zero, and by virtue of neglecting longitudinal membrane current one gets

$$2\pi \int_0^\infty J_z\rho\, d\rho = -2\pi\sigma_i \int_0^a \frac{\partial\Phi^i}{\partial z}\,\rho\, d\rho - 2\pi\sigma_o \int_a^\infty \frac{\partial\Phi^o}{\partial z}\,\rho\, d\rho = 0 \quad (5.137)$$

In terms of the definitions of $I_l{}^o$ and $I_l{}^i$ given in Eqs. (5.131) and (5.132) one gets

$$I_l{}^o = -I_l{}^i$$

showing that Eq. (5.124) is valid.

It remains now to investigate how well Eqs. (5.125) and (5.126) are satisfied by the three-dimensional model. Upon examination, one notes that in order for the actual three-dimensional axon to satisfy Eqs. (5.125) and (5.126) it is necessary for a proportionality to exist between total current $I_l{}^o$ and the electric field along both the outer and the inner membrane surface. Since the electric field and *current density* are linearly related by Ohm's law, this is equivalent to the requirement that the total current and the current density at the membrane surface be proportional, a condition that intuitively is not likely to be satisfied by a complex current-density field. Thus, if we define the effective cross-sectional area $A^o$ and $A^i$ by

$$I_l{}^o(z) = -A^o(z)\sigma_o \frac{\partial \Phi_s{}^o}{\partial z} \tag{5.138}$$

$$I_l{}^i(z) = -A^i(z)\sigma_i \frac{\partial \Phi_s{}^i}{\partial z} \tag{5.139}$$

satisfaction of Eqs. (5.125) and (5.126) is equivalent to the requirement that $A^o(z)$ and $A^i(z)$ be independent of $z$.

To investigate Eqs. (5.125) and (5.126) we note that the specification of $\Phi_s{}^o$ for the external region and $\Phi_s{}^i$ for the internal region uniquely specifies the potential everywhere in the respective regions. It is then possible, based on these fields, to compute the current densities everywhere and hence the total currents according to Eqs. (5.131) and (5.132). One may then proceed to compare the electric field at the membrane surface with the corresponding total currents and in this way examine $r_2$ and $r_1$ [or, equivalently, $A^o(z)$ and $A^i(z)$] for constancy with respect to $z$.

The procedure is to utilize (5.116) and the definition in (5.131) [differentiating (5.116) under the integral sign and interchanging the order of integration], and the result is

$$I_l{}^o(z) = j\sigma_o \int_{-\infty}^{\infty} \frac{kF^o(k)}{K_0(|k|a)} e^{-jkz} \left[ \int_a^{\infty} \rho K_0(|k|\rho)\, d\rho \right] dk \tag{5.140}$$

where $F^o(k)$ is given by (5.114), and the superscript designates the external region. The integration in $\rho$ is straightforward, and one obtains

$$\int_a^{\infty} \rho K_0(|k|\rho)\, d\rho = \left[ \frac{\rho K_1(|k|\rho)}{|k|} \right]_a^{\infty} = -\frac{a K_1(|k|a)}{|k|} \tag{5.141}$$

If the identity

$$-\frac{aK_1(|k|a)}{|k|} = \frac{a^2K_0(|k|a)}{2} - \frac{a^2K_2(|k|a)}{2} \qquad (5.142)$$

is used, then Eq. (5.140) with the result of (5.141) can be expressed as the sum of two integrals, namely,

$$I_l{}^o(z) = \frac{a^2\sigma_o}{2} \int_{-\infty}^{\infty} jkF^o(k)e^{-jkz}\, dk - \frac{a^2\sigma_o}{2} \int_{-\infty}^{\infty} jkF^o(k)\frac{K_2(|k|a)}{K_0(|k|a)}\, e^{-jkz}\, dk$$

$$(5.143)$$

The first integral can be identified if the derivative with respect to $z$ is taken of both sides of Eq. (5.113) and the definition of $F^o(k)$ given in Eq. (5.115) is used. The result is

$$I_l{}^o(z) = -\pi a^2\sigma_o \frac{\partial \Phi_s{}^o(z)}{\partial z} - \frac{a^2\sigma_o}{2} \int_{-\infty}^{\infty} jkF^o(k)\frac{K_2(|k|a)e^{-jkz}}{K_0(|k|a)}\, dk \qquad (5.144)$$

The first term on the right-hand side of Eq. (5.144) is precisely in the proportional form necessary for Eq. (5.125) to be valid. It remains to examine the influence of the integral expression in Eq. (5.144) on the current-field relationship. Because of the complexity of the integral this cannot be done, in general. Instead, this question is considered by means of specific action potentials.

Before proceeding with this evaluation, we consider a repetition of the above analysis but applied to the axoplasmic region. In this case, the appropriate eigenfunction expressions are

$$A'(k)I_0(k\rho)e^{jkz} + B'(k)I_0(k\rho)e^{-jkz}$$

where $I_0$ is the modified Bessel function of the first kind. This form is chosen since it alone is well behaved at the origin and at the same time permits the complex exponential $z$ variation. By an entirely analogous development, one can show that the potential anywhere within the internal region is

$$\Phi^i(\rho,z) = \frac{1}{2\pi} \int_{-\infty}^{\infty} F^i(k)\frac{I_0(|k|\rho)}{I_0(|k|a)}\, e^{-jkz}\, dk \qquad (5.145)$$

where

$$F^i(k) = \int_{-\infty}^{\infty} \Phi_s{}^i(z)e^{jkz}\, dz = C'(k)I_0(|k|a) \qquad (5.146)$$

We now make use of Eq. (5.132) and, as before, differentiate under the integral sign [in Eq. (5.145)] and interchange the order of integration [in Eq. (5.132)]. This results in

$$I_l{}^i(z) = j\sigma \int_{-\infty}^{\infty} \frac{kF^i(k)}{I_0(|k|a)}\, e^{-jkz}\left[\int_0^a \rho I_0(|k|\rho)\, d\rho\right] dk \qquad (5.147)$$

The $\rho$ integration gives

$$\int_0^a \rho I_0(|k|\rho)\, d\rho = \left[\frac{a \rho I_1(|k|\rho)}{|k|}\right]_0 = \frac{aI_1(|k|a)}{|k|} \tag{5.148}$$

Except for the difference in sign, this is the dual to Eq. (5.141); and the result, corresponding to Eq. (5.144), is now clearly

$$I_l^i(z) = \pi a^2 \sigma_i \frac{\partial \Phi_s^i(z)}{\partial z} + \frac{a^2 \sigma_i}{2} \int_{-\infty}^{\infty} jk F^i(k) \frac{I_2(|k|a)}{I_0(|k|a)} e^{-jkz}\, dk \tag{5.149}$$

As before, the result consists of a proportional term and an additional integral. For this case, the coefficient of the first term $\sigma_i \pi a^2$ is precisely the classical internal resistance per unit length $r_1$. Thus, in this case the integral term constitutes a correction to the classical cable expression (5.126).

As noted, the evaluation of the cable equation has been reduced to a consideration of the integrals in Eqs. (5.144) and (5.149). Since this cannot be done in general, calculations utilizing typical squid data were performed. This work follows the method at the end of Sec. 5.9 except that three gaussian curves, each with different amplitude, mean, and standard deviation, are superposed to synthesize quite closely to real action-potential waveforms. As a test of the validity of (5.125) and (5.126) the quantities $r_1$ and $r_2$ were determined from

$$r_1(z) = -\frac{1}{I_l^i(z)} \frac{\partial \Phi_s^i(z)}{\partial z} \tag{5.150}$$

$$r_2(z) = -\frac{1}{I_l^o(z)} \frac{\partial \Phi_s^o(z)}{\partial z} \tag{5.151}$$

utilizing (5.144) and (5.149) with synthetic data shown in Fig. 5.21. For the aforementioned equations to be satisfied, $r_1(z)$ and $r_2(z)$ should be constants (independent of $z$). Table 5.3 lists computed values, and one notes satisfaction for $r_1$ but not $r_2$. One also notes that $r_1 \gg r_2$, as anticipated earlier.

Table 5.3 Typical values of longitudinal resistances per unit length $r_1$ and $r_2$

| $z$, cm | $r_2$, k$\Omega$/cm | $r_1$, k$\Omega$/cm |
|---|---|---|
| 0.0 | 0.06 | 15.91 |
| 0.2 | 0.70 | 15.91 |
| 0.4 | 0.62 | 15.91 |
| 0.6 | 0.48 | 15.91 |
| 0.8 | 0.27 | 15.91 |
| 1.0 | 0.13 | 15.91 |
| 1.2 | 0.23 | 15.91 |
| 1.4 | 0.11 | 15.91 |

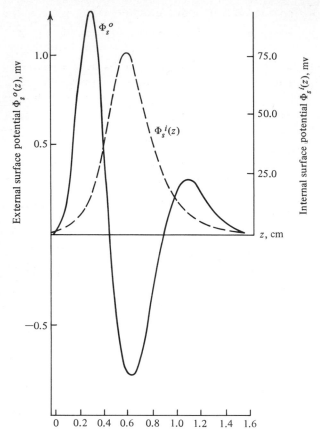

**Fig. 5.21**  Plot of external (solid line) and internal (broken line) surface-potential distribution vs. axial distance $z$ (in centimeters).  [*J. Clark and R. Plonsey, A Mathematical Evaluation of the Core Conductor Model, Biophys. J.*, **6**:95 (1966).]

In conclusion, it is clear that the linear-core-conductor model is not satisfactory for external conditions but should adequately describe internal effects.   In particular, no direct relationship exists between the longitudinal electric field at the surface ($-\partial \Phi_s{}^o / \partial z$) and the total longitudinal current $I_i{}^o$, as assumed in the classical theory.   On the other hand, other evidence indicates that the axon behavior is not seriously affected by neglecting potential variations in the external medium altogether by taking $r_2 = 0$.   This procedure cannot be followed, of course, if the axon field is itself the subject of study.

## 5.11 TRANSMEMBRANE POTENTIAL AS A SOURCE FUNCTION

The results in Table 5.3 are based on the synthetic data in Fig. 5.21. Very accurate data exist for transmembrane potentials, but only poor measurements of $\Phi_s{}^o$ are available. This function, however, is a second-order quantity; because of its low magnitude $\Phi_s{}^i$ is approximately equal to the transmembrane potential. It is, however, possible to relate both $\Phi_s{}^o$ and $\Phi_s{}^i$ rigorously to the transmembrane potential $V(z)$.† We do this by imposing the condition of continuity of normal component of current across the axon membrane.

An expression for the transmembrane current per unit length is given in (5.136). Combining this with (5.116) leads to

$$i_m{}^o = \sigma_o a \int_{-\infty}^{\infty} |k| F^o(k) \frac{K_1(|k|a)}{K_0(|k|a)} e^{-jkz} \, dk \qquad (5.152a)$$

In a similar way we can obtain

$$i_m{}^i = -\sigma_i a \int_{-\infty}^{\infty} |k| F^i(k) \frac{I_1(|k|a)}{I_0(|k|a)} e^{-jkz} \, dk \qquad (5.152b)$$

Since $i_m{}^i = i_m{}^o$, we have

$$-\sigma_0 F^o(k) \frac{K_1(|k|a)}{K_0(|k|a)} = \sigma_i F^i(k) \frac{I_1(|k|a)}{I_0(|k|a)} \qquad (5.153)$$

Now

$$V(z) = \frac{1}{2\pi} \int_{-\infty}^{\infty} [F^i(k) - F^o(k)] e^{-jkz} \, dk$$

$$= \frac{1}{2\pi} \int_{-\infty}^{\infty} \alpha(|k|a) F^o(k) e^{-jkz} \, dk \qquad (5.154)$$

where

$$\alpha(|k|a) \equiv -\left[ \frac{\sigma_o K_1(|k|a) I_0(|k|a)}{\sigma_i K_0(|k|a) I_1(|k|a)} + 1 \right] \qquad (5.155)$$

The inverse transform of (5.154) yields

$$V(k) = \int_{-\infty}^{\infty} V(z) e^{jkz} \, dk = F^o(k) \alpha(|k|a)$$

Thus, given the transmembrane potential $V(k)$, $F^o(k)$ and hence $\Phi_s{}^o$ can be determined. In particular, the external potential field is related to the transmembrane potential by

$$\Phi^o(\rho,z) = \frac{1}{2\pi} \int_{-\infty}^{\infty} \frac{V(k) K_0(|k|\rho)}{\alpha(|k|a) K_0(|k|a)} e^{-jkz} \, dk \qquad (5.156)$$

---

† D. B. Geselowitz, Comment on the Core Conductor Model, *Biophys. J.*, **6**:691 (1966).

Clark[1] utilized Eq. (5.156) to calculate the field distribution surrounding an unmyelinated cylindrical nerve fiber in a uniform conducting region of infinite extent.   The experimentally determined transmembrane potential $V(z)$ of the crayfish lateral giant axon, as obtained by Watanabe and Grundfest,[2] was used.   An analytic expression for $V(z)$ was obtained by summing three gaussian curves with appropriate amplitude, mean, and standard deviation for best fit.   A plot of $V(z)$ and the computed surface potential $\Phi_s{}^o(z)$ are given in Fig. 5.22.   One notes that $\Phi_s{}^o(z)$ has the expected triphasic form.   Furthermore, the magnitude of $\Phi_s{}^o(z)$ is relatively small and serves as a further justification for approximating the extracellular medium as isopotential.   A field plot for this example is given in Fig. 5.23.   The general form corresponds to that computed by Lorente de Nó and illustrated in Fig. 5.13.

To the extent that the above example can be generalized, the transformation of (5.66) to (5.68) appears justifiable as long as the external

[1] J. Clark, "Bioelectric Field Interaction between Adjacent Nerve Fibers," Ph.D. dissertation, Case Western Reserve University, November, 1967.
[2] A. Watanabe and H. Grundfest, Impulse Propagation of the Crayfish Lateral Giant Axon, *J. Gen. Physiol.*, **45**:267 (1961).

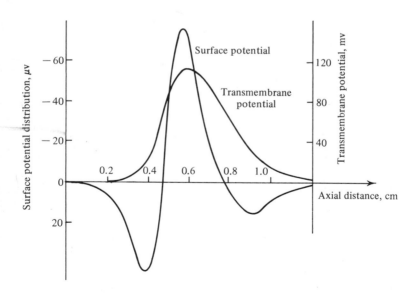

**Fig. 5.22**   Transmembrane-potential variation from direct measurement and surface-potential distribution computed therefrom, for the crayfish lateral giant axon. (*J. Clark, "Bioelectric Field Interaction between Adjacent Nerve Fibers," Ph.D. thesis, Case Western Reserve University, November,* 1967.)

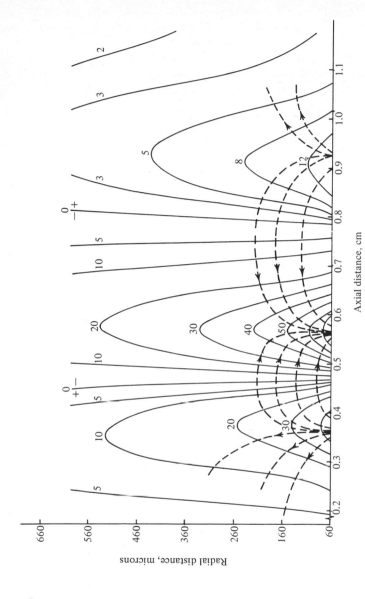

**Fig. 5.23** Calculated field distribution of a crayfish lateral giant axon in a uniform conducting medium of infinite extent based on measured transmembrane potential. The axon radius is 60 microns. Isopotential lines are solid, current flow lines are dashed; potentials are in microvolts. (*After J. Clark, "Bioelectric Field Interaction between Adjacent Nerve Fibers," Ph.D. thesis, Case Western Reserve University, November, 1967.*)

medium has a characteristic dimension that is large compared with the fiber radius (i.e., is extensive). Such a condition would appear to ensure negligible potential variations in the external medium (i.e., a short-circuiting effect). A further consequence is the validation of the linear-core-conductor model, provided the external resistance is set equal to zero, a procedure already commented on. One prediction of this model is a linear relationship between transmembrane current and the second derivative of the surface potential with respect to the axial variable [see (5.84)]. An investigation of this equation utilizing the aforementioned rigorous example was performed by Clark and Plonsey,[1] and excellent results are noted. These conclusions are of great importance in the analysis of volume-conductor fields of action currents.

If an inactive fiber is placed in the conducting medium parallel to the active fiber, action currents will enter and leave the inactive fiber. It is of interest to determine the transmembrane voltage which results thereby. This induced voltage will alter the excitation threshold and, if sufficiently high, could result in ephaptic stimulation of the inactive fiber. An analytic expression for the desired transmembrane potential can be developed by utilizing (5.156) as a source function.

Referring to Fig. 5.24, we see that an expression for the potential

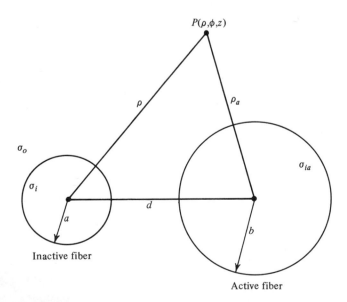

**Fig. 5.24**   Geometry for fiber-interaction problem.

within the axoplasm of the inactive fiber has the form

$$\Phi^i(\rho,\phi,z) = \sum_{n=0}^{\infty} \cos n\phi \int_{-\infty}^{\infty} A_n(k) I_n(|k|\rho) e^{-jkz}\, dk \qquad (5.157)$$

which corresponds to (5.145) except that a $\phi$ dependence must now be included. For the extracellular region

$$\Phi^o(\rho,\phi,z) = \Phi_a{}^o(\rho_s,z) + \sum_{n=0}^{\infty} \cos n\phi \int_{-\infty}^{\infty} B_n(k) K_n(|k|\rho) e^{-jkz}\, dk \quad (5.158)$$

where $\Phi_a{}^o$ is the potential field of the active fiber alone. That is, $\Phi_a{}^o$ is precisely the expression (5.156) but with $\rho$ replaced by $\rho_a$ to correspond with the geometry and nomenclature defined in Fig. 5.23. Note that the coordinate system origin is at the center of the *inactive* fiber.

Now one can use the Bessel function addition theorem[1]

$$K_0(|k|\rho_a) = \sum_{n=0}^{\infty} (2 - \delta_n{}^0) \cos n\phi K_n(|k|d) I_n(|k|\rho) \qquad \text{for } \rho < d$$

$$= \sum_{n=0}^{\infty} (2 - \delta_n{}^0) \cos n\phi I_n(|k|d) K_n(|k|\rho) \qquad \text{for } \rho > d$$

where $\delta_n{}^0 = 1$ for $n = 0$ and zero for $n \neq 0$. The result is that (5.158) can be written

$$\Phi^o(\rho,\phi,z) = \sum_{n=0}^{\infty} \cos n\phi \int_{-\infty}^{\infty} [P_n(k) I_n(|k|\rho) + B_n(k) K_n(|k|\rho)] e^{-jkz}\, dk$$
$$\text{for } \rho < d \quad (5.159a)$$

$$\Phi^o(\rho,\phi,z) = \sum_{n=0}^{\infty} \cos n\phi \int_{-\infty}^{\infty} \left[ \frac{I_n(|k|d)}{K_n(|k|d)} P_n(k) + B_n(k) \right]$$
$$\times K_n(|k|\rho) e^{-jkz}\, dk \qquad \text{for } \rho > d \quad (5.159b)$$

where

$$P_n(k) = \frac{(2 - \delta_n{}^0) V(k) K_n(|k|d)}{\alpha(|k|b) K_0(|k|b)}$$

The undetermined functions $A(k)$ and $B(k)$ can be found from the boundary conditions at the membrane interface of the inactive fiber (at $\rho = a$). Here we have continuity of normal current density

$$-\sigma_i \frac{\partial \Phi^i}{\partial \rho}\bigg|_{\rho=a} = -\sigma_o \frac{\partial \Phi^o}{\partial \rho}\bigg|_{\rho=a} = J_m \qquad (5.160)$$

---

[1] A. Gray and G. B. Mathews, "A Treatise on Bessel Functions and their Application to Physics," Dover Publications, Inc., New York, 1966.

and if we assume the simple $RC$ membrane of Fig. 3.18, the following impedance relationship must hold:

$$J_m = \sigma_m \Phi_m + C_m \frac{\partial \Phi_m}{\partial t} \tag{5.161}$$

where

$$\Phi_m(\phi,z) = \Phi_i(a,\phi,z) - \Phi_o(a,\phi,z)$$

Equations (5.160) and (5.161) are sufficient to permit the determination of $A(k)$ and $B(k)$. In the application of (5.161) the time derivative can be related to spatial variations (which are known) by noting that in view of propagation all field qualities have a $(z + vt)$ dependence, where $v$ is the propagation velocity. Consequently,

$$\frac{\partial \Phi_m}{\partial t} = v \frac{\partial \Phi_m}{\partial z}$$

An analytic form for the spatial derivative of $\Phi_m$ with respect to $z$ [consider (5.157) and (5.159)] can be readily obtained.

Using results obtained as outlined above, Clark[1] calculated numerical solutions for the field arising from an active fiber in the extracellular region and in the inactive fiber axoplasm due to an active fiber source. The basic data, those of transmembrane potential of the active fiber, were chosen as noted earlier (shown in Fig. 5.22). Other values utilized (refer to Fig. 5.24) were

$$\sigma_i = \sigma_{ia} = 0.0106 \text{ mho/cm} \qquad a = 40 \text{ microns}$$

$$\sigma_o = 0.05 \text{ mho/cm} \qquad b = 60 \text{ microns}$$

$$C_m = 0.61 \text{ } \mu f/cm^2 \qquad d = 160 \text{ microns}$$

$$\sigma_m = 0.000322 \text{ mho/cm}^2$$

Table 5.4 lists the transmembrane potential $\Phi_m$ as a function of $\phi$ at $z = 0.625$ cm, while a field plot in a longitudinal plane is given in Fig. 5.25.

The magnitudes of potential in the external medium, as given in Fig. 5.25, and the transmembrane potential induced in the inactive fiber, as shown in Table 5.4, are relatively low. For the more realistic condition that the fibers are located within a nerve trunk these potentials can be expected to be greatly increased. That is, the potential external to the active fiber should be proportional to the active transmembrane potential times the ratio of effective external resistance (per unit length) to axoplasmic resistance (per unit length). The latter ratio could

[1] J. Clark, op. cit.

Table 5.4  $\Phi_m$ versus $\phi$
calculated at $z = 0.625$ cm*

| $\phi$, deg | $\Phi_m$, $\mu v$ |
|---|---|
| 0 | $-19.1$ |
| 30 | $-17.5$ |
| 60 | $-13.8$ |
| 90 | $-9.6$ |
| 120 | $-6.3$ |
| 150 | $-4.5$ |
| 180 | $-3.9$ |

\* J. Clark, "Bioelectric Field Interaction between Adjacent Nerve Fibers," Ph.D. thesis, Case Western Reserve University, November, 1967.

increase perhaps one hundredfold when the external medium is confined by a nerve-trunk sheath.   By an extension of the previous mathematics it is possible to consider this model analytically.   The case of a single fiber in a nerve trunk has been treated by Clark,[1] and a potential enhancement in the order of 100 has been found.

It is interesting to consider the inactive transmembrane potential as a function of $r_m$ and $C_m$.   Such a study shows that variations of $r_m$ about the physiological range produce very small changes; the controlling parameter is $C_m$.[†]   In other words, corresponding to the effective temporal frequencies of the action potential, the capacitive reactance of the membrane is much smaller than its resistance and hence the passive membrane behavior is mainly dependent on the magnitude of $C_m$.

## 5.12  INHOMOGENEOUS MEDIA[2]

The relationship between potential field and bioelectric sources developed in Sec. 5.6 was based on an assumption that the medium is of uniform conductivity.   In this section we shall derive appropriate expressions for the case where the conductivity is discontinuous at one or more surfaces.

[1] J. Clark, op. cit.
[†] Ibid.
[2] The material in this section is based on D. B. Geselowitz, On Bioelectric Potentials in an Inhomogeneous Volume Conductor, Biophys. J., 7:1 (1967).

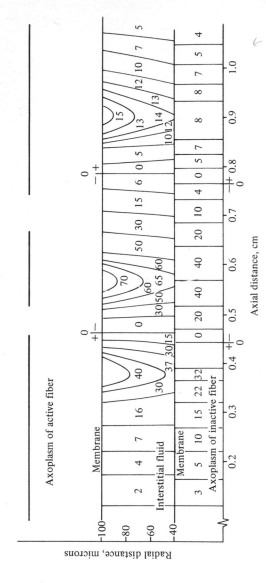

**Fig. 5.25** Calculated field distribution for an active fiber of radius 60 microns and an inactive fiber of radius 40 microns with a center-to-center spacing of 160 microns. Designated potentials are in microvolts. (*J. Clark, "Bioelectric Field Interaction between Adjacent Nerve Fibers," Ph.D. thesis, Case Western Reserve University, November, 1967.*)

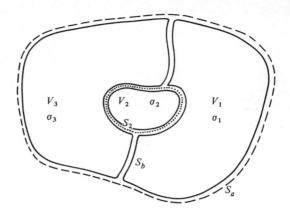

**Fig. 5.26** Geometry for inhomogeneous Green's theorem.

To introduce this problem, we first consider a specific example, which is illustrated in Fig. 5.26. The region $V$ is composed of three internal subunits $V_1$, $V_2$, $V_3$ of conductivity $\sigma_1$, $\sigma_2$, $\sigma_3$, respectively. Now the divergence (or Gauss' theorem) asserts that

$$\int_V \nabla \cdot \mathbf{A} \, dV = \int_S \mathbf{A} \cdot d\mathbf{S} \tag{5.162}$$

provided that $\mathbf{A}$ is finite and continuous in $V$. We shall now define

$$\mathbf{A} = \psi\sigma\nabla\phi$$

where $\psi$ and $\phi$ are scalar fields that are finite, continuous, and twice differentiable in $V$ (Fig. 5.26), and $\sigma$ takes on the value $\sigma_1$ in $V_1$, $\sigma_2$ in $V_2$, and $\sigma_3$ in $V_3$. We may apply (5.162) separately in each of the subregions, so that

$$\int_{V_1} \nabla \cdot (\psi\sigma_1\nabla\phi) \, dV = \int_{S_1} \psi\sigma_1\nabla\phi \cdot d\mathbf{S} \tag{5.163a}$$

$$\int_{V_2} \nabla \cdot (\psi\sigma_2\nabla\phi) \, dV = \int_{S_2} \psi\sigma_2\nabla\phi \cdot d\mathbf{S} \tag{5.163b}$$

$$\int_{V_3} \nabla \cdot (\psi\sigma_3\nabla\phi) \, dV = \int_{S_3} \psi\sigma_3\nabla\phi \cdot d\mathbf{S} \tag{5.163c}$$

If Eqs. (5.163) are summed, then we get

$$\int_V \nabla \cdot (\psi\sigma\nabla\phi) \, dV = \sum_{i=1}^{2} \int_{S_i} \psi\sigma \frac{\partial\phi}{\partial n_i} dS_i$$

$$+ \sum_{j=1}^{3} \int_{S_j} \left( \psi'\sigma' \frac{\partial\phi_j'}{\partial n_j'} + \psi''\sigma'' \frac{\partial\phi_j''}{\partial n_j''} \right) dS_j \tag{5.164}$$

where the surfaces $S_i$ are those that bound the volume $V$,† while the surfaces $S_j$ are internal surfaces which in all cases are described twice. An example of the former is $S_a$, while the latter is $S_b$. For surfaces $S_j$ the prime notation refers to the fields and conductivity on one side, while the double prime refers to these variables on the other side. The direction of the normal is always into the surface (outward from the subregion). In the notation of (5.164) $\sigma$ denotes the conductivity within $V$ and may be viewed as a piecewise continuous function.

We now duplicate the development leading to (5.164) but with $\psi$ and $\phi$ interchanged. This gives

$$\int_V \boldsymbol{\nabla} \cdot (\phi\sigma\boldsymbol{\nabla}\psi) \, dV = \sum_{i=1}^{2} \int_{S_i} \phi\sigma \frac{\partial\psi}{\partial n_i} \, dS_i$$
$$+ \sum_{j=1}^{3} \int_{S_j} \left( \phi'\sigma' \frac{\partial\psi_j'}{\partial n_j'} + \phi''\sigma'' \frac{\partial\psi_j''}{\partial n_j''} \right) dS_j \quad (5.165)$$

Now if we subtract (5.165) from (5.164), a generalized form of Green's theorem results which is applicable when $\sigma$ is discontinuous. At this point we shall generalize the result to an arbitrary number of subvolumes with $m$ bounding component surfaces and $q$ internal surface sections. The final result is[1]

$$\int_V [\boldsymbol{\nabla} \cdot (\psi\sigma\boldsymbol{\nabla}\phi) - \boldsymbol{\nabla} \cdot (\phi\sigma\boldsymbol{\nabla}\psi)] \, dV = \sum_{i=1}^{m} \int_{S_i} \sigma \left( \psi \frac{\partial\phi}{\partial n_i} - \phi \frac{\partial\psi}{\partial n_i} \right) dS_i$$
$$+ \sum_{j=1}^{q} \int_{S_j} \left[ \sigma' \left( \psi' \frac{\partial\phi_j'}{\partial n_j'} - \phi' \frac{\partial\psi_j'}{\partial n_j'} \right) + \sigma'' \left( \psi'' \frac{\partial\phi_j''}{\partial n_j''} - \phi'' \frac{\partial\psi_j''}{\partial n_j''} \right) \right] dS_j$$
$$(5.166)$$

It is now possible to follow the procedure in Sec. 5.6 but utilizing the more general expression (5.166). We are interested in the potential at an arbitrary point $P(x',y',z')$ in a region $V_p$ of conductivity $\sigma_p$. We define

$$\psi = \frac{1}{r} = \frac{1}{\sqrt{(x - x')^2 + (y - y')^2 + (z - z')^2}} \quad (5.167)$$

† Although in Fig. 5.26 the bounding surface is external and singly connected, it is possible for this surface to be multiply connected as well. Thus, in Fig. 5.26 we could apply (5.164) to the volume $V = V_1 + V_3$, in which case $S_i$ would include the surface $S_2$.

[1] W. R. Smythe, "Static and Dynamic Electricity," 2d ed., p. 53, McGraw-Hill Book Company, New York, 1950.

and let $\phi = \Phi$, where $\Phi$ is the desired potential field; that is,

$$\nabla^2 \Phi = -\frac{I_v}{\sigma} \tag{5.168}$$

Furthermore, at the interface between regions of different conductivity the normal component of current $(-\sigma\, \partial\Phi/\partial n)$ must be continuous, as noted in (5.16), as must the potential itself,[1] hence

$$\sigma' \frac{\partial \Phi'}{\partial n'_j} + \sigma'' \frac{\partial \Phi''}{\partial n''_j} = 0 \tag{5.169a}$$

$$\Phi'_j = \Phi''_j \tag{5.169b}$$

The volume integral in (5.166) can be simplified since for each subregion

$$\nabla \cdot (\psi\sigma\nabla\Phi) - \nabla \cdot (\Phi\sigma\nabla\psi) = \sigma(\psi\nabla^2\Phi - \Phi\nabla^2\psi) \\ + \sigma(\nabla\psi \cdot \nabla\Phi - \nabla\Phi \cdot \nabla\psi) = \sigma(\psi\nabla^2\Phi - \Phi\nabla^2\psi) \tag{5.170}$$

and this is also true for the summation over all subregions, i.e., true for $V$. By using (5.167), (5.168), and (5.170) in (5.166), the left-hand side becomes

$$-\int_V \frac{I_v\, dV}{r} - \int_{V_\epsilon} \sigma\Phi\nabla^2\left(\frac{1}{r}\right) dV \\ = -\int_V \frac{I_v\, dV}{r} - \sigma_p\Phi(P) \int_{V_\epsilon} \nabla \cdot \left[\nabla\left(\frac{1}{r}\right)\right] dV \tag{5.171}$$

where $V_\epsilon$ is an infinitesimal volume surrounding $P$ [outside this region $\nabla^2(1/r) = 0$, and there is no additional contribution to the volume integral]. Since $\Phi$ is well behaved and essentially unchanging in the infinitesimal volume $V_\epsilon$, it is removed from the integral with the value $\Phi(P)$. We also note that

$$\int_{V_\epsilon} \nabla \cdot \left[\nabla\left(\frac{1}{r}\right)\right] dV = \int_{S_\epsilon} \nabla\left(\frac{1}{r}\right) \cdot d\mathbf{S} = -\frac{4\pi r^2}{r^2} = -4\pi \tag{5.172}$$

We now have, utilizing (5.167), (5.168), (5.169), (5.171), and (5.172), the following:

$$-\int_V \frac{I_v}{r}\, dV + 4\pi\sigma_p\Phi(P) = \sum_i \int_{S_i} \sigma \left[\frac{1}{r}\frac{\partial\Phi}{\partial n_i} - \Phi\frac{\partial(1/r)}{\partial n_i}\right] dS_i \\ + \sum_j \int_{S_j} \Phi \frac{\partial(1/r)}{\partial n_j}(\sigma''_j - \sigma'_j)\, dS_j$$

choosing $n_j$ to be defined from the primed to the double-primed region

[1] This follows from the physical requirement that at most a layer of charge can accumulate at a conductive interface. The property of a charge layer is that the potential must be continuous but the normal derivative discontinuous.

(thus $n_j = n_j' = -n_j''$). Solving for $\Phi(P)$, we finally obtain

$$\Phi(P) = \frac{1}{4\pi\sigma_p} \int_V \frac{I_v}{r}\, dV + \frac{1}{4\pi\sigma_p} \sum_i \int_{S_i} \sigma \left[ \frac{1}{r} \frac{\partial \Phi}{\partial n_i} - \Phi \frac{\partial(1/r)}{\partial n_i} \right] dS_i$$

$$+ \frac{1}{4\pi\sigma_p} \sum_j \int_{S_j} \Phi \frac{\partial(1/r)}{\partial n_j} (\sigma_j'' - \sigma_j')\, dS_j \quad (5.173)$$

This result is a generalization of (5.92). Note that the true (or primary) sources $I_v$ contribute to the potential at $P$ as if the entire volume were uniform with the conductivity that exists at $P$ (that is, $\sigma_p$). The surfaces $S_i$ bounding $V$ contribute to the potential as a consequence of sources which lie outside $V$. If $V$ is chosen to be all of space, then $S_i$ is the surface at infinity and the integral can be dropped by the argument advanced in connection with (5.89). The surface integrals over $S_j$ represent the effect of the discontinuity in conductivity and are expressed in the form of equivalent layer sources.

Usually, in electrophysiological systems, the region under study is bounded by a nonconducting region of infinite extent. In electrocardiography, for example, the body is in air. In nerve physiology the preparation is surrounded by air, etc. In such cases we can specify the surfaces in the second summation of (5.173) to be the interface with the outer nonconducting region, where $\sigma'' = 0$. We may separately designate this surface as $S_0$ and obtain

$$\Phi(P) = \frac{1}{4\pi\sigma_p} \int_V \frac{I_v}{r}\, dV - \frac{1}{4\pi\sigma_p} \int_{S_0} \sigma\Phi \frac{\partial(1/r)}{\partial n}\, dS_0$$

$$+ \frac{1}{4\pi\sigma_p} \sum_j \int_{S_j} \Phi \frac{\partial(1/r)}{\partial n_j} (\sigma_j'' - \sigma_j')\, dS_j \quad (5.174)$$

where $\sigma$ may be piecewise continuous, as before.

Where the sources are conceived to arise from membrane activity, then $I_v \equiv 0$. On the other hand, there will be $k$ membrane surfaces across which the condition of activity produces a discontinuity in potential along with a continuity of the normal component of current. Mathematically,

$$\Phi_k' \neq \Phi_k'' \quad (5.175a)$$

$$\sigma_k' \frac{\partial \Phi_k'}{\partial n_k'} + \sigma_k'' \frac{\partial \Phi_k''}{\partial n_k''} = 0 \quad (5.175b)$$

replaces (5.169) for such surfaces. The result is that at these surfaces the second summation on the right-hand side of (5.166) gives

$$\sum_k \int_{S_k} (\sigma_k''\Phi_k'' - \sigma_k'\Phi_k') \frac{\partial(1/r)}{\partial n_k}\, dS_k \quad (5.176)$$

where $n_k$ is positive when directed from the primed to the double-primed region, which we choose as from the interior to the exterior of the cell, respectively. Under this consideration (5.166) becomes

$$\Phi(P) = \frac{1}{4\pi\sigma_p} \sum_k \int_{S_k} (\sigma_{ko}\Phi_{ko} - \sigma_{ki}\Phi_{ki}) \frac{\partial(1/r)}{\partial n_k} \, dS_k$$

$$+ \frac{1}{4\pi\sigma_p} \sum_j \int_{S_i} \Phi(\sigma_j'' - \sigma_j') \frac{\partial(1/r)}{\partial n_j} \, dS_j - \frac{1}{4\pi\sigma_p} \int_{S_0} \sigma\Phi \frac{\partial(1/r)}{\partial n} \, dS_0$$

$$(5.177)$$

where the subscript $o$ (for cell exterior) replaces the double prime and $i$ (for cell interior) the single prime. The integrals over $S_k$ represent the primary-source contribution, which appears in the form of an equivalent double layer. This result could also have been obtained by utilizing (5.69), which can be cited as a justification of (5.175). Discontinuities in conductivity. which are responsible for the zero and $j$ surfaces are the basis for the secondary (induced) sources and disappear when $\sigma_j'' = \sigma_j'$. The primary source will not vanish, of course, when $\sigma_k' = \sigma_k''$. Note that all sources are referred to a homogeneous conducting medium whose conductivity is that which exists at $P$.

If the region under consideration contains cells in an inactive (passive) condition, then the interface between cell interior and exterior requires the following constraining equations [in place of either (5.169) or (5.175)]:

$$\sigma_k' \frac{\partial\Phi_k'}{\partial n_k'} + \sigma_k'' \frac{\partial\Phi_k''}{\partial n_k''} = 0 \tag{5.178}$$

$$\sigma_k' \frac{\partial\Phi_k'}{\partial n_k'} = r_m(\Phi_k'' - \Phi_k') + C_m \frac{\partial}{\partial t}(\Phi_k'' - \Phi_k') \tag{5.179}$$

Equation (5.178) expresses the requirement that the normal component of current be continuous across the membrane. Equation (5.179) relates the transmembrane current to the transmembrane voltage, assuming a linear parallel resistance-capacitance representation (as discussed in Sec. 3.12).

If in place of $\phi = \Phi$ in (5.166) we choose $\phi = \Phi/\sigma$, where $\Phi$ is given by (5.168), then

$$\int_V \left[ -\frac{I_v}{r\sigma} - \Phi\nabla^2 \left(\frac{1}{r}\right) \right] dV$$

$$= -\sum_j \int_{S_i} \frac{1}{r} (E_n' - E_n'') \, dS_j + \int_{S_0} \frac{1}{r} E_n'' \, dS_0 \tag{5.180}$$

where $E_n'$ and $E_n''$ are the normal components of the electric field at an interface approached from the primed and double-primed sides, respectively, with positive normal from primed to double-primed regions as before. That is, $E_n' = -\partial\Phi'/\partial n'$ and $E_n'' = \partial\Phi''/\partial n''$. If we define

$$E_j = \tfrac{1}{2}(E_n' + E_n'')_j = \tfrac{1}{2}E_n'\left(1 + \frac{\sigma_j'}{\sigma_j''}\right)$$

then

$$(E_n' - E_n'')_j = 2E_j\frac{\sigma_j'' - \sigma_j'}{\sigma_j'' + \sigma_j'} \tag{5.181}$$

Accordingly, we finally have

$$\Phi(P) = \frac{1}{4\pi}\int_V \frac{I_v}{\sigma r}\,dV$$
$$- \frac{1}{4\pi}\sum_j \int_{S_i} \frac{2E_j}{r}\frac{\sigma_j'' - \sigma_j'}{\sigma_j'' + \sigma_j'}\,dS_j + \frac{1}{4\pi}\int_{S_o} \frac{2E_0}{r}\,dS_0 \tag{5.182}$$

where $E_0$ is the outward normally directed field within the bounding nonconducting medium at the interface.

A comparison of (5.174) and (5.182) reveals differences in both the volume integral and the surface integral. For (5.174) the volume integral evaluates the potential produced by the sources as if they were in an unbounded uniform medium of conductivity $\sigma_p$. For (5.182) each element of source contributes as if the medium were uniform and infinite in extent but at the conductivity in which the source element (rather than the field point) exists. Thus in the volume integral in (5.182) $\sigma$ appears under the integral sign since for the inhomogeneous medium it will be a piecewise continuous function of position. The contribution from the volume integral in (5.182) will therefore differ from that in (5.174). For the homogeneous case both results become identical, of course.

In comparing the surface integrals of (5.174) with (5.182) we note that they express the effect of the conductive discontinuities by means of equivalent sources: current double layers in (5.174), which are located in a medium of conductivity $\sigma_p$, and single layers in (5.182). In the latter case the single-layer sources $\omega_j$ have the strength (current per unit area) of

$$\omega_j = 2\sigma_k E_j\frac{\sigma_j'' - \sigma_j'}{\sigma_j'' + \sigma_j'} \tag{5.183}$$

referred to an arbitrary medium of uniform conductivity $\sigma_k$.

From a physical standpoint, it is known that the steady-state boundary condition expressed by (5.169a) requires the existence of an induced surface charge layer at the conductive interface.[1] The second

---

[1] R. Plonsey and R. Collin, "Principles and Applications of Electromagnetic Fields," p. 179, McGraw-Hill Book Company, New York, 1961.

integral in (5.182) contributes a potential component $\Phi_c$, given by

$$\Phi_c = \frac{1}{4\pi} \sum_j \int_{S_i} \frac{2E_j}{r} \frac{\sigma_j'' - \sigma_j'}{\sigma_j'' + \sigma_j'} \, dS_j$$

$$= \frac{1}{4\pi\epsilon} \sum_j \int_{S_i} \frac{2\epsilon E_j}{r} \frac{\sigma_j'' - \sigma_j'}{\sigma_j'' + \sigma_j'} \, dS_j \tag{5.184}$$

in a medium of uniform permittivity $\epsilon$. From this expression one can identify the quantity $2\epsilon E_j(\sigma_j'' - \sigma_j')/(\sigma_j'' + \sigma_j')$ as the induced surface charge density. This result can be confirmed as follows. Let the surface charge be $\rho_{sj}$; then a secondary field normal to the surface (directed away from the surface on both sides) has the magnitude $\rho_{sj}/2\epsilon$. The total field is the sum of the applied and induced fields and is augmented on one side and reduced on the other side of the surface charge. Application of (5.169a) then gives

$$\sigma' \left( E_j + \frac{\rho_{sj}}{2\epsilon} \right) = \sigma'' \left( E_j - \frac{\rho_{sj}}{2\epsilon} \right) \tag{5.185}$$

where $E_j$ denotes the applied field. Solving for $\rho_{sj}$ yields

$$\rho_{sj} = 2\epsilon E_j \frac{\sigma'' - \sigma'}{\sigma'' + \sigma'} \tag{5.186}$$

confirming the original assertion, as well as the identification of $E_j$ as the applied field. Thus the surface-integral terms in (5.182) can be related (by the proportionality factor $\epsilon$) to the actual charge density as given by (5.186) provided that the permittivity of the medium, $\epsilon$, is everywhere the same. The form of Eq. (5.180) could thus have been written immediately, based on these principles; indeed, this is the approach used by Gelernter and Swihart.[1]

If the permittivity is not uniform but also takes the value $\epsilon'$ in conductive medium $\sigma'$ and $\epsilon''$ in $\sigma''$, then a more complicated expression for the charge density $\rho_{sj}$ results. The normal component of electric field due to $\rho_{sj}$ is $\rho_{sj}/2\epsilon'$ and $\rho_{sj}/2\epsilon''$ in the primed and double-primed medium, while the applied electric displacement $D_j$ is continuous across the interface. Application of (5.169a) thus gives

$$\sigma' \left( \frac{D_j + \rho_{sj}}{2\epsilon'} \right) = \sigma'' \left( \frac{D_j - \rho_{sj}}{2\epsilon''} \right) \tag{5.187}$$

and this leads to

$$\rho_{sj} = 2D_j \frac{\sigma'' \epsilon_j' - \sigma' \epsilon_j''}{\sigma'' \epsilon_j' + \sigma' \epsilon_j''} \tag{5.188}$$

[1] H. L. Gelernter and J. C. Swihart, A Mathematical-Physical Model of the Genesis of the Electrocardiogram, *Biophys. J.*, **4**:285 (1964).

From the definition of $E_j$, we have

$$E_j = \left( \frac{D_j}{\epsilon_j'} + \frac{\rho_{sj}}{2\epsilon_j'} \right) + \left( \frac{D_j}{\epsilon_j''} - \frac{\rho_{sj}}{2\epsilon_j''} \right)$$

so that we get

$$\rho_{sj} = 2E_j \frac{\sigma_j'' \epsilon_j' - \sigma_j' \epsilon_j''}{\sigma_j'' + \sigma_j'} \tag{5.189}$$

Thus under these conditions each integrand in the surface integrals of (5.182) must be multiplied by $(\sigma_j'' \epsilon_j' - \sigma_j' \epsilon_j'')/(\sigma_j'' - \sigma_j')$ to evaluate the total charge density at the respective interface. This charge density includes bound (arising from the dielectric discontinuity) charge as well as "true" charge.

The above result shows that the actual charge distribution depends on the dielectric properties of the medium. On the other hand, the results given by (5.174) and (5.182) are independent of the dielectric permittivity and depend only on the conductive properties of the medium. Because of this, the sources in (5.174) and even in (5.182) can be viewed as effective (current) sources with reference to the physical sources being unnecessary for a complete mathematical formulation. One can think of the medium as a network of resistors and capacitors; the steady-state solution to the application of a current step, say, depends only on the resistance network; however, the capacitors determine the transient solution and their accumulated charge.

If, as before, the sources are considered as arising from membrane activity alone, then $I_v \equiv 0$ and the active membrane surfaces $S_k$ are characterized by (5.175). Substituting these conditions into (5.166), but with $\psi = 1/r$ and $\phi = \Phi/\sigma$, results in the addition of the following surface integrals:

$$\sum_k \int_{S_k} \left[ \frac{\sigma_k'}{r} \frac{\partial \Phi_k'}{\partial n_k} \left( \frac{1}{\sigma_k'} - \frac{1}{\sigma_k''} \right) + (\Phi_k'' - \Phi_k') \frac{\partial (1/r)}{\partial n_k} \right] dS_k$$

In the notation that primed regions correspond to the interior of the individual cell, while the double-primed regions represent the exterior, and positive normal $n_k$ is directed outward from the cell, the membrane current density $J_m$ is

$$J_m = -\sigma_k' \frac{\partial \Phi_k'}{\partial n_k}$$

Since the transmembrane potential $V = \Phi_k' - \Phi_k''$, the above expression becomes

$$\sum_k \int_{S_k} \left[ \frac{J_m}{r} \left( \frac{1}{\sigma_j''} - \frac{1}{\sigma_k'} \right) - V \frac{\partial (1/r)}{\partial n_k} \right] dS_k$$

and (5.180) may be written

$$4\pi\Phi(P) = \sum_k \int_{S_k} \left[ \frac{J_m}{r} \left( \frac{1}{\sigma_{ko}} - \frac{1}{\sigma_{ki}} \right) - V \frac{\partial(1/r)}{\partial n_k} \right] dS_k$$

$$- \sum_j \int_{S_j} \frac{2E_j}{r} \frac{\sigma_j'' - \sigma_j'}{\sigma_j'' + \sigma_j'} dS_j + \int_{S_0} \frac{2E_0}{r} dS_0 \quad (5.190)$$

where, as before, $\sigma_{ko}$ is the conductivity exterior to the cell, $\sigma_k''$, and $\sigma_{ki}$ that interior to the cell, $\sigma_k'$. The integrals $S_k$ result from primary-source contributions, while the remainder are the consequence of secondary sources which arise at discontinuities in conductivity. Note that the potential from the primary sources is that which would exist in a uniform infinite medium and does not depend on the conductivity at $P$. Furthermore, both equivalent single- and double-layer sources are required to represent the primary source in this representation. Should $\sigma_{ko} = \sigma_{ki}$, then a more accurate form for the first term in (5.190) is required which takes into account the separate inner and outer membrane surfaces. For this case the integrand for $S_k$ becomes

$$\left( \frac{m_d J_m}{r} - V \right) \nabla \left( \frac{1}{r} \right) \cdot d\mathbf{S}_k$$

where $m_d$ is the membrane thickness. In this connection $J_m$ can be interpreted as the dipole moment per unit volume of the primary (impressed) source as defined in Sec. 5.3; that is, $\mathbf{J}_i = J_m \mathbf{n}_k$.

## 5.13  INTEGRAL-EQUATION APPLICATIONS

The equations that have been developed in the previous section can be applied to electrophysiological problems on both a macroscopic and a microscopic scale. As an example of the former we have the determination of current distributions in the human torso due to electric generators in the heart. This constitutes the electrocardiographic problem (to be considered in detail in the next chapter) where, since the lungs, fat layer, and vertebrae have different conductivities, the influence of these inhomogeneities must be considered. A comparable problem is one where the impedance of the torso is measured by using external electrodes. In one such system an apparent correlation of impedance with cardiac output results.[1] In order to investigate this phenomenon it is necessary to determine the current flow field within the inhomogeneous torso, taking into account fluctuations in geometry and conductivity with cardiac output.

---

[1] R. D. Allison, Stroke Volume, Cardiac Output, and Impedance Measurement, *Proc. 19th Ann. Conf. Eng. Med. Biol.*, **19**:191 (1966).

On the cellular level we might be interested in the transmembrane potential induced in a passive cell or group of cells because of a stimulating current. In this case the axoplasmic, interstitial, and external media may all have different conductivities. Furthermore, each membrane (cellular or sheath) requires representation by some impedance description. Appropriate boundary conditions are given in Sec. 5.12 for active and passive membranes and at phase boundaries between regions of different conductivities [see (5.169), (5.175), (5.178), and (5.179)].

As a simple example of the application of the formulations in the previous section to the problem of computation of impedance (hence relating impedance to shape and conductivity), consider Fig. 5.27, which depicts a single arbitrarily shaped region which is stimulated by surface electrodes. For the volume itself we have no sources; consequently, only a single surface integral remains in (5.173) [see also (5.92)], which is

$$\Phi(P) = \frac{1}{4\pi} \int_{S_0} -\Phi \frac{\partial(1/r)}{\partial n}\, dS + \frac{1}{4\pi} \int_{S_0} \frac{1}{r} \frac{\partial \Phi}{\partial n}\, dS \tag{5.191}$$

The boundary conditions which must be applied are:

$\partial\Phi/\partial n = 0$ over the surface not contacted by the electrodes
$\Phi = V$ over electrode 1 and $\Phi = 0$ over electrode 2

Now, as described by Pilkington, Metz, and Barr,[1] if each electrode

[1] T. C. Pilkington, W. C. Metz, and R. C. Barr, Calculation of Resistance of Three-dimensional Configurations, *Proc. IEEE*, **53**:307 (1965).

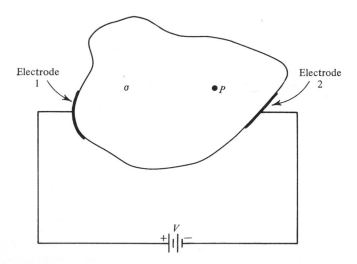

**Fig. 5.27** Integral-equation determination of the impedance of an irregularly shaped body.

surface is divided into $L$ elements, and the remaining surface into $M$ elements, the expression for potential can be evaluated at the center of each element (just inside the conducting region), with the result that $2L + M$ equations can be written. If the surface integrals are approximated by a finite summation over the same surface elements, we have $2L$ unknown gradients over the electrodes and $M$ unknown potentials over the remaining surface. Since $(2L + M)$ equations for $(2L + M)$ unknowns are available, solutions for the potential and current fields can be found. With this result the total impedance, among other quantities, can be calculated. Of course, accuracy depends on the fineness with which the surfaces are divided and the accuracy with which the geometry is specified. Matrix-inversion difficulties also increase as the number of elements goes up. In this discussion it was assumed, but it is not necessary, that each electrode surface is divided into an equal number of elements. Furthermore, the region need not be homogeneous. For the inhomogeneous case additional surface integrals arise in (5.173). Approximating them by finite summations introduces further unknowns; however, a like number of additional equations can be written.

In Fig. 5.28 a finite cylindrical nerve fiber, with internal conductance equal to that in the external volume conductor, is assumed to be stimulated by a sinusoidal exciting field. This field is denoted the applied field $\Phi_a$. The total potential field is found by the superposition of the applied field and a secondary field arising from the potential discontinuity at the membrane. The latter is found from (5.177), and the resultant expression is

$$\Phi = \Phi_a + \frac{1}{4\pi} \int_{S_m} (\Phi_o - \Phi_i) \frac{\partial(1/r)}{\partial n}\, dS \tag{5.192}$$

where $S_m$ is the membrane surface. The unknown and desired function is the transmembrane potential $V_m = \Phi_i - \Phi_o$. In terms of $V_m$ (5.192) becomes

$$\Phi = \Phi_a + \frac{1}{4\pi} \int_{S_m} V_m\, d\Omega \tag{5.193}$$

A matrix formulation for $V_m$ can be achieved by taking the normal derivative of the total potential just outside the cell membrane, calculating the membrane current, and relating this to the transmembrane potential by means of the membrane impedance. For an $RC$ membrane (5.160) and (5.161) can be applied, resulting in an integral equation for the (unknown) transmembrane potential. If the nerve is divided into $N$ elements, then $N$ equations in $N$ unknown transmembrane potential samples result. Thus, at the $k$th element we have

$$\left.\frac{\partial\Phi}{\partial\rho}\right|_k = \left.\frac{\partial\Phi_a}{\partial\rho}\right|_k + \frac{1}{4\pi} E_\rho\Big|_k = -V_{mk}\left(\frac{\sigma_m + j\omega C_m}{\sigma}\right) \tag{5.194}$$

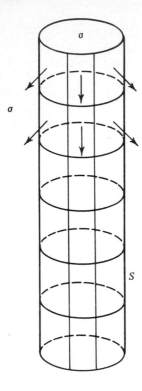

**Fig. 5.28** Geometry for the matrix solution of the stimulation of an inactive cylindrical axon by a polarizing source.

where $E_\rho$ is the electric field in the radial direction due to the integral term in (5.193) and $V_{mk}$ is the transmembrane potential for the $k$th element. [Field quantities in (5.194) must be thought of as complex phasors.] Matrix inversion of (5.194) permits solving for the $V_{mk}$. If solutions are found at a number of different angular frequencies $\omega$, one can utilize a Fourier summation to find the transmembrane potential due to a step or a spike.

Although this example utilized a cylindrical cell, the method clearly does not depend on any particular shape. Actually, for the circular cylindrical cell the approach given in Sec. 5.11 is probably preferable. The technique noted here is known as the integral-equation method of solution of Laplace's equation.[1] This approach seems advantageous, in general, since a three-dimensional problem is considered by a two-dimensional formulation. On the other hand, powerful numerical-analysis techniques are available for a direct consideration of Laplace's equation utilizing a difference approximation. A comparison of both

[1] Some general considerations of this technique in applied electromagnetic theory may be found in R. F. Harrington, Matrix Methods for Field Problems, *Proc. IEEE,* **55**:136 (1967).

methods may be found in a paper by Terry and Plonsey,[1] while details of the difference-equation approach can be found in Varga.[2]

## BIBLIOGRAPHY

Lorente de Nó, R.: "A Study of Nerve Physiology," Rockefeller Institute Series, vols. 131 and 132, New York, 1947.

Morse, P. M., and H. Feshbach: "Methods of Theoretical Physics," McGraw-Hill Book Company, New York, 1953.

Plonsey, R., and R. Collin: "Principles and Applications of Electromagnetic Fields," McGraw-Hill Book Company, New York, 1961.

Smythe, W. R.: "Static and Dynamic Electricity," 2d ed., McGraw-Hill Book Company, New York, 1950.

Stratton, J. A.: "Electromagnetic Theory," McGraw-Hill Book Company, New York, 1941.

[1] F. Terry and R. Plonsey, Comparison of Iterative Techniques for Solving Internal Neumann Problems, *Proc. 20th Ann. Conf. Eng. Med. Biol.*, **20**:215 (1967).

[2] R. Varga, "Matrix Iterative Analysis," Prentice-Hall, Inc., Englewood Cliffs, N.J., 1962.

# 6
# Electrocardiography

## 6.1 INTRODUCTION

This chapter is concerned with an electrophysiological system of considerable interest, that of the heart. This topic has been chosen to illustrate some applications of the earlier work since it provides one of the best understood and successfully modeled bioelectric systems. As a consequence of its level of development, we shall also be able to consider several further applications of mathematics and engineering to biology. We begin with a very brief description of the basic mechanical and electrical properties of the heart. Following this, the electrophysiology of heart muscle will be considered briefly. The remainder, and major, emphasis of the chapter will then center on the heart as a bioelectric source in an inhomogeneous volume conductor. This material will include some background on conventional electrocardiography in addition to basic theoretical concepts.

## 6.2  MECHANICAL ACTIVITY OF THE HEART

Figure 6.1 illustrates the major anatomical parts of interest.  Blood is returned to the heart through the inferior and superior venae cavae where it enters the right atria (RA).  The blood then flows into the right ventricle (RV), from which it is forced into the lungs, oxygenated, and then returned to the left atria (LA).  From here, it enters the left ventricle (LV), whereupon the fresh blood is then forced back into the system through the aorta.

The physical activity of the heart is rhythmic and consists of the following recognized intervals:

1. *Diastasis.*  This corresponds to the end of the main pressure pulse in the system.  The atrioventricular valves (tricuspid and mitral valves of Fig. 6.1) have been open for some time and all chambers

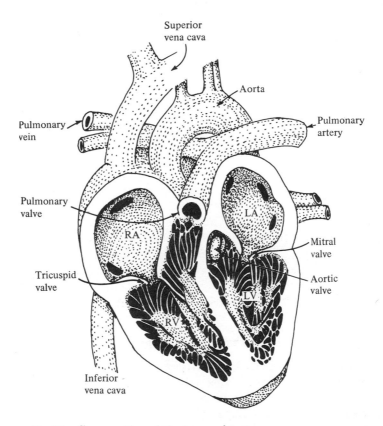

**Fig. 6.1**  Cross section of the human heart.

have filled with blood. This condition may be thought of as the initial or resting state of the heart.

2. *Atrial contraction.* Diastasis ends with the beginning of atrial contraction. This causes only a relatively slight rise in intra-atrial pressure. The ventricular volume and pressure increase as a result of the entrance of blood from the atria.

3. *Ventricular isometric contraction.* The first phase of ventricular contraction corresponds to an increase in pressure but not in volume. The increase in pressure closes the AV valves, hence preventing an initial change in volume. At the outset of ventricular contraction, the pressure in the discharge vessel of the right ventricle (pulmonary artery) is 7 mm Hg, while that in the aorta (the artery carrying blood from the left ventricle) is 80 mm. The aortic and pulmonary valves, illustrated in Fig. 6.1, remain closed until a sufficient pressure is built up.

4. *Ventricular ejection.* When the ventricular pressure builds up to a value which exceeds that in the aorta, the aortic valve opens and the left ventricle rapidly discharges blood through the aorta into the system. A peak pressure of about 125 mm Hg is reached. After the major portion of ejection, the ventricular muscle fibers are shortened and can no longer contract forcefully. The ventricular and aortic pressures begin to drop.

5. *Isometric relaxation.* When ventricular pressure falls sufficiently, the aortic and pulmonary pressures exceed that within the chambers and the aortic and pulmonary valves close. At this point, ejection ceases. As the ventricle continues to relax, the pressure falls still further; but while the pressure exceeds that in the atria no exit is provided for trapped blood, so that the relaxation is characterized as isometric.

6. *Ventricular filling.* When the ventricular pressure drops below that of the atria, the AV valves open and rapid ventricular filling begins. This phase is followed by a slowing down of filling as the ventricle reaches a maximum diastolic size. This is the period of diastasis, and it is the completion of one cycle to be followed by the next cycle upon the initiation of atrial systole.

## 6.3  ELECTRICAL ACTIVITY OF THE HEART

The mechanical activity described above is accompanied by electrical activity in the heart. This activity begins in the sinus venosus or *SA node*, which is consequently known as the *pacemaker*. Cells in this region have the characteristic that the resting potential is not maintained but continually diminishes until the cell "fires." Upon recovery, the process is repeated, so that a periodic excitation results. The intra-

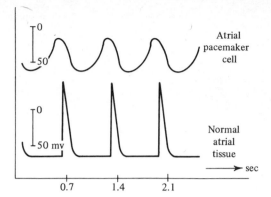

**Fig. 6.2** Activity of a single atrial pacemaker and a single cell of atrial tissue. [*T. C. West, G. Falk, and P. Cervono, J. Pharm. Exptl. Therap.*, **117**: 245 (1956).]

cellular recording of a pacemaker cell is shown in Fig. 6.2, along with a normal atrial cell for comparison.

As noted, the heart rate is basically determined by the pacemaker activity in the SA node, which operates, in a sense, as a free-running multivibrator. The rate, however, is modified by the competing effects of the parasympathetic and sympathetic nerves. The effect of the vagus nerve is one of slowing the heart, while the sympathetic nerves increase the rate. Cutting both nerves to the heart results in an increased rate, so that the vagal effects appear to be dominant under resting conditions. It should be noted that cardiac output depends on both heart rate and stroke volume, so that an increase in heart rate, for example, does not necessarily result in an increased heart output.

The activity initiated by the SA node spreads through the muscle of the atria at about 1 m/sec. For the human, about 80 msec is required for the complete activation of the atria. Toward the end of this period, the electrical activity reaches the *AV node*, which is the sole muscular connection between the atria and ventricles. The propagation in the AV node is very slow, about 0.1 m/sec. Upon passage through this node, the excitation travels at about 2 m/sec through the specialized *right* and *left bundles* and then through the arborized *Purkinje fibers*. This system serves to initiate electrical activity in the ventricular musculature. An illustration of the system is shown in Fig. 6.3.

The heart muscle (myocardium) behaves in many ways as if it were a single cell. Excitation spreads at about 0.5 m/sec and causes mechanical contractions in an efficient, synchronous way. While electron microscopy shows the myocardium to be made up of many independent elements, it behaves functionally, at least, as a *syncytium*. Cardiac muscle is divided into its many elements by vertical and horizontal double membranes. The latter are actually irregular tonguelike processes. The boundaries they form between muscle "domains" are known

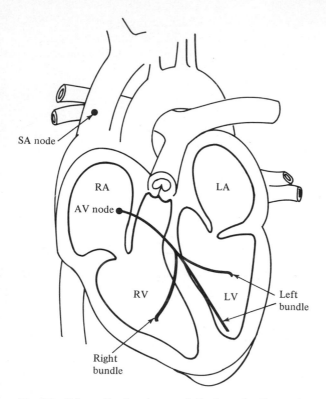

**Fig. 6.3**  Schematic showing specialized conduction system
of the heart.

as *intercalated disks*, and constitute the basis for discarding the concept
of the syncytium in the anatomical sense.    This is illustrated in Fig. 6.4.

Studies of heart excitation have been carried out in an effort to
elucidate the pathways of activation.    In general, these have involved
insertion of thin bipolar electrodes into the beating heart of a dog and
measuring the difference of potential.    The result, in general, is a monophasic wave.    This is interpreted as arising from the passage of a wave
(double-layer) of activity, and the time of passage at the midpoint of
the electrode pair is the time when the signal reaches its peak.    Through
measurement at many points throughout the heart and with the ECG
as a time reference, one can establish the location of the activation wave-
fronts at successive instants of time.    The result of such a study for ventricular excitation is shown in Fig. 6.5, which is due to Scher and Young.[1]

[1] A. M. Scher and A. C. Young, Ventricular Depolarization and the Genesis of QRS,
*Ann. N.Y. Acad. Sci.*, **65**:768 (1957).

It can be seen that the activation pattern is fairly complex even for the normal heart. The Purkinje system is responsible for the almost simultaneous initiation of activity at several points in the myocardium.

## 6.4 CELLULAR ACTIVITY

A typical ventricular transmembrane action potential is illustrated in Fig. 6.6. The initial rapid upstroke is labeled phase 0 and is similar to that observed in other muscle and in nerve. A phase of early repolarization is labeled phase 1. This is followed by a period of prolonged and slow repolarization (phase 2) which is referred to as the "plateau." A terminal rapid repolarization phase (referred to as phase 3) completes the cycle, and phase 4 represents the diastolic period of the heart. One notes from Fig. 6.6 that cardiac ventricular action potentials differ from those of other cells in the prolonged duration of the recovery phase.

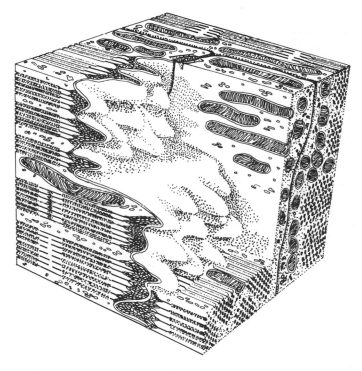

**Fig. 6.4** Reconstruction of submicroscopic anatomy of cardiac muscle of mouse, showing intercalated disk. [*F. S. Sjöstrand, E. Anderson-Cedergran, and M. M. Dewey, Ultrastructure Res.*, **1**:271 (1958).]

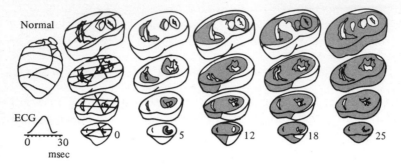

**Fig. 6.5** Pathway of normal ventricular excitation in dog as discerned by noting the extent of depolarization at 0.5, 12, 18, and 25 msec after beginning of the QRS complex.   The small drawing of the heart indicates positions of planes in which records were taken.   Lead II electrocardiogram is labeled to indicate the total duration of electrical activity. [*A. M. Scher, Electrical Correlates of the Cardiac Cycle, in T. C. Ruch and H. D. Patton (eds.), "Physiology and Biophysics," W. B. Saunders Company, Philadelphia, 1965.*]

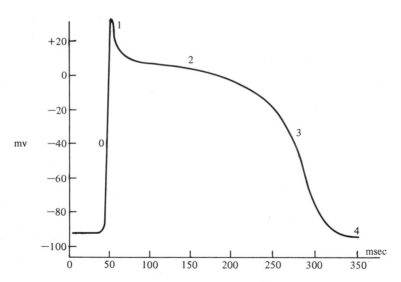

**Fig. 6.6** Transmembrane action potential for ventricular fiber.   (*After B. F. Hoffman and P. F. Cranefield, "Electrophysiology of the Heart," McGraw-Hill Book Company, New York, 1960.   By permission of the publishers.*)

For atrial pacemaker fibers phase 4 is characterized by a slow *depolarization* which eventually reaches threshold. In such fibers the upstroke (phase 0) is normally longer in duration than for nonpacemaker cells. In addition, there is little or no plateau, but rather a continuous uniform depolarization characteristic of phase 3. An illustration of the atrial pacemaker transmembrane potential is given in Fig. 6.2.

The passive electrical properties of Purkinje cardiac fibers turn out to be similar to those of striated muscle and require a two-time-constant model. By using the current-step and foot-of-action-potential techniques, the values of $C_m = 2.4$ $\mu f/cm^2$, $r_m = 1,800$ ohm-cm$^2$, $C_s = 7$ $\mu f/cm^2$, $r_m = 300$ ohm-cm$^2$ have been measured by Fozzard[1] (see Fig. 3.14 for nomenclature). As noted in Chap. 3, the current-step (dc) method yields the net parallel capacitance $(C_m + C_s)$, while the action-potential "foot" depends on the $r_s C_s$ time constant, and thereby serves to determine $C_s$.

While important differences between nerve and cardiac muscle exist, they are characterized by similar ionic features (high internal to external potassium concentration ratio and low internal to external sodium and chloride concentration ratios). The mechanism of activation is also the same for both, namely, increase in permeability to sodium. Furthermore, the resting potential is of the same order of magnitude, and the sodium-potassium pump appears to be similar. Consequently, the underlying mechanisms which account for the respective electrical activity are similar. However, the cardiac action potential does differ from nerve in several respects. First, the plateau is very much longer in duration than one obtains from nerve. Secondly, the pacemaker phenomenon occurs in nerve only under special conditions. And finally, a measurement of membrane impedance throughout the cardiac action potential reveals, in addition to the drop following activation, an increase during repolarization which sometimes exceeds that measured at rest. Except for the last two phenomena it is possible, by suitable modification, to utilize the Hodgkin-Huxley equations to account for measured behavior. That is, through suitable modifications both pacemaker activity and prolonged action potentials can be realized.[2] Factors other than strictly electrochemical ones, however, may also contribute to the genesis of the plateau potential; a reduction in the size of the surrounding myocardium has been observed to decrease the duration of the plateau.[3]

[1] H. A. Fozzard, Membrane Capacity of the Cardiac Purkinje Fiber, *J. Physiol.*, **182**:255 (1966).

[2] D. Noble, A Modification of the Hodgkin-Huxley Equations Applicable to Purkinje Fiber Action and Pacemaker Potentials, *J. Physiol.*, **16**:371 (1966).

[3] A. Ebara, Plateau Potential of Oyster Myocardium, *Japan. J. Physiol.*, **16**:371 (1966).

One difficulty yet to be resolved is the interpretation of voltage-clamp measurements on the Purkinje fiber. The problem arises because the membrane behaves as a two-time-constant network. A consequence is that the series $r_s C_s$ element causes a prolonged capacitive current which cannot be separated from the ionic currents as readily as could be done for the single parallel capacitance $C_m$. Computations of the initial (maximum) depolarization rate and conduction velocity from a Hodgkin-Huxley analysis are successful when a single parallel capacitance is assumed with a value around 3 $\mu f/cm^2$; this value is appropriate for describing the initial short-time-constant behavior.

The absence of satisfactory voltage-clamp techniques and data has hampered the study of the electrophysiological properties of cardiac membrane. The direct measurement of transmembrane potentials has, however, permitted the development of some understanding of the ionic basis for cardiac excitability. For example, Weidmann[1] developed a modified voltage-clamp technique for studying the sodium current system in Purkinje fibers. Two closely spaced intracellular electrodes were used, with one controlling the current of the second to maintain a fixed voltage in the first. After 50 msec this "local" voltage clamp was released and the tissue stimulated. The resultant *rate* of rise of action potential was taken as a measure of $g_{Na}$, somewhat similar to the two-step experiments of Hodgkin-Huxley that serve to evaluate the inactivation parameter $h$. As illustrated in Fig. 6.7, the resultant curves are similar in shape to those obtained for the squid axon. On the basis of these results, Weidmann concluded that $\alpha_h$ and $\beta_h$ were of the same order of magnitude and vary with membrane potential in the same way as for the squid. In absence of adequate data the activation system ($m$) has also been taken as similar to that in nerve.

The behavior of the potassium current in cardiac fibers is quite different from that in nerve fibers. The value of potassium conductance for transmembrane potentials near the potassium Nernst potential appears to be fairly well described by the constant-field equation (3.135), namely,[2]

$$(g_K)_{V_K} = \frac{P_K F^3 V_K}{(RT)^2} \frac{[C_K]_i [C_K]_o}{[C_K]_o - [C_K]_i} \tag{6.1}$$

where $V_K = (RT/F) \ln ([C_K]_o/[C_K]_i)$. When the external potassium concentration is increased, thereby depolarizing the membrane, (6.1) correctly predicts an increase in conductance. On the other hand, an outward electrochemical gradient causing an outward flow of current pro-

---

[1] S. Weidmann, The Effect of the Cardiac Membrane Potential on the Rapid Availability of the Sodium Carrying System, *J. Physiol.*, **127**:213 (1955).

[2] Equation (6.1) is obtained from (3.135) by evaluating $dJ_K/dV$ at $V_K$.

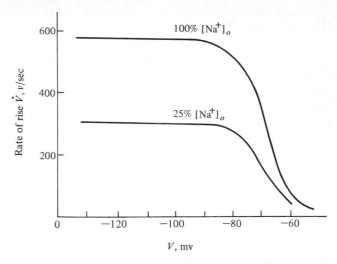

**Fig. 6.7**  Relationship between clamp potential $V$ and maximum rate of rise of action potential.  The membrane is held at $V$ for 50 msec, following which an action potential is elicited.  Reducing $[Na^+]_o$ to 25 percent of full strength reduces the magnitude but not the shape of the curve.  [*After S. Weidmann, J. Physiol.*, **127**:213 (1955).]

duces a reduction in conductance, and the reverse for inward currents. This is opposite the normal rectification predicted by the constant-field equation.  An empirical relationship[1]

$$g_K = (g_K)_{V_K} f_1(V - V_K) \tag{6.2}$$

may be utilized to include the effect of anomalous rectification.  Additional evidence suggests that $g_K$ is also time- and voltage-dependent, so that (6.2) should be extended to

$$g_K = (g_K)_{V_K} f_1(V - V_K) + f_2(V,t) \tag{6.3}$$

This permits taking into account, for example, the increase in conductance which occurs for depolarizing potentials in excess of 30 mv.  The time-dependent component is related to the dependence of $g_K$ on the parameter $n$ in the squid axon.

A modification of the Hodgkin-Huxley equations for Purkinje fibers was formulated by Noble.[2]  He utilized a sodium system that was

---

[1] D. Noble, A Modification of the Hodgkin-Huxley Equations Applicable to Purkinje Fiber Action and Pacemaker Potentials, *J. Physiol.*, **160**:317 (1962).

[2] *Ibid.*

similar to that for the squid axon for $(m)$ and a modified $(h)$ to incorporate the results of Weidmann such as described by Fig. 6.7. The anomalous K rectification was considered by means of (6.1) and (6.2). The time dependence of $f_2$ in (6.3) was assumed described by a slow $n$-type process, while the voltage dependence was chosen to account for the sigmoid steady-state potassium current-voltage relation. With these changes the main features of measured action potentials could be duplicated (e.g., initial spike, plateau, fast terminal repolarization). In addition, pacemaker activity could be simulated by a small decrease in $g_K$ or $[C_K]_o$. The main features of measured impedance variations (fall during spike, rapid recovery at the beginning of plateau, rise during plateau until in excess of resting value, fall during phase 3, and slow rise during pacemaker potential) were also duplicated.[1]

### 6.5 THE ACTIVATION PROCESS

The cardiac cells are irregular in shape, but have a nominal dimension of 15 by 15 by 100 microns. These cells are stacked together somewhat like bricks, as illustrated in Fig. 6.8. The plasma membranes of cells

[1] The increase in impedance during the plateau coincides with an increase in ionic conductances $g_{Na}$ and $g_K$. This seeming inconsistency is explained by the fact that the impedance depends on the slope conductances $(di_{Na}/dV)$ and $(di_K/dV)$ rather than the chord conductances $g_{Na}$ and $g_K$. Present state of knowledge does not preclude the possibility that $g_{Na}$ and $g_K$ do, in fact, decrease during the plateau.

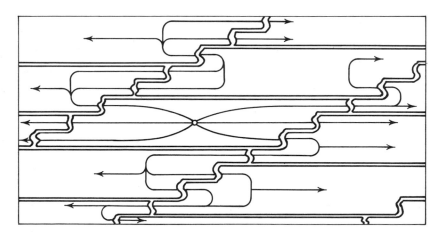

**Fig. 6.8** Simplified and formalized diagram illustrating the electrical structure of a portion of a rat atrial trabecula. Lines with arrowheads indicate the major pathways of current flow from an electrode in the intracellular fluid. (*J. W. Woodbury, Cellular Electrophysiology of the Heart, in "Handbook of Physiology," sec. 2, p. 237. American Physiological Society, Washington, D.C., 1962.*)

which adjoin end to end form the intercalated disk structure which is pictured in Fig. 6.4. Within each cell the myofibrils attach at the intercalated disk and run parallel to the axis of the cell. The disk interrupts the myofibrils only along the $Z$ line; in general, it crosses a cell in a stepwise manner, running parallel to the fibrils an integral number of sarcomeric levels between steps.

Besides the normal intercellular gap of 200 Å between adjoining plasma membranes, three additional types of contact have been observed. These are the desmosomes, myofibrillar insertion plaques, and the quintuple-layered membrane junctions.[1] The last-mentioned regions have been referred to by Dewey and Barr[2] as the "nexus." This type of contact refers to a fusing of adjacent membranes along their outer leaflets, and it is believed that this constitutes a low-resistance intercellular connection. The nexuses occur in the intersarcoplasmic regions of the intercalated disks. In general, they run parallel to the fiber axis, often for an entire sarcomere, while those occurring normal to the axis involve much smaller areas. An illustration of the nexal junction is given in Fig. 6.9.

On a larger scale, cardiac cells are packed into bundles with a diameter that is six, or fewer, cells across. While the extracellular space within the bundles is very limited, much space exists between the bundles. Thus, any cardiac cell is within three cell "diameters" from a large extracellular space. The bundles merge and divide and tend to form much larger bundles, with cross-sectional dimension in the order of millimeters, which are called *trabeculae*. These, in turn, branch and re-form (*anastamose*), thus constituting the structure of the myocardium.

Activity initiated at a cell in the myocardium will spread to adjoining cells until all become activated. The spread of activity from a depolarizing region of the plasma membrane of a cell to the remainder of the membrane is accomplished by the local current flow in exactly the same way as discussed for nerve. But there is some question as to how activation is transmitted intercellularly. Most of the evidence favors the view that the nexus in the intercalated disk is a region of high electrical conductivity which permits a sufficiently high current density to be created by an active cell in some of its neighbors. This local current is adequate to depolarize the adjacent cells in the same way that activity spreads contiguously along the membrane of the same cell.

---

[1] A. R. Muir, Further Observations on the Cellular Structure of Cardiac Muscle, *J. Anat. London*, **99**:27 (1965).

[2] M. M. Dewey and L. Barr, Intercellular Connection Between Smooth Muscle Cells: The Nexus, *Science*, **137**:670 (1962); see also L. Barr, M. M. Dewey, and W. Berger, Propagation of Action Potentials and the Structure of the Nexus in Cardiac Muscle, *J. Gen. Physiol.*, **48**:797 (1965).

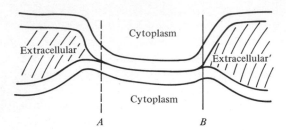

**Fig. 6.9** Fusion of plasma membrane to form a
*nexus* (*A* to *B*). [*Copyright* 1962 *by the Am.*
*Assoc. for Advancement of Science.   M. M. Dewey*
*and L. Barr, Science,* **137**:670 (1962).]

Evidence for electric transmission between cells is provided by an
experiment conducted by Barr, Dewey, and Berger.[1] Using a thin
bundle of frog atrial muscle in a Ringer's solution, they were able to block
electric transmission across a 0.5-mm sucrose gap which established two
isolated regions.   By making an electric connection between the latter
regions, conduction was restored.   Only transmission of an electrical
nature is consistent with the results of this experiment.

In another approach Weidmann[2] studied the diffusion of radiopotassium [$^{42}$K] across the intercalated disks of sheep ventricular fibers.
Bundles of fibers were pulled through a hole in a partition.   One half was
then charged by radio potassium and the other half washed by inactive
Tyrode solution.   The washing was designed to remove interstitial [$^{42}$K];
what remained was presumably intracellular.   After 6 hr, the steady-
state intracellular distribution of [$^{42}$K] was measured and an average space
constant of 1.55 mm (about 15 cell lengths) was determined.   This
result could only mean that the resistance of the intercalated disk is low,
the average resistance being estimated as 3 ohm-cm$^2$.   On this basis,
propagation of the cardiac action potential by local circuit currents
becomes theoretically possible.

An important experiment was performed by Woodbury and Crill[3]
using the rat atrial trabecula.   They plotted equipotential lines on the
surface of the thin tissue when current was applied via an intracellular

[1] *Ibid.*

[2] S. Weidmann, The Diffusion of Radiopotassium across the Intercalated Discs of
Mammalian Cardiac Muscle, *J. Physiol.,* **187**:323 (1966).

[3] W. E. Crill, in J. W. Woodbury, Cellular Electrophysiology of the Heart, in "Handbook of Physiology," sec. 2, p. 237, American Physiological Society, Washington,
D.C., 1962.

electrode located at the origin.   (The result is plotted in Fig. 6.10.)
Since the trabecula was only about 75 microns thick, while being 500 to
700 microns across, the resultant potential field should have a two-
dimensional character.   That is, if we let the $z$ axis be normal to the
surface, then since the surface extent is very large compared with the
thickness, we expect the potential variations with respect to $z$ to be
negligible.   This means that the potential field is a function of polar
coordinates $\rho$, $\phi$, and can be characterized as arising from a sheet of
cardiac cells connected together at the intercalated disks.   If the latter
are of low resistance, the potential field should appear in its gross behavior, at least, as arising from a single flat cell of the same thickness
with an upper and a lower membrane surface.

For such a cell lying in an extracellular medium of large extent,

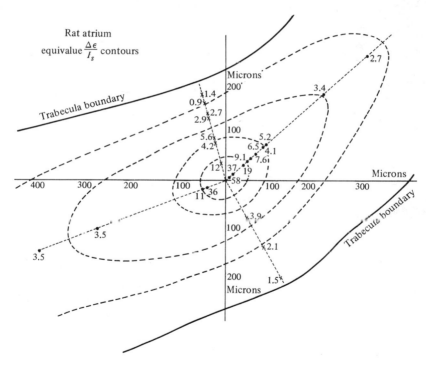

**Fig. 6.10**  Equipotential contour map of a rat atrial trabecula, due to intracellular applied current at the origin.   Numbers are the ratio of change in
potential, $\Delta\epsilon$, to applied current $I_s$.   From inside to outside contours are 10, 5,
3.5, and 2.5 mv/$\mu$a.   (*W. E. Crill, in J. W. Woodbury, Cellular Electrophysiology
of the Heart, in "Handbook of Physiology," sec. 2, p. 237, American Physiological
Society, Washington, D.C., 1962.*)

where the external resistance can be neglected in comparison with the intracellular resistance, Ohm's law can be applied to give "cable equations" appropriate to the two-dimensional geometry.   We get

$$\frac{\partial V}{\partial \rho} = \frac{I_1(\rho)}{\sigma 2\pi \rho d} \tag{6.4}$$

where $V$ is the transmembrane potential (inside with respect to outside), $I_1$ is the total internal current at the radius $\rho$ (positive being radially inward), $\sigma$ is the cytoplasmic conductivity, and $d$ is the cell thickness. If we let $i_m$ be the outward transmembrane current per unit area, then

$$i_m = \frac{1}{2\pi \rho} \frac{\partial I_1}{\partial \rho} \tag{6.5}$$

to satisfy continuity of current.   From (6.4) and (6.5) we get

$$\frac{\partial^2 V}{\partial \rho^2} = \frac{-I_1}{\sigma 2\pi \rho^2 d} + \frac{\partial I_1/\partial \rho}{\sigma 2\pi \rho d} \tag{6.6}$$

$$\frac{\partial^2 V}{\partial \rho^2} = -\frac{\sigma 2\pi \rho d}{\sigma 2\pi \rho^2 d} \frac{\partial V}{\partial \rho} + \frac{2\pi \rho i_m}{\sigma 2\pi \rho d} \tag{6.7}$$

Finally, we get

$$\frac{\partial^2 V}{\partial \rho^2} + \frac{1}{\rho} \frac{\partial V}{\partial \rho} = \frac{i_m}{\sigma d} \tag{6.8}$$

If we utilize the simple one-time-constant membrane network, then $i_m$ and $V$ can be related to the resistance $r_m$ and capacitance $C_m$ (for a unit area) as

$$\frac{i_m}{2} = \frac{V}{r_m} + C_m \frac{\partial V}{\partial t} \tag{6.9}$$

where the factor of 2 takes into account that transmembrane current flows equally through the upper and lower membrane surfaces.   Then (6.8) can be written

$$\lambda^2 \left( \frac{\partial^2 V}{\partial \rho^2} + \frac{1}{\rho} \frac{\partial V}{\partial \rho} \right) - V - \tau \frac{\partial V}{\partial t} = 0 \tag{6.10}$$

where

$$\lambda = \sqrt{\frac{\sigma d r_m}{2}} \quad \text{and} \quad \tau = r_m C_m \tag{6.11}$$

Note that $1/\sigma d$ is the resistance of a section of cytoplasm of unit width and unit length (with thickness $d$), and if this be designated $r_c$, then $\lambda = \sqrt{r_m/2r_c}$.   Under steady-state conditions, $\partial/\partial t = 0$, (6.10) becomes Bessel's equation[1]

$$\frac{d^2V}{d\rho^2} + \frac{1}{\rho}\frac{dV}{d\rho} - \frac{V}{\lambda^2} = 0$$

and the solution is

$$V(\rho) = AK_0\left(\frac{\rho}{\lambda}\right) + BI_0\left(\frac{\rho}{\lambda}\right) \qquad (6.12)$$

where $I_0$ and $K_0$ are the zero-order modified Bessel functions of the first and the second kind, respectively. Because $I_0$ increases indefinitely with increasing $\rho$, we may set $B = 0$ for a region that is essentially unbounded. Note that $K_0$ has the appropriate singularity at $\rho \to 0$ which corresponds to the assumption of a line source of current at that location; the source strength is represented by the constant $A$.

If the data available in Fig. 6.10 are used to plot potential vs. distance along the fiber axis from the source, the points in Fig. 6.11 result. The solid curve shown is $V_0K_0(\rho/\lambda)$, and it is seen that a very close fit is obtained. The space constant $\lambda$ comes out approximately 170 microns. Measurements at right angles to the fiber also fit a Bessel function but have a space constant which is slightly more than half. This anisotropy might be due to the more circuitous path followed by current flow normal to the fiber axis if one considers that intercellular current can flow only across the intercalated disks. This idea is suggested by the sketch of current flow lines in Fig. 6.8.

For the equivalent heart cell of 15 microns thickness and with

---

[1] An alternative derivation of this result is based on the assumption that the ventricular syncytial network is a square lattice of cylindrical fibers with zero resistance at the nodes. Then starting at an origin and numbering the nodes along a linear path, one notes that corresponding to node $n$ there are $(8n - 4) \approx 8n$ ways of connecting to node $(n - 1)$. Consequently, if $r_i$ be the resistance per unit length of the fiber and $r_m$ the membrane resistance for a unit length, the effective longitudinal resistance between nodes $(n - 1)$ and $(n)$ is $r_i/8n$ while the effective leakage resistance is $r_m/8n$. If the internodal distance is designated as $d$ and the distance along a linear fiber path from the origin is $\rho$, we can approximate the discrete model by a continuum, giving

$$\frac{dV}{d\rho} = \frac{Ir_id}{8\rho} \qquad \frac{dI}{d\rho} = \frac{8V\rho}{r_md}$$

If one now takes a derivative of the first equation with respect to $\rho$ and utilizes the second equation, the result is

$$\frac{d^2V}{d\rho^2} + \frac{1}{\rho}\frac{dV}{d\rho} - \frac{1}{\lambda^2}V = 0$$

with $\lambda = \sqrt{r_m/r_i}$. Note that $\lambda$ is essentially that of (6.11). This model and the derivation of the Bessel equation were given by I. Tanaka and Y. Sasaki, On the Electrotonic Spread in Cardiac Muscle of the Mouse, *J. Gen. Physiol.*, **49**:1089 (1966).

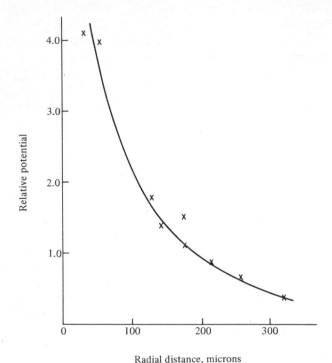

**Fig. 6.11** Spatial decrement of electrotonic potential in a rat atrial trabecula. Ordinate: change in membrane potential per unit of applied current. Abscissa: distance from intracellular current-applying electrode along the radius parallel to the trabecula edge. Crosses are measured points, and the solid curve is theoretical. (*W. E. Crill and J. W. Woodbury in J. W. Woodbury, Cellular Electrophysiology of the Heart, in "Handbook of Physiology," sec. 2, p. 237, American Physiological Society, Washington, D.C., 1962.*)

membrane resistance of 1,000 ohm-cm$^2$ and internal myoplasm conductivity of 0.01 mho/cm, Eq. (6.11) yields a theoretical space constant of 880 microns. This exceeds the measured value by 5. One possible explanation is that the intercalated disk has a resistance which effectively increases the internal resistance by 5 squared. Since the internal longitudinal resistance of a cell 15 by 15 by 100 microns would be 0.445 × 10$^6$ ohms, then the intercalated disk resistance per cell cross section should be 11.1 × 10$^6$ ohms or 25 ohm-cm$^2$. This exceeds the value of 3 ohm-cm$^2$ obtained by Weidmann and suggests that an alternative explanation be sought. Woodbury and Crill[1] and Woodbury and Gordon[2] note that because the trabecula contains bundles of closely packed fibers, none of which is more than three fibers distant from a large extracellular space, the effective membrane area separating internal and external media may substantially exceed the upper and lower surface areas of the idealized two-dimensional model. In particular, paths from internal to external media through the side walls of the individual cells and through the intercalated disks may be significant, and these would reduce the effective membrane resistance. If one assumes the intercalated disk with 3 ohm cm², then a geometric factor of eightfold reduction in resistance, taking as a lower limit $r_m = 125$ ohm-cm², would yield a space constant of 170 microns.

The evidence for electrical spread of activity based on the two-dimensional model requires that the intercalated disk resistance be in the order of 1 ohm-cm².† Thus, a value of 25 ohm-cm² would be too high, while 3 ohm-cm² would probably be satisfactory. One can interpret the two-dimensional data as confirming the existence of high-resistance intercalated disks. This point of view has been developed by Sperelakis et al.[3] Measurements made on isolated frog ventricle by Tarr and Sperelakis[4] give values of 12 mΩ which are consistent with the high-resistance intercalated disk computed above. Tarr and Sperelakis also were unable to confirm electrotonic coupling between adjacent cells when current was injected intracellularly into one cell and recorded from an adjoining cell with a second microelectrode. These results are therefore in direct conflict with those of Woodbury and Crill. Sperelakis suggests that the explanation for the spread of activity is the presence of a chemical mediator between cells (similar perhaps to the mechanism of a neuromuscular junction), but no direct evidence has been discovered.

The preponderance of evidence, however, favors the assumption that the intercalated disk has a low resistance. Tanaka and Sasaki[5] compared the electrotonic spread measured in the ventricle of an adult mouse with the theoretical values obtained from their lattice model, which assumes zero resistance at the intercalated disk, and obtained good

[1] *Loc. cit.*

[2] J. W. Woodbury and A. M. Gordon, The Electrical Equivalent Circuit of Heart Muscle, *J. Cellular Comp. Physiol.*, **66**:35 (1966).

† Weodbury and Crill, *loc. cit.*

[3] N. Sperelakis, T. Hoshiko, and R. M. Berne, Nonsyncytial Nature of Cardiac Muscle: Membrane Resistance of Single Cells, *Am. J. Physiol.*, **198**(3):531 (1960).

[4] M. Tarr and M. Sperelakis, Weak Electrotonic Interaction between Contiguous Cardiac Cells, *Am. J. Physiol.*, **207**:691 (1964).

[5] *Loc. cit.*

agreement.   For $r_m = 1,000$ ohm-cm², $\sigma = 0.01$ mho/cm, and with the
fiber diameter as 10 microns, $\lambda$ comes out to be 500 microns.   By choosing data from measurements within 100 microns of the current source, in
which region their model consists of four linear fibers, and with the linear
fiber theory as a basis, a value of $\lambda = 70$ microns is obtained.   But if
measurements at a sufficiently great distance (1 and 1.5 mm) are utilized
and the Bessel function behavior assumed, then computed values of $\lambda$ lie
between 400 and 700 microns.   Thus, a good agreement with the two-
dimensional theory is obtained.   If all internal resistance is lumped into
the intercalated disk, then an upper bound of 2 ohm-cm² is computed
which is still a very low value.   For an examination of potential behavior
near the polarizing electrode, the specific geometry of the electrode and
the muscle fibers must be included.   By extrapolating back to the origin
a good agreement between the measured and theoretical input resistance
is obtained.

In addition to the above experiments on atrial and ventricular
syncytial structures, experiments have been performed on Purkinje
fibers[1] and papillary muscle.[2]   Exponential electrotonic spread which is
obtained fits the linear cable theory and further substantiates the assumption of high conductivity in the intercalated disk.   The values for space
constant, furthermore, check reasonably well with predictions from cable
theory.

## 6.6  BIOELECTRIC SOURCES IN THE HEART

The mathematical basis for describing the bioelectric sources in the heart
during activity is contained in (5.177) or (5.190).   These expressions
formulate the potential field in terms of surface integrals over each cardiac
cell.   Considering (5.177), one notes that the cellular source is given as a
double layer over the enclosing plasma membrane with a strength
$\tau(S) = \sigma_{ko}\Phi_{ko} - \sigma_{ki}\Phi_{ki}$.   For a cell at rest $\tau(S)$ will be constant,[3] and the
contribution to the potential from the $k$th such cell, $\Phi_k$, is given by

$$\Phi_k(P) = \frac{1}{4\pi\sigma_p} \int_{S_k} \tau(S) \frac{\partial(1/r)}{\partial n_k} dS \tag{6.13}$$

[1] S. Weidmann, The Electrical Constants of Purkinje Fibers, *J. Physiol.*, **118**:348
(1952).

[2] A. Kamiyama and K. Matsuda, Electrophysiological Properties of Canine Ventricular Fiber, *Japan. Physiol.*, **16**:407 (1966).

[3] This is essentially the case, but not absolutely true if neighboring active cells set up
local currents, in which case $\tau$ will be caused to deviate from constancy by the nonuniform current flow across the high-impedance membrane.   However, most current
is confined to the intercellular region.

and this evaluates to zero, as demonstrated in connection with the discussion of (5.73).

For illustrative purposes let us consider that the ventricular action potential is trapezoidal, as illustrated in Fig. 6.12. Under these conditions $\tau(S)$ will be a constant over each cell surface during its rest phase and during the plateau. Accordingly, a contribution to the external potential field from a given cell occurs during the "upstroke" or "downstroke." In Sec. 6.7 we shall see that in the resulting signal, the electrocardiogram, the integrated effect of the "upstrokes" is the QRS and that of the "downstrokes" is the T wave.

Actually what is important in evaluating $\tau(S)$ is the *instantaneous* distribution of potential over the membrane (inner and outer surface), so that Fig. 6.12 by itself is not useful as the basis for specifying $\tau(S)$. However, there is abundant evidence that activity spreads uniformly through the myocardium in a wavelike manner (at least in the outer two-thirds of the wall), so that time and space variations can be interrelated by means of the "propagation" velocity. This is a three-dimensional equivalent of the propagated action potential of a linear fiber which was discussed in Chap. 4. The thickness of this advancing wave may be estimated from a mean propagation velocity of 0.5 m/sec for myocardial cells. Multiplying 0.5 m/sec by the rise time of 1 msec yields an active region of 500 microns. Thus, the activation wavefront can be thought of as being 500 microns wide; note that it encompasses five or more cardiac cells.

On the basis of the above discussion, we are assured that if the $k$th cardiac cell is within the active wavewidth, then $\tau(S)$ will not be uniform

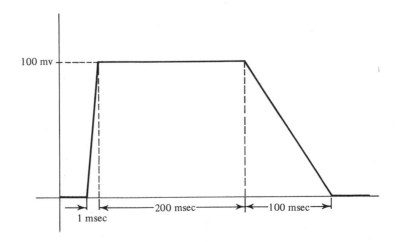

Fig. 6.12   Idealized ventricular transmembrane potential.

and (6.13) must yield a finite contribution.  Since the individual cell is small compared with most dimensions of interest, the double layer $\tau(S)$, which is defined on the surface of a cell, can be summed vectorially to obtain a net dipole for the cell.  In this way the actual sources in the wavewidth can be replaced by an averaged dipole moment per unit volume.  The latter, which we designate as $\mathbf{J}_i$, is averaged from the physical source entities in the same way as the mathematical *charge density* is related to a physical cloud of point charges.  Some additional discussion of this formulation is given in Sec. 5.3.

If the source density $\mathbf{J}_i$ is uniform over a surface normal to the direction of propagation, which is usually assumed, then the field set up by the activation wave as a whole can be obtained from the solid-angle formula of (5.97).  If a pair of electrodes are placed in the heart along a line normal to an advancing wavefront, then the resulting potential difference corresponds to the difference in solid angle subtended.  In this case, the wave thickness must be considered, and one could think of the advancing wave as made up of several dipole layers as defined by $\mathbf{J}_i$.  One expects, in general, no signal until the wavefront begins passing the electrode reached first.  The signal will then increase linearly until the entire wave is contained between the electrodes.  A plateau follows corresponding to the propagation of the wave between the electrodes, and a linear decline to zero signals passage of the wave through the second electrode.  If the electrodes are separated by a distance which is less than the wavewidth, then a triangular signal should result, and the peak amplitude will then be inversely proportional to the electrode spacing.

The predicted bipolar voltage noted above is confirmed by actual measurements in the outer ventricular wall.[1]  In Fig. 6.13, the voltage is shown as registered between a fixed electrode 11 and electrodes 10, 9, 8, and 7, each spaced 2 mm from the other.  It can be observed that as the electrode-pair spacing increases, the width of the wave does also,

[1] D. Durrer et al., Spread of Activation in the Left Ventricular Wall of the Dog.  III, *Am. Heart J.*, **48**:13 (1954).

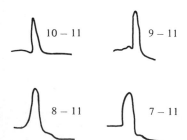

**Fig. 6.13**  Bipolar recordings between electrodes at sites numbered 7, 8, 9, 10, 11, along a line normal to the ventricular wall with spacings of 2 mm between electrodes.  [*D. Durrer, L. H. Van der Tweel, and J. R. Blickman, Am. Heart J.*, **48**:13 (1954).]

whereas the amplitude remains constant. This result implies that the wavewidth is less than 2 mm. However, since the same phenomenon is observed with electrodes spaced 1 mm apart, the wavewidth must be less than 1 mm; the previous estimate of 0.5 mm is thus roughly corroborated. Recent findings of Vander Ark and Reynolds[1] put the width at 0.9 ± 0.12 mm.

For bipolar recording from the inner ventricular layers, Durrer et al. found that as the electrode spacing increased, the voltage rather than the duration of the response increased. Their interpretation of this result is that the inner layers are excited more or less synchronously. If one considers that the inner one-third layer of the heart is activated at the same time and if one resolves the resultant volume source $\mathbf{J}_i$ into uniform strata, then by the solid-angle formalism, the main contribution to the bipolar potential is from laminae which lie between the electrodes. Thus, increase in electrode spacing would increase the measured amplitude. It is not possible to predict, analytically, the waveform to be expected since this depends on the temporal behavior of $\mathbf{J}_i$. An explanation of the simultaneous excitation of the inner wall of the heart is that it results from almost synchronous activity of the conduction (Purkinje) system.

A similar study utilizing bipolar electrodes spaced 1 mm apart in the outer third of the left ventricular wall of a dog heart was conducted by Solomon et al.[2] The waveform which they recorded, illustrated in Fig. 6.14, was triangular. If one divides the measured potential difference by the spacing, the result is an average electric field. Thus, if the electrodes lie on the $x$ axis with one at the origin and the second at $x = d$, then

$$E_x\left(\frac{d}{2}\right) \approx \frac{\Phi(0) - \Phi(d)}{d} \tag{6.14}$$

Now, if we assume propagation in the $x$ direction with a velocity $\theta$, then

$$\frac{\partial E_x}{\partial x} \approx \frac{1}{\theta d}\frac{\partial}{\partial t}[\Phi(0) - \Phi(d)] \tag{6.15}$$

Now, if the field is assumed to be uniform in the transverse plane, then

[1] C. R. Vander Ark and E. W. Reynolds, Experimental Study of Propagated Electrical Activity of the Heart, *Circulation*, **36**:II-255 (1967).

[2] J. C. Solomon, R. H. Selvester, W. L. Kirk, Jr., and R. B. Pearson, New Theoretical Model of the Electromotive Surface in the Heart, *Proc. 19th Ann. Conf. Eng. Med. Biol.*, **8**:133 (1966).

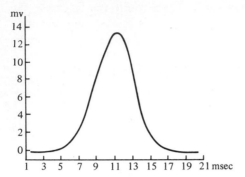

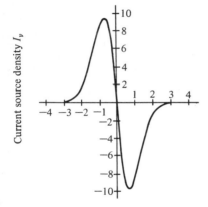

**Fig. 6.14**  Shape of electromotive activation surface in the heart.  The upper curve represents weighted bipolar recordings (an average of 222 recordings in the outer third of ventricular wall is shown); both $E_x$ and $J_i$ have the same shape. The lower curve contains the computed source density $I_v$ corresponding to the measured points and the theoretical function above. (*From J. C. Solomon, R. B. Pearson, W. L. Kirk, Jr., and R. H. Selvester, Model of the Electromotive Activity in the Heart, personal communication.*)

the effective source density $I_v$ [see (5.30)], is given by

$$I_v(x) = -\sigma \nabla^2 \Phi = -\sigma \frac{\partial E_x}{\partial x} \approx -\left(\frac{\sigma}{\theta d}\right) \frac{\partial}{\partial t} [\Phi(0) - \Phi(d)] \qquad (6.16)$$

A plot of the relative strength of $I_v$ as evaluated by Solomon et al. is given in Fig. 6.14.  One concludes from these measurements that $J_i$, which is proportional to $E_x$ by virtue of (5.41), rather than being uniform as previously assumed, varies in amplitude roughly as an error function.

One might have predicted this result from an extension of (5.76) which shows that the spatial distribution of double-layer strength is proportional to the spatial derivative of the potential.  For the propagating wave, the latter is proportional to the temporal derivative.  The derivative of the leading edge of the ventricular action potential, illustrated in Fig. 6.6, conforms to the double-layer distribution given in Fig. 6.14.

## 6.7 INTRODUCTION TO THE ELECTROCARDIOGRAM

As a result of the electrical activity of the cardiac muscle, an electric field is established in the conducting region of the body surrounding the heart. The potentials that arise at the surface of the body, resulting from cardiac activity, are known as the electrocardiogram (ECG). We shall be concerned with both theoretical and empirical relations between the heart activity and the related surface potentials.

Normally, ECG genesis is considered in two parts. The first consists in specifying, in some way, the current sources in the heart. For example, this could be accomplished by giving the space-time distribution of $J_i$, the dipole moment per unit volume. Since the sources are distributed, we may seek simplifications to make the mathematical analysis easier. A crude approximation is to simply find the vector summation of all dipole elements $J_i \, dV$. That is, we form the vector $p = \int_v J_i \, dV$, and let this single dipole represent the heart electrically. We shall later introduce a rigorous analysis where $p$ is the leading term of an infinite multipole expansion. For the present, we assume that the electrical activity of the heart can be represented by the dipole $p$. Interestingly, this rather crude approximation turns out to be amazingly good. Almost all clinical electrocardiography and vectorcardiography are based on the notion that the heart may be represented by a single dipole, the *equivalent heart vector*.

The second portion of the ECG problem is to obtain surface potentials due to the effective dipole (heart) source. In this analysis, the body is normally assumed to be linear, homogeneous, uniform, and isotropic. These are not particularly good assumptions, but, again, satisfactory results are normally obtained. We shall consider some effects of inhomogeneities in a later section.

## 6.8 STANDARD LEADS OF EINTHOVEN

The earliest work in electrocardiography was undertaken by Einthoven. He developed a system of *lead* (electrode) placement at the extremities of the body on the assumption that this would enhance the validity of the dipole heart model. We now proceed to a consideration of this system of leads, which are called the *standard* or *limb leads*. Despite the proved inadequacies of this system from a theoretical point of view, this constitutes the most common clinical system today.

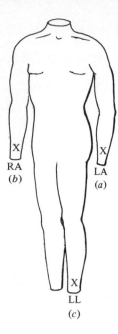

RA
(b)

LA
(a)

LL
(c)

**Fig. 6.15** Standard (or limb) leads of Einthoven. Each lead (electrode) is located by an X.

We define the "standard" leads $V_I$, $V_{II}$, $V_{III}$ as follows[1] (see Fig. 6.15):

$$V_I = V_{ab} \qquad \text{(lead 1)} \tag{6.17}$$

$$V_{II} = V_{cb} \qquad \text{(lead 2)} \tag{6.18}$$

$$V_{III} = V_{ca} \qquad \text{(lead 3)} \tag{6.19}$$

In the above, $a$ corresponds to the wrist of the left arm (LA), $b$ to the wrist of the right arm (RA), and $c$ to the lower portion of the left leg (LL). From Kirchhoff's voltage law, we have

$$V_{ca} + V_{ab} = V_{cb}$$

hence

$$V_{III} + V_I = V_{II} \tag{6.20}$$

Equation (6.20) can be represented by the following "vector" relations in the *Einthoven equilateral triangle*, illustrated in Fig. 6.16. The origin is at the center of the triangle, while the origin for each lead is the projection on the corresponding side. Positive directions are taken as indicated on the figure and correspond to a reversal of the double-subscript notation (by convention). If the lead voltages are plotted along

[1] The double-subscript notation is read as $V_{ab} = \Phi(a) - \Phi(b)$ (that is, the potential of $a$ minus the potential at $b$).

the sides of the triangle, each from its respective origin, then the projections of their termini define the unique vector **V** as illustrated in Fig. 6.16. We can show that all three projections must meet in a common point. To do this let us assume the existence of the resultant vector **V**. As a consequence, the scalar leads must be

$$[\text{Lead II}] = V_{\text{II}} = -V \cos(120° - \alpha) = \frac{V}{2}\cos\alpha - \frac{\sqrt{3}}{2}V\sin\alpha$$

$$(6.21)$$

$$[\text{Lead III}] = V_{\text{III}} = -V\sin(\alpha + 30°)$$

$$= -\left(\frac{V}{2}\cos\alpha + \frac{\sqrt{3}}{2}\sin\alpha\right) \quad (6.22)$$

$$[\text{Lead I}] = V_{\text{I}} = V\cos\alpha \qquad\qquad (6.23)$$

We see that

$$V_{\text{III}} + V_{\text{I}} = \frac{V}{2}\cos\alpha - \frac{\sqrt{3}}{2}V\sin\alpha = V_{\text{II}} \qquad (6.24)$$

Hence, the construction establishes a relationship which satisfies (6.20).

From Fig. 6.16, it is seen that the vector **V** contains all the information of the three separate lead components. Its direction plotted at characteristic instants in the cardiac cycle has been found to have diagnostic significance. If positive $\alpha$ is measured clockwise from the hori-

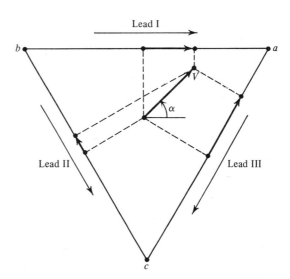

**Fig. 6.16** Einthoven triangle. Positive leads are represented by directed segments as shown for leads I, II, and III.

zontal (the usual ECG convention), then it is called the *electrical axis*. In the normal ECG $0 < \alpha < 90°$. In the Fig. 6.16, $\alpha$ would be described as negative, a condition referred to as *left-axis deviation*. For $\alpha > 90°$, the condition is *right-axis deviation*.

A typical ECG "scalar" lead is shown in Fig. 6.17. The significant quantities are the heights of the P, Q, R, S, and T waves, the durations of each wave, and the time interval between waves. The temporal reference point is usually chosen as the peak of the R wave, and the duration of the cardiac cycle is the RR interval. This is roughly 0.8 sec.

The electrical axis noted above is actually calculated from the effective lead voltages $V_1$, $V_2$, $V_3$. The latter are determined as follows:

$$V_1 = Q_1 + R_1 + S_1$$
$$V_2 = Q_2 + R_2 + S_2$$
$$V_3 = Q_3 + R_3 + S_3$$

where $Q_i$, $R_i$, and $S_i$ are peak values of the Q, R, and S waves in lead $i$. Note that since $Q_i$, $R_i$, and $S_i$ are taken as peak values, they are not necessarily simultaneous. Thus, the Einthoven construction may not be unambiguous. Other remarks concerning the "tracing" in Fig. 6.17 are:

1. The P wave corresponds to atrial activity.
2. The QRS (complex) is the result of ventricular activity.

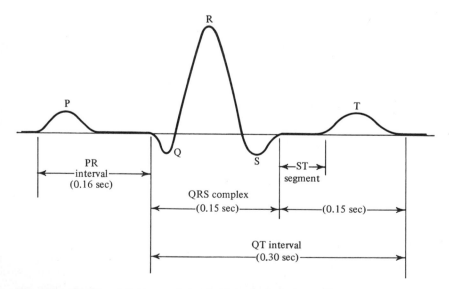

**Fig. 6.17**   Significant features of standard (scalar) electrocardiogram.

3. The T wave corresponds to ventricular repolarization. Atrial repolarization is masked by the QRS.
4. The PR interval is a measure of AV conduction time, and disorders of conduction are related to this interval.
5. The base line is established by the TP segment of the wave. The relative level of the ST segment is an important diagnostic measure. Normal ST segments are at the base line, while coronary insufficiency results in a depressed segment.
6. The QT interval gives the total duration of the ventricular systole. It should be less than half the preceding RR interval.

### PRECORDIAL LEADS

An important reference lead has been arbitrarily established by connecting the "limb leads" (that is, RA, LA, and LL) together, each through a 5,000-ohm resistance. The reference formed is known as the *Wilson central terminal*. If, with respect to this, the "exploring electrode" is placed over the chest at each of six standard locations, the *precordial leads* $V_1$, $V_2$, $V_3$, $V_4$, $V_5$, and $V_6$ are established. Figure 6.18 shows a normal record for each precordial (chest) lead and also indicates the chest position for each lead.

The central terminal of Wilson was devised so that it would be "indifferent" to the electrical activity in the heart. This meant that any difference of potential between an exploring lead and the central terminal could only result from activity in the vicinity of the exploring (unipolar) electrode. An analytical basis for interpreting the relation-

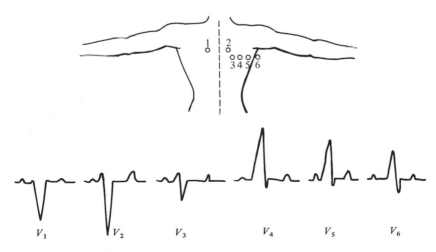

**Fig. 6.18** Precordial leads. Location and typical records.

ship between bioelectric activity and location of electrodes which measure that activity was given in the previous chapter. If one electrode is indeed remote to the region where sources are located, then the measured potential difference will depend solely on the second electrode and its relative position. The Wilson central terminal involves points on the body which are as remote from the heart as possible, so that it serves satisfactorily. By choosing an average over the extremity potentials, it was believed that it was even less sensitive to source variations. Its main advantage is that the reference is well defined and, therefore, consistent results, from patient to patient, can be obtained.

In addition to the precordial leads, *augmented unipolar limb leads* are used, where two limb leads are connected through 5,000 ohms each to form a reference, and the relative potential of the remaining limb lead is measured. If the positive terminal is the right arm, the lead is

$$aV_R$$

The left arm (with respect to RA and LL tied through 5 kΩ each) is lead

$$aV_L$$

Similarly, the left leg gives lead

$$aV_F$$

The three standard leads plus $V_1$, $V_2$, $V_3$, $V_4$, $V_5$, $V_6$ and $aV_R$, $aV_L$, $aV_F$ constitute *the standard 12-lead ECG*.

## CENTRIC–DIPOLE MODEL

For a highly idealized torso that is represented as a uniform conducting sphere, and with a heart model that consists of a central dipole (see Fig. 6.19), the surface potentials can be readily computed. We have for

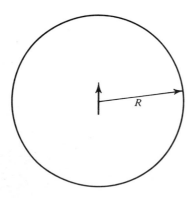

**Fig. 6.19**  Centric dipole in conducting sphere.

the potential $\Phi$,

$$\nabla^2 \Phi = 0 \tag{6.25}$$

$$\left. \frac{\partial \Phi}{\partial r} \right|_{r=R} = 0 \tag{6.26}$$

Solutions to (6.25) in spherical coordinates, with $\partial/\partial\phi = 0$ due to symmetry, are in the form of

$$\Phi = \sum_n [a_n r^n P_n (\cos\theta) + b_n r^{-(n+1)} P_n (\cos\theta)] \tag{6.27}$$

Since the potential near the origin must vary as $1/r^2$, we choose $n = 1$. [Other terms in positive powers of $r$ are not excluded on this ground, but one can confirm, when applying (6.26), that since $b_n = 0$ $(n \neq 1)$ it is necessary that $a_n = 0$ $(n \neq 1)$.] Consequently, since $P_1 (\cos\theta) = \cos\theta$,

$$\Phi = ar\cos\theta + \frac{b}{r^2}\cos\theta \tag{6.28}$$

$$\left. \frac{\partial \Phi}{\partial r} \right|_{r=R} = a\cos\theta - \frac{2b}{R^3}\cos\theta = 0 \tag{6.29}$$

Hence $a = 2b/R^3$.

$$\Phi = \frac{b}{R^2}\cos\theta \left[ \frac{2r}{R} + \left(\frac{R}{r}\right)^2 \right] \tag{6.30}$$

The potential over the surface is $(3b/R^2)\cos\theta$ and is three times what it would be if the body were infinite in extent.

If we adopt the fiction that the points $\theta = 30°$, $\phi = 0°$; $\theta = 30°$, $\phi = 180°$; $\theta = 180°$ are LA, RA, and LL, respectively, then the potentials produced by a central dipole of arbitrary orientation[1] and magnitude are precisely those given by the Einthoven triangle. This can be confirmed by reference to Fig. 6.20, where the case of a "unit" dipole vector $(3b/R^2 = 1)$ at an angle $\alpha$ with the horizontal is described. Applying (6.30) to evaluate the potential at the "limb" locations, one obtains

$$\Phi_{\text{LA}} = \cos(\alpha - 30°) = \frac{\sqrt{3}}{2}\cos\alpha + \frac{1}{2}\sin\alpha \tag{6.31a}$$

$$\Phi_{\text{RA}} = \cos(150° - \alpha) = -\frac{\sqrt{3}}{2}\cos\alpha + \frac{1}{2}\sin\alpha \tag{6.31b}$$

$$\Phi_{\text{LL}} = \cos(90° + \alpha) = -\sin\alpha \tag{6.31c}$$

---

[1] The Einthoven system tacitly assumes that the "heart dipole" lies in the frontal plane, here defined by $\phi = 0°$, $180°$.

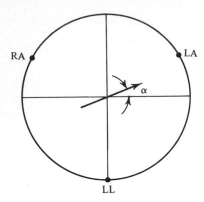

**Fig. 6.20** Spherical-model simulation of the Einthoven system.

For a vector amplitude $V/\sqrt{3}$,

$$V_{\mathrm{I}} = V \cos \alpha \tag{6.32a}$$

$$V_{\mathrm{II}} = \frac{V}{2} \cos \alpha - \frac{\sqrt{3}}{2} V \sin \alpha \tag{6.32b}$$

$$V_{\mathrm{III}} = -\frac{V}{2} \cos \alpha - \frac{\sqrt{3}}{2} V \sin \alpha \tag{6.32c}$$

Comparison of Eqs. (6.32) and (6.21) to (6.23) confirms that the centric spherical model conforms to the Einthoven triangle relationships.

The Einthoven triangle of Fig. 6.16 has been introduced solely on the basis that the lead voltages can be correctly represented as the projections of a vector $V$ at an appropriate angle $\alpha$. Specifically, the geometry was chosen so that (6.21) to (6.23) are necessarily satisfied. No further implication in the meaning of the vector $V$ can be given. However, (6.32) permits $V$ to be thought of as a heart vector in the highly oversimplified spherical-torso centric-dipole model.

On the basis of the Einthoven system, the Wilson central terminal potential can be evaluated. We first note that for an essentially infinite input-impedance recording system, no net current flows out of the central terminal, so that if its potential is designated $\Phi_{\mathrm{CT}}$ we have

$$\frac{\Phi_{\mathrm{LA}} - \Phi_{\mathrm{CT}}}{5,000} + \frac{\Phi_{\mathrm{RA}} - \Phi_{\mathrm{CT}}}{5,000} + \frac{\Phi_{\mathrm{LL}} - \Phi_{\mathrm{CT}}}{5,000} = 0 \tag{6.33}$$

from the Kirchhoff node equation. We therefore confirm

$$\Phi_{\mathrm{CT}} = \frac{\Phi_{\mathrm{LA}} + \Phi_{\mathrm{RA}} + \Phi_{\mathrm{LL}}}{3} \tag{6.34}$$

or $\Phi_{CT}$ is the mean of the extremity potentials.   Using (6.31), we have

$$\Phi_{CT} = \frac{\sqrt{3}}{2}\cos\alpha + \frac{\sin\alpha}{2} - \frac{\sqrt{3}}{2}\cos\alpha + \frac{\sin\alpha}{2} - \sin\alpha = 0 \quad (6.35)$$

which is the basis for the Wilson central terminal design.

The lead formed by pairing the right-arm electrode with the central terminal is designated $V_R$, that is,

$$V_R = \Phi_{RA} - \Phi_{CT} = \Phi_{RA} \tag{6.36a}$$

according to the highly idealized model above.   Similarly,

$$V_L = \Phi_{LA} - \Phi_{CT} = \Phi_{LA} \tag{6.36b}$$
$$V_F = \Phi_{LL} - \Phi_{CT} = \Phi_{LL} \tag{6.36c}$$

By using Kirchhoff's laws, it is not too hard to show that

$$aV_R = 1.5V_R \tag{6.37a}$$
$$aV_L = 1.5V_L \tag{6.37b}$$
$$aV_F = 1.5V_F \tag{6.37c}$$

One should note that this result is actually not dependent on the idealized properties of the Wilson central terminal given above.

Based on the geometrical relationships contained in (6.31) and (6.32), Fig. 6.21 can be constructed.   For a given heart dipole located at the origin and inclined at an angle $\alpha$, the projection on any of the labeled axes corresponds to the voltage that would be measured in that particular lead to the extent that the approximations are satisfied.   The use of such geometrical relationships is quite helpful, and the following section considers their development for realistic torso geometry and for a heart dipole not restricted to the frontal plane.

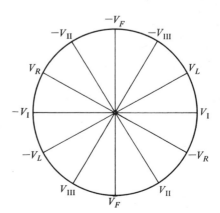

**Fig. 6.21**  ECG reference frame for the idealized (Einthoven) system.

Analytical solutions are available for eccentric dipoles in the sphere and also in the spheroid. Such models correspond somewhat more closely to the actual geometry of the heart in the body. For the eccentric dipole in a sphere Frank[1] has derived a closed-form solution for the surface potential. The result is expressed in the equation

$$\Phi(R,\theta,\phi) = \frac{p}{4\pi\sigma Rr}\left\{\cos\psi\left[\frac{1-f^2}{(1+f^2-2f\mu)^{\frac{3}{2}}}-1\right]\right.$$
$$\left.-\frac{\sin\psi\cos\phi}{\sin\theta}\left[\frac{3f-3f^2\mu+f^3-\mu}{(1+f^2-2f\mu)^{\frac{3}{2}}}+\mu\right]\right\}$$

where $p$ is the dipole-moment magnitude, $r$ is the radial position of the dipole, $f = r/R$, $\mu = \cos\theta$, and $\psi$ is the angle made by the dipole with the polar $(z)$ axis. The radius of the sphere is $R$, and a standard spherical coordinate system is utilized. For $f \to 0$ (note that $r \to 0$ also) the above expression approaches (6.30) in the limit.

## 6.9  LEAD–VECTOR CONCEPT

A more rational approach to the relationship between the heart-dipole source and the resultant surface potentials is embodied in the lead-vector concept introduced by Burger and Van Milaan.[2] The initial assumptions are:

1. Body is heterogeneous, resistive, and linear.
2. Electrical activity of the heart can be represented by a dipole fixed in position but variable in magnitude and orientation.
3. Actual body shape is to be used.

Suppose the heart-dipole current source has a vector magnitude **p**. In view of the linearity of the medium, as assumed above, the potential at an arbitrary surface point $Q$, relative to a chosen reference, must be proportional to the magnitude of **p**. The proportionality must exist also with respect to the components of **p**. If the rectangular components are $p_x$, $p_y$, $p_z$, then the potential at $Q$, $\Phi_Q$, can be expressed as

$$\Phi_Q = c_x p_x + c_y p_y + c_z p_z \tag{6.38}$$

where $c_x$, $c_y$, $c_z$ are proportionality constants which are valid for the conditions specified. A somewhat simpler notation results if $c_x$, $c_y$, $c_z$ are considered to be components of a vector **c** which is known as the *lead vector*.

[1] E. Frank, Electric Potential Produced by Two Point Current Sources in a Homogeneous Conducting Sphere, *J. Appl. Phys.*, **23**:1225 (1952).

[2] H. C. Burger and J. B. Van Milaan, Heart Vector and Leads. Parts I, II, III, *Brit. Heart J.*, **8**:157 (1946), **9**:154 (1947), **10**:229 (1948).

This gives

$$\Phi_Q = \mathbf{c} \cdot \mathbf{p} \qquad (6.39)$$

The value of $\mathbf{c}$ depends on the location of $Q$ and the equivalent heart dipole, the body shape, the electrical characteristics of the body tissues, etc. If three different surface points $Q_i$ are chosen and their respective values of $c_i$ are determined, then the dipole moment $\mathbf{p}$ can be found from a specific simultaneous potential measurement at each point. For example, we might measure

$$V_{\mathrm{I}} = \mathbf{A} \cdot \mathbf{p} \qquad (6.40a)$$

$$V_{\mathrm{II}} = \mathbf{B} \cdot \mathbf{p} \qquad (6.40b)$$

$$V_{\mathrm{III}} = \mathbf{C} \cdot \mathbf{p} \qquad (6.40c)$$

$$V_{\mathrm{BR}} = \mathbf{D} \cdot \mathbf{p} \qquad (6.40d)$$

where $V_{\mathrm{BR}}$ is the back to right arm lead. The value of $\mathbf{p}$ can be found since (6.40) includes three independent linear equations. One can confirm the algebraic solution for $p_x$, namely,

$$p_x = \frac{1}{\Delta} [V_{\mathrm{I}}(B_y D_z - B_z D_y) + V_{\mathrm{II}}(A_z D_y - A_y D_z)$$
$$+ V_{\mathrm{BR}}(A_y B_z - A_z B_y)]$$

where

$$\Delta = A_x(B_y D_z - B_z D_y) + A_y(B_z D_x - B_x D_z)$$
$$+ A_z(B_x D_y - B_y D_x) = \mathbf{A} \cdot \mathbf{B} \times \mathbf{D}$$

In obtaining this result, use was made of $V_{\mathrm{I}} + V_{\mathrm{III}} = V_{\mathrm{II}}$, which leads to $\mathbf{A} + \mathbf{C} = \mathbf{B}$ in (6.40).

In current ECG practice, the positive $x$ direction is chosen as right to left, positive $z$ is front to back, while positive $y$ is top to bottom. The coordinate system is thus an orthogonal right-handed system. The frontal plane corresponds to $xy$, the horizontal plane is $xz$, while the saggital plane is $yz$.

Frank[1] determined the coefficients $\mathbf{A}, \mathbf{B}, \mathbf{C}, \mathbf{D}$ of (6.40) from measurements in an electrolytic-tank model of the human torso. He established the following relative relationships for the scalar leads:

$$V_{\mathrm{I}} = 76p_x - 27p_y + 14p_z \qquad (6.41a)$$

$$V_{\mathrm{II}} = 30p_x + 146p_y - 16p_z \qquad (6.41b)$$

utilizing a torso model based on an average male shape. This result is significantly different from the values obtained for a central dipole in a

[1] E. Frank, General Theory of Heart-vector Projection, *Circulation Res.*, **2**:258 (1954).

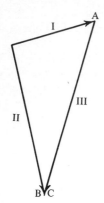

**Fig. 6.22** "Burger triangle." Lead vectors as determined from a torso model that includes spine and lungs. [*After H. C. Burger and J. B. Van Milaan, Brit. Heart J.*, **9**:154 (1947).]

spherical body, where, utilizing (6.21) to (6.23) and the nomenclature of (6.39), we have

$$V_{\mathrm{I}} = 2\sqrt{3}\,p_x \qquad\qquad\qquad (6.42a)$$

$$V_{\mathrm{II}} = \sqrt{3}\,p_x + 3p_y \qquad\qquad\qquad (6.42b)$$

This result involves no $z$ dependence since the Einthoven model is restricted to the frontal plane.

A plot of the lead vectors **A**, **B**, and **C** determined from electrolytic-tank measurements by Burger and Van Milaan[1] is shown in Fig. 6.22. The figure represents the frontal plane only; measurements, which included $z$ components (and correspond closely to Frank's results), are

$$V_{\mathrm{I}} = 65p_x - 21p_y + 17p_z \qquad\qquad (6.43a)$$

$$V_{\mathrm{II}} = 25p_x + 120p_y - 15p_z \qquad\qquad (6.43b)$$

Note that $\mathbf{A} + \mathbf{C} = \mathbf{B}$ in Fig. 6.22, a result that follows from (6.20) and (6.40). In the Einthoven theory, **A**, **B**, **C** forms an equilateral triangle with **A** horizontal. If the three-dimensional character is considered [i.e., by using (6.43)], then the Burger triangle would rotate with **B** moving "out of" and **A** "into" the frontal plane, and would, therefore, differ even more significantly from the simple Einthoven theory.

Frank defines a surface known as an *image surface* or *space*. Each point on the latter corresponds to a point on the body surface. Its property is that a vector connecting the reference origin to a point on the image surface is the monopolar lead vector for the corresponding physical point. Thus, if the difference of potential between two body points (any arbitrary lead) is desired, the vector joining the corresponding points in image space is the appropriate lead vector.

[1] *Loc. cit.*

## 6.10 LEAD FIELDS AND RECIPROCITY

For a given electrode pair on the body surface, we could map the behavior of the lead vector as a function of position of the heart dipole, thus generating a *lead-vector field*. In view of the assumption of homogeneity it might be expected that this field would be mathematically "regular." A development that makes evident the dependent behavior of the lead-vector field is the *lead field* of McFee and Johnston.[1] The lead field arises from general reciprocity relations, and we shall proceed by developing these first.

Consider an arbitrary volume $V$ bounded by the surface $S$, as illustrated in Fig. 6.23. Let $\Phi_1$ and $\Phi_2$ be any two scalar fields defined in $V$. Then if $\sigma$ is the conductivity within $V$, which we assume a regular function of position, the following vector identities must be satisfied:

$$\nabla \cdot \Phi_1(\sigma\nabla\Phi_2) = \Phi_1\nabla \cdot (\sigma\nabla\Phi_2) + \sigma\nabla\Phi_1 \cdot \nabla\Phi_2 \qquad (6.44)$$

$$\nabla \cdot \Phi_2(\sigma\nabla\Phi_1) = \Phi_2\nabla \cdot (\sigma\nabla\Phi_1) + \sigma\nabla\Phi_1 \cdot \nabla\Phi_2 \qquad (6.45)$$

If we subtract (6.45) from (6.44), integrate term by term over the volume $V$, and use the divergence theorem, we obtain

$$\int_S \Phi_1(\sigma\nabla\Phi_2) \cdot d\mathbf{S} - \int_S \Phi_2(\sigma\nabla\Phi_1) \cdot d\mathbf{S} = \int_V [\Phi_1\nabla \cdot (\sigma\nabla\Phi_2) - \Phi_2\nabla \cdot (\sigma\nabla\Phi_1)] \, dV \qquad (6.46)$$

Now let us assume that $\Phi_1$ is the scalar potential in $V$ due to sources within $V$ specified by $I_v = -\nabla \cdot \mathbf{J}_i$ [see (5.41)]. We shall further assume that $\Phi_2$ arises solely from currents caused to flow across the surface $S$; we designate the latter by $K$ (amperes per unit area). We require

$$\int K \, dS = 0$$

since, as noted in (5.28), the current is solenoidal. The following con-

[1] R. McFee and F. D. Johnston, Electrocardiographic Leads, *Circulation*, **8**:564 (1953), **9**:255 (1954), **9**:868 (1954).

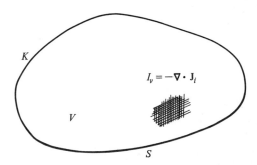

**Fig. 6.23** "Torso" volume geometry.

straints must then be satisfied by $\Phi_1$ and $\Phi_2$:

$$\nabla \cdot (\sigma \nabla \Phi_1) = -I_v \qquad \text{[using (5.35)]} \tag{6.47}$$

$$\sigma \nabla \Phi_2 \cdot d\mathbf{S} = K\, dS \qquad \text{(taking positive } K \text{ as } \textit{inflow}) \tag{6.48}$$

When $I_v$ is the source, no currents cross the boundary surface; hence

$$\nabla \Phi_1 \cdot d\mathbf{S} = 0 \tag{6.49}$$

For the source $K$ at the surface, the current must be solenoidal everywhere in $V$; hence

$$\nabla \cdot (\sigma \nabla \Phi_2) = 0 \tag{6.50}$$

By using (6.47) to (6.50), (6.46) becomes

$$\int_S \Phi_1 K\, dS = \int_V \Phi_2 I_v\, dV \tag{6.51}$$

which is the desired form of the reciprocity theorem.

Consider a specific $I_v$ which consists of a point current source of magnitude $I_0$ at $\mathbf{r}_1$ and a point sink of equal magnitude at $\mathbf{r}_2$, where $\mathbf{r}_1$ and $\mathbf{r}_2$ are the respective position vectors. Thus

$$I_v = [\delta_v(\mathbf{r}_1 - \mathbf{r}) - \delta_v(\mathbf{r}_2 - \mathbf{r})]I_0 \tag{6.52}$$

and $\delta_v(r)$ is a three-dimensional delta function. Consider further that $\Phi_2$ arises from a specific distribution $K$ which is that of a surface electrode $a$ that provides an inflow of a unit current, while electrode $b$ has an outflow of a unit current. If the position vector of $a$ is $\mathbf{r}_a$ and that of $b$ is $\mathbf{r}_b$, then for very small electrodes, we have

$$K = \delta_s(\mathbf{r}_a - \mathbf{r}) - \delta_s(\mathbf{r}_b - \mathbf{r}) \tag{6.53}$$

where $\delta_s$ is a two-dimensional delta function. If (6.52) and (6.53) are substituted into (6.51), then we get

$$I_0[\Phi_2(\mathbf{r}_1) - \Phi_2(\mathbf{r}_2)] = \Phi_1(\mathbf{r}_a) - \Phi_1(\mathbf{r}_b) \tag{6.54}$$

This result asserts that the difference of potential between two surface points $a$ and $b$ (that is, the lead voltage) due to a unit source and sink at 1 and 2 ($I_0 = 1$) must be equal to the difference of potential that would be measured between the source points (1 and 2) if a unit of current were applied at the leads $a$ and $b$. The latter (equivalent) condition is referred to as *reciprocal excitation*.

If the separation of $r_1$ and $r_2$ approaches zero, the source described by (6.52) becomes a dipole with moment $I_0(\mathbf{r}_1 - \mathbf{r}_2)$. (It is tacitly assumed that $I_0 \to \infty$ as $|\mathbf{r}_2 - \mathbf{r}_1| \to 0$ so that the moment is finite.) Now, by a Taylor series expansion

$$\Phi_2(\mathbf{r}_1) = \Phi_2(\mathbf{r}_2) + \nabla \Phi_2 \cdot (\mathbf{r}_1 - \mathbf{r}_2) + \cdots \tag{6.55}$$

Thus, if we consider $(\mathbf{r}_2 - \mathbf{r}_1) \to 0$, then the higher terms in (6.55) can be neglected. Under these conditions the resultant dipole has a moment $I_0(\mathbf{r}_1 - \mathbf{r}_2) = \mathbf{m}_0$, and the potential $\Phi_1(\mathbf{r}_a) - \Phi_1(\mathbf{r}_b) = V_{ab}$ due to $\mathbf{m}_0$ is

$$V_{ab} = \boldsymbol{\nabla}\Phi_2 \cdot \mathbf{m}_0 \qquad (6.56)$$

using (6.54) and (6.55). In (6.56), $\Phi_2$ is the potential due to "reciprocally energizing" the pickup leads $ab$. The scalar field $\Phi_2$ has been designated the "lead field" by McFee and Johnston,[1] who first studied it. By comparison with earlier nomenclature we see that $\boldsymbol{\nabla}\Phi_2$ is the lead vector (field), while $\mathbf{m}_0$ corresponds to the "heart vector" and $V_{ab}$ is the lead voltage. This result clarifies the nature of physically realizable lead-vector fields, namely, that they must be derivable as the gradient of a scalar field which satisfies Laplace's equation. This is an important limitation and will be discussed further in a subsequent section.

Since the actual bioelectric sources in the heart can be characterized as a current dipole moment per unit volume, $\mathbf{J}_i$, an appreciation of the effect of a particular lead in responding to such sources is apparent in the lead-field formulation. As an example, consider the electrode system in Fig. 6.24. The lead field due to electrode pair $a$ and $b$ is sketched, along with the corresponding equipotential surfaces (dotted). From this sketch, we see that $\boldsymbol{\nabla}\Phi_2$ in the region of the heart is approximately constant and in the $x$ direction. Accordingly, the lead $ab$ will record the linearly superposed contributions from the $x$ component dipoles which are active in the heart. This result is obtained by generalizing (6.56) to the case of a volume distribution of dipoles, $\mathbf{J}_1$, namely,

$$V_{ab} = \int_V \boldsymbol{\nabla}\Phi_2 \cdot \mathbf{J}_i \, dV \qquad (6.57)$$

By utilizing the lead-field concept, one can consider a lead system that is made up of more than two electrodes which are connected together

[1] *Ibid.*

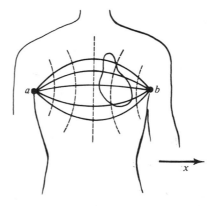

**Fig. 6.24** Lead field (for reciprocally energizing $a$ and $b$). Dotted curves are isopotentials.

by a suitable resistance network with a two-terminal output.  If we let the output pair of the network correspond to the points $a$ and $b$ of the previous discussion, and consider the resistance network as part of the volume conductor, then (6.56) continues to be satisfied.  By manipulating the number and orientation of electrodes, and choosing the network of resistances appropriately, one can realize more closely lead fields with desired configuration.  Thus for an $x$ lead a still more uniform lead-vector field than illustrated in Fig. 6.24 can be obtained by use of multiple leads.  As an example, consider, say, 10 equispaced electrodes placed on the right side of the torso, each connected through a resistor to a common output terminal (that is, $a$), and a like system on the left side (yielding terminal $b$).  Then if the $ab$ lead is reciprocally energized, current flow lines (the lead-vector field) should be much more uniform and directed more precisely along the horizontal.  This approach, with application to electrocardiography, is discussed in some detail in a paper by Brody and Romans.[1]

The lead field is also useful in indicating the *relative* contribution to the total lead voltage from each dipole element of a distribution of sources.  In Fig. 6.24, sources near the electrodes would be more strongly emphasized because of the higher values of gradient.  Also, of course, the orientation of the dipole must be such as to have a component in the direction of $\nabla\Phi$.  Further discussions of reciprocity and lead fields are given by Brody et al.[2] and by Plonsey.[3]

## 6.11  ORTHOGONAL LEAD SYSTEM

In Sec. 6.9 it was shown that the lead vectors of the conventional limb leads actually form a skewed scalene triangle.  Furthermore, this triangle lies mainly, though not exactly, in the frontal plane.  If the interpretation of lead voltages is to be put on a quantitative and comparative basis, a clear first step would be to set up three leads, each of which responds to an orthogonal component of the "heart vector."  More specifically, we desire to set up three lead fields, each of which is uniform in the heart region and where the gradients are mutually orthogonal.  (One should verify that such a lead field is physically realizable.)  Such a lead

[1] D. A. Brody and W. E. Romans, A Model Which Demonstrates the Quantitative Relationship between the Electromotive Forces of the Heart and the Extremity Leads, *Am. Heart J.*, **45**:263 (February, 1953).

[2] D. A. Brody, J. C. Bradshaw, and J. W. Evans, A Theoretical Basis for Determining Heart-lead Relationships of the Equivalent Cardiac Multipole, *IRE Trans. Biomed. Electron.*, **BME8**:139 (1961).

[3] R. Plonsey, Reciprocity Applied to Volume Conductors and the ECG, *IEEE Trans. Biomed. Electron.*, **BME10**:9 (January, 1963).

system is known as an orthogonal lead system and directly measures the $x$, $y$, and $z$ components of the "heart vector." More exactly, it measures the linear superposition of the distributed $x$, $y$, and $z$ components of $\mathbf{J}_i\, dV$, as is evident from (6.57).

In the design of an orthogonal lead system a human torso model (a plastic model filled with electrolyte) is utilized so that the lead voltages can be calibrated with respect to active artificial dipole sources. Active dipole sources, oriented in either the $x$, $y$, or $z$ direction, are placed, say, at the approximate anatomical center of the heart within the torso model. One now seeks a lead which responds to the $x$ dipole but not $y$ or $z$. This corresponds to realizing $c_y = c_z = 0$ in (6.39). In three-dimensional space, this lead vector is then in the $x$ direction and yields the $x$ component of the heart vector. Similarly, $y$ and $z$ leads are found that respond to the $y$ and $z$ components of the heart dipole only. For a given permissible error there is no single solution, and other criteria must be used to select an optimum system. For example, since the activity of the heart consists of a distribution of dipole moments, the lead system, as evaluated in the torso model, should continue to be satisfactory if the active dipole is displaced from the anatomical center to epicardial positions. Such a system will then produce a heart vector that corresponds to the vector sum of individual dipole elements; thus it has the property that the lead field is uniform in the region of the heart, as desired.

Because of the linearity of the system, it is not necessary that the $x$ component, say, be obtained from a single electrode pair. It is possible to combine the potentials at three or more points on the body surface through resistance networks, such that across the output terminals the voltage is proportional to $p_x$ and independent of $p_y$ and $p_z$. This mechanism was noted in the discussion on lead fields. Its utilization permits the selection of electrode locations that more nearly satisfy such criteria as convenience in applying electrodes, well-defined location with respect to anatomical landmarks, minimized sensitivity to imprecise electrode positioning and/or to body shape, and uniformity and correctness of geometry of the lead field. By suitable resistance networks it is also possible to adjust the lead-vector magnitudes to be equal. This gives rise to what is designated as the *corrected orthogonal lead system*, a number of which have been designed, notably the Frank,[1] SVEC III,[2] and McFee and Parungao.[3]

[1] E. Frank, An Accurate Clinically Practical System for Spatial Vectorcardiography, *Circulation*, **13**:737 (1956).

[2] O. H. Schmitt and E. Simonson, The Present Status of Vectorcardiography, *Arch. Internal Med.*, **96**:574 (1955).

[3] R. McFee and A. Parungao, An Orthogonal Lead System for Clinical Electrocardiography, *Am. Heart J.*, **62**:93 (1961).

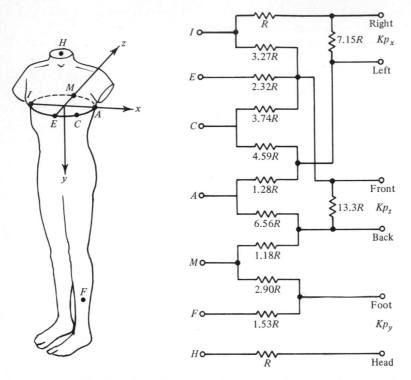

**Fig. 6.25** The Frank lead system with its associated resistor network.
[*E. Frank, Ann. N.Y. Acad. Sci.,* **65**:980 (1957).]

The Frank system is shown in Fig. 6.25. Note that the number of electrodes which are required is seven rather than a theoretical minimum of four. In the SVEC III, fourteen electrodes are required.

   A comparison of the electrode systems of Frank, SVEC II, SVEC III, Wilson-Burch, and Duchosal-Sulzer was made by Schmitt.[1] An analysis of the variation of lead voltage as a consequence of displacement of an artificial dipole to the corners of a 5-cm cube (i.e., roughly the volume of the heart) centered at the heart was made, utilizing an electrolytic torso model. As has been noted, the translation of the dipole should cause no lead-voltage change in an ideal lead system. Both Frank and SVEC III systems perform quite satisfactorily in this respect. The standard deviation from the mean of the SVEC III is 10.2, 3.0, and 4.0 percent for $x$, $y$, and $z$ leads. For the Frank system, the figures are 15.2, 6.9, and 16.5

---

[1] O. H. Schmitt, Lead Vectors and Transfer Impedance, *Ann. N.Y. Acad. Sci.,* **65**:1902 (1957).

percent, respectively.   This compares with 20.4, 12.3, and 14.2 percent for the limb leads I, II, and III, respectively.

A comparison of the Frank, SVEC III, and McFee-Parungao systems was performed by Brody and Arzbaecher,[1] based on the lead-field approach.   Lead-field maps in the principal planes of the torso due to the appropriate lead should be uniform, and it is relatively easy to evaluate the performance actually achieved.   The plots for the frontal plane of the SVEC III and the McFee-Parungao system are shown in Fig. 6.26.   It can be seen that the SVEC III introduces distortion in the lower left-hand portion of the field.   Although the Frank, SVEC III, and McFee-Parungao systems are different in performance, they are relatively equivalent.   Other orthogonal systems such as the cube and the tetrahedron introduce considerable distortion.

In contrast to the standard scalar leads, the outputs of the $x$, $y$, and $z$ orthogonal leads are independent.   For convenience in visualizing the three signals, they may be combined "vectorially" to form the heart vector.   One can discuss the path traced by the tip of the heart vector since it contains the same information as the component signals.   The three lead signals themselves are referred to as *scalar records*, while consideration of their composition into a vector constitutes the subject of vectorcardiography.   The trajectory of the vector tip is referred to as the *vector loop*, and since the ECG is cyclical, the vector loop is closed.

[1] D. A. Brody and R. C. Arzbaecher, A Comparative Analysis of Several Corrected Vector-cardiographic Leads, *Circulation*, **39**:533 (1964).

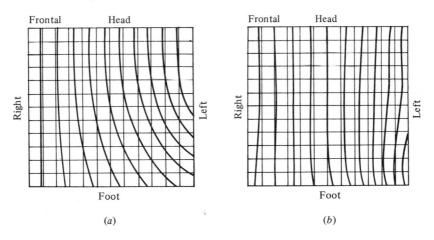

**Fig. 6.26**   Lead-field maps for the frontal plane.   (*a*) SVEC III system; (*b*) McFee-Parungao system.   [*D. A. Brody and R. C. Arzbaecher, Circulation, **39**:533 (1964).*]

The main interest is in the QRS (ventricular activation), and this also, in general, forms a closed loop. The vector loop is considered by means of its projection on the horizontal, frontal, and sagittal planes.

One interesting aspect of vectorcardiography is that, having determined the heart vector, one can project this on any lead vector, thereby synthesizing other lead outputs. For example, Frank,[1] by determining the lead vectors of each of the precordial leads, "resolved" orthogonal-lead data into precordial estimates. In Fig. 6.27, some of these results are shown, and it is seen that they are qualitatively satisfactory. They would be precisely correct if the electrical activity of the heart were exactly represented by a dipole.

From a clinical standpoint, Pipberger[2] has shown that essentially all (clinical) information contained in the 12-lead standard system can also be found in the corrected orthogonal lead system, provided lead resolution is used in doubtful cases. (In the study conducted, 7.3 percent of cases

[1] E. Frank, Determination of the Electrical Center of Ventricular Depolarization in the Human Heart, *Am. Heart J.*, **49**:670 (1955).

[2] H. Pipberger et al., Correlation of Clinical Information in the Standard 12-lead ECG and in Corrected Orthogonal 3-lead ECG, *Am. Heart J.*, **61**:34 (January, 1961).

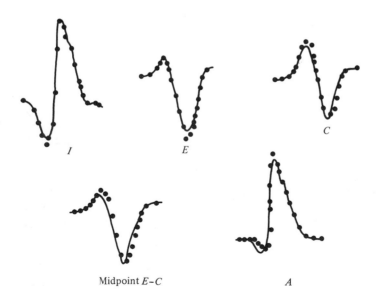

**Fig. 6.27** Solid curve: potential variations at designated electrode locations (refer to Fig. 6.25). Points: obtained by lead resolution utilizing the output of the Frank system and a lead vector obtained from a homogeneous torso model. [*From E. Frank, Am. Heart J.,* **49**:670 (1955).]

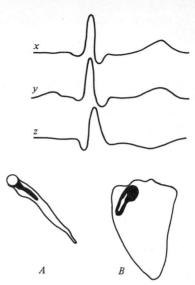

**Fig. 6.28** Scalar orthogonal leads $x$, $y$, and $z$, as labeled. Loop $A$ is a frontal-plane projection of the spatial loop. Loop $B$ is a "polar projection." [*From H. V. Pipberger and T. N. Carter, Circulation*, **25**:827 (1962).]

were in this category.) The use of lead resolution appears to improve the diagnostic ability of the clinician since the records are displayed in a standard form. Thus, even though no information is created by lead resolution, the ability to recognize patterns appears to be enhanced by this operation.

Even further removed from the original standard system is the resolution of the vector ECG on special spatial axes. For example, the QRS vector loop in most cases lies approximately in a plane in space. The projection of the vector loop on this plane constitutes a *polar projection*. The projection puts the vector ECG in a form which is independent of the standard $x$, $y$, and $z$ axes and rather more closely related to the physical "axis" of the heart itself. That is, since the plane of the vector loop should rotate with the heart, the polar ECG projection should be approximately independent of the orientation of the (physical) heart in the chest cavity. Since the physical orientation of normal healthy hearts is a variable, this procedure, at least in principle, removes this factor from the ECG. Any skewness of the resultant heart vector is then solely an electrical phenomenon and a reflection of the electrophysiology of the heart itself. An example of the resultant scalar records and vector loop projections is illustrated in Fig. 6.28.

On the basis of these ideas, McFee et al.[1] define a *normalized electro-*

---

[1] R. McFee, R. S. Wilkinson, and J. A. Abildskov, On the Normalization of the Electrical Orientation of the Heart and the Representation of Electrical Axis by Means of an Axis Map, *Am. Heart J.*, **62**:391 (1961). Also see *Am. Heart J.*, **65**:220 (1963).

*cardiogram* by a specific procedure of coordinate axes rotations.   These are chosen so that deflections in one lead are a maximum, are minimum in a second (this is normal to the plane of the loop), and are equally negative and positive in the third lead.   It should be emphasized that the justification of these techniques is based on the validity of the dipole model.


## 6.12  THE "DIPOLE HYPOTHESIS"

The heart model which has been assumed in the previous sections is that of a dipole fixed in position but free to change its direction and magnitude. This formulation is the underlying basis for the orthogonal lead systems, for axis rotation, and indeed for vectorcardiography.   We consider in this section some of the evidence relating to the adequacy of representing the electrical activity in the heart by a simple dipole.

Some evidence that the dipole representation is a satisfactory one is seen in the evaluation of resolved leads.   If, indeed, arbitrary leads are obtainable as linear combinations of only the $x$, $y$, and $z$ components of an orthogonal lead system, then the dipole basis of cardiac activity is established.   Typical of such computations are the potentials shown in Fig. 6.27.   Agreement, as judged visually, is quite good.   Peak-to-peak discrepancies, as determined in a study of 19 normals by Okada et al.[1] as a percentage of the average of the actual and synthesized signal, were 23 percent with an SD of ±7 percent.

A more comprehensive evaluation can be made by plotting equipotentials on the surface of the torso and comparing such patterns with those produced on the surface of a torso model with true dipole excitations.   The earliest such study was performed by Frank,[2] and Fig. 6.29 illustrates the comparison at the 10-msec instant of the QRS.   The comparison was found to be equally favorable at successive instants during the ventricular activation.

Work by more recent experimenters has modified this picture and suggests, first, that Frank's subject was unusually dipolar; second, that his equipotential lines were insufficiently close together; and third, that deviations from dipolarity are more likely to occur for subjects with heart disease.   In Fig. 6.30 an example of a nondipolar record obtained by Taccardi[3] is shown.

[1] R. H. Okada, P. Langner, and S. A. Briller, Synthesis of Precordial Potentials from the SVEC III Vectorcardiographic System, *Circulation Res.*, **7**:185 (1959).

[2] E. Frank, Absolute Quantitative Comparison of Instantaneous QRS Equipotentials on a Normal Subject with Dipole Potentials on a Homogeneous Torso Model, *Circulation Res.*, **3**:243 (1955).

[3] B. Taccardi, Distribution of Heart Potentials on the Thoracic Surface of Normal Human Subjects, *Circulation Res.*, **12**:341 (1963).

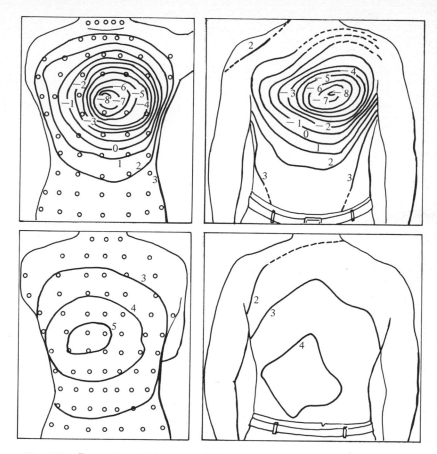

**Fig. 6.29**  Comparison of an equipotential map for a dipole in a homogeneous torso model (left) and a human subject (right) at $t = 10$ msec following the beginning of the normal QRS. The potential difference between adjacent equipotentials is 0.25 mv.  [*E. Frank, Circulation Res.,* **3**:243 (1955).]

Surface-potential maps of humans and dogs have been made by Taccardi,[1] Boineau et al.,[2] Horan et al.,[3] and Spach et al.[4]  These display

[1] *Ibid.*

[2] J. P. Boineau, M. S. Spach, T. C. Pilkington, and R. C. Barr, Relationship between Body Surface Potential and Ventricular Excitation in the Dog, *Circulation Res.,* **19**:489 (1966).

[3] L. Horan, N. C. Flowers, and D. A. Brody, Body Surface Potential Distribution, *Circulation Res.,* **13**:373 (1963).

[4] M. S. Spach et al., Body Surface Isopotential Maps in Normal Children Ages 4 to 14 Years, *Am. Heart J.,* **72**:640 (1966).

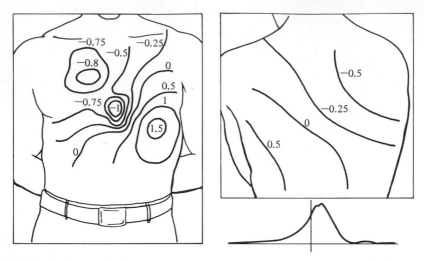

**Fig. 6.30** Surface-potential map of a human subject at the instant during the QRS which is indicated. Potential values are in millivolts. [*B. Taccardi, Circulation Res.*, **12**:341 (1963).]

patterns of changing complexity that appear dipolar at certain times during the QRS (usually early and late) and nondipolar at other times. A measure of nondipolarity is the appearance of more than one maximum or minimum (or both). Figure 6.30 is illustrative of the appearance of such "multipolarity." All investigations show evidence of multipolarity during some portion of the QRS.

The above study of the dipolarity of the heart is a fairly direct one. Another approach, which is somewhat inferential, is that of "cancellation." To explain this technique we note that, from (6.39),

$$V_1(t) = c_{11}p_1(t) + c_{12}p_2(t) + c_{13}p_3(t)$$
$$V_2(t) = c_{21}p_1(t) + c_{22}p_2(t) + c_{23}p_3(t) \qquad (6.58)$$
$$V_3(t) = c_{31}p_1(t) + c_{32}p_2(t) + c_{33}p_3(t)$$

where $p_1$, $p_2$, $p_3$ are the three orthogonal components of the assumed heart vector; $V_1$, $V_2$, $V_3$ are the lead voltages at three arbitrary (and we assume independent) leads; and $c_{ij}$ are geometrical constants. Now the above equations can be used to solve for $p_1$, $p_2$, $p_3$ in terms of $V_1$, $V_2$, $V_3$, giving

$$p_1(t) = \alpha_{11}V_1(t) + \alpha_{12}V_2(t) + \alpha_{13}V_3(t)$$
$$p_2(t) = \alpha_{21}V_1(t) + \alpha_{22}V_2(t) + \alpha_{23}V_3(t) \qquad (6.59)$$
$$p_3(t) = \alpha_{31}V_1(t) + \alpha_{32}V_2(t) + \alpha_{33}V_3(t)$$

If we now consider an arbitrary fourth lead $V_4(t)$, then from (6.39)

$$V_4(t) = c_{41}p_1(t) + c_{42}p_2(t) + c_{43}p_3(t) \tag{6.60}$$

and, substituting (6.59) for $p_1$, $p_2$, and $p_3$, we have

$$V_4(t) = A_1V_1(t) + A_2V_2(t) + A_3V_3(t) \tag{6.61}$$

where $A_1$, $A_2$, and $A_3$ are geometrical constants and independent of time. Thus a necessary condition for the electrical activity of the heart to be accounted for by a dipole generator is that any arbitrary lead be a linear contribution of any other three linearly independent leads. Note that (6.61) is true at any instant of time; it is an identity.

In the "cancellation" experiments three arbitrary leads are led into amplifiers with variable gain, summed, and subtracted from a fourth arbitrary lead. If the difference can be made essentially zero through an appropriate adjustment of the gain controls, then cancellation has been effected and the results considered a substantiation of the dipole hypothesis. Good cancellation results were reported by Schmitt et al.,[1] although they noted that cancellation was less satisfactory for abnormals. However, work—notably by Brody and Copeland,[2] Morton et al.,[3] and McFee[4]— has shown cancellation to be a very imprecise technique and that good cancellation can be effected even in studies where sources are clearly non-dipolar. In particular, as pointed out by McFee, one can always choose the constants [$A_1$, $A_2$, $A_3$ in (6.61)] such that exact cancellation occurs at three instants of time.

Another indirect method for discussing the dipole heart model is the application of factor analysis to ECG waveforms recorded on the torso. As contained in (6.39), a necessary condition for dipole behavior is that an arbitrary lead be related to the $x$, $y$, and $z$ components of the heart vector by a linear relationship. That is, the dipole condition requires

$$V_k(t) = k_xp_x(t) + k_yp_y(t) + k_zp_z(t) \tag{6.62}$$

where $k_x$, $k_y$, $k_z$ are constants that depend only on the dipole location, the reference-potential point, and the surface point $k$. By recording at a great many sites ($i = 1, 2, \ldots, I$) one can investigate whether each $V_i$

[1] O. H. Schmitt, R. B. Levine, and E. Simonson, Electrocardiographic Mirror Pattern Studies, *Am. Heart J.*, **45**:416, 500, 655 (1953).

[2] D. A. Brody and G. D. Copeland, Electrocardiographic Cancellation, *Am. Heart J.*, **56**:381 (1958).

[3] R. F. Morton, W. E. Romans, and D. A. Brody, Cancellation of Esophageal Electrocardiograms, *Circulation*, **15**:897 (1957).

[4] R. McFee, On the Interpretation of Cancellation Experiments, *Am. Heart J.*, **59**:433 (1960).

is, indeed, a linear combination of the three independent time-varying signals, as required by (6.62).

If the potential at any $i$th electrode is sampled in time ($t = t_1$, $t_2$, . . . $t_N$), then the voltage matrix of potentials at $i$, at instants $t$, can be formulated by

$$V = AX \tag{6.63}$$

where $V$ is an $(I \times N)$ matrix, $X$ is an $(I \times N)$ matrix of factors, and $A$ is an $(I \times I)$ matrix of distributive coefficients. If each waveform were independent, the factors $X$ would be unique and (6.63) would be orthogonal. In experiments conducted by Horan et al.,[1] using 180 surface points ($I = 180$), diagonalization of the transforming matrix $A^{-1}$ (that is, $X = A^{-1}V$) yielded only eight eigenvalues that were significantly different from zero. That is, only the first eight rows of $A^{-1}$ need to be considered, so that only eight rows of $X$ are significant. In other words, the surface potentials could, essentially, be recovered from eight principal factors. They found that in man the mean-square error using only one principal factor was 46.8 percent; two, 20.6 percent; three, 14.6 percent; four, 11.9 percent; five, 9.1 percent; six, 6.9 percent; seven, 3.6 percent; and eight, 0.3 percent. On the basis of these results one could conclude that surface potentials are more complex than can be accounted for by a dipole model, but that the latter is a fair approximation. These results are somewhat more conservative than those obtained by Scher et al.;[2] however, the latter work was based on a smaller number of surface electrocardiograms (8 to 36).

A mathematically direct means of evaluation of the dipole content of a distributed source utilizes multipole theory. Several investigations have taken this approach. These results will be discussed in the following section. It will be seen that this work also yields conclusions consistent with that presented in this section.

## 6.13  MULTIPOLE THEORY

It is clear from experiments described in Sec. 6.12 that the dipole representation of the electrical activity in the heart is, in general, too crude. The actual heart sources, as we have defined them, are specified by a dipole-moment density function $\mathbf{J}_i$. This function has the dimensions of

[1] L. Horan, N. C. Flowers, and D. A. Brody, Principal Factor Waveforms of the Thoracic QRS Complex, *Circulation Res.*, **15**:131 (1964).

[2] A. M. Scher, A. C. Young, and W. M. Meredith, Factor Analysis of the Electrocardiogram, *Circulation Res.*, **7**:519 (1960).

a dipole moment per unit volume and is space- and time-dependent. There may be occasions when it will be useful to fully characterize the heart activity by $\mathbf{J}_i$. On the other hand, this is a very complex formulation. In view of the fact that the relatively simple dipole model is often a good approximation to the actual source, it is reasonable to seek an approach that retains the dipole as a leading term but that provides additional (correction) components. Such an approach is embodied in the multipole representation. We shall see that the multipole theory applies to any arbitrary source distribution. Furthermore, since it is essentially equivalent to the actual source distribution, it provides a useful mechanism for evaluating general properties of electrocardiographic sources.

We begin by considering an infinite region of uniform conductivity $\sigma$ which contains an arbitrary distribution of current sources $\mathbf{J}_i$. An origin is set up at an arbitrary point in the vicinity of the current sources, and an expression for the resulting potential field at a distant point is sought. We first consider an element of source at $\xi, \eta, \zeta$ and a (fixed) field point at $P(x,y,z)$, as illustrated in Fig. 6.31. The current-source density $I_v$ is related to the dipole-moment density by (5.41), that is, $I_v = -\nabla \cdot \mathbf{J}_i$, so that

$$d\Phi_p = \frac{1}{4\pi\sigma} \frac{I_v \, dV}{R} = \frac{1}{4\pi\sigma} \frac{I_v \, dV}{\sqrt{(x-\xi)^2 + (y-\eta)^2 + (z-\zeta)^2}} \tag{6.64}$$

where $d\Phi_p$ is the contribution to the total potential at $P$ from element $I_v \, dV$. The denominator of (6.64) can be replaced by a Taylor series

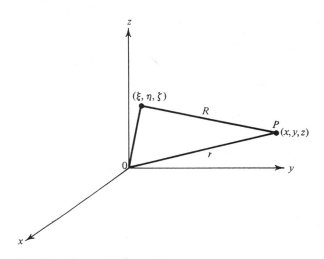

**Fig. 6.31**  Source field geometry.

expansion about the origin in powers of $(\xi,\eta,\zeta)$ given by

$$
\frac{1}{R} = \frac{1}{\sqrt{(x-\xi)^2 + (y-\eta)^2 + (z-\zeta)^2}} = \left\{ \frac{1}{r} - \left[ \xi \frac{\partial}{\partial x}\left(\frac{1}{r}\right) \right. \right.
$$
$$
+ \eta \frac{\partial}{\partial y}\left(\frac{1}{r}\right) + \zeta \frac{\partial}{\partial z}\left(\frac{1}{r}\right) \Big] + \frac{1}{2}\left[ \xi^2 \frac{\partial^2}{\partial x^2}\left(\frac{1}{r}\right) + \eta^2 \frac{\partial^2}{\partial y^2}\left(\frac{1}{r}\right) \right.
$$
$$
+ \zeta^2 \frac{\partial^2}{\partial z^2}\left(\frac{1}{r}\right) + 2\xi\eta \frac{\partial^2}{\partial x\, \partial y}\left(\frac{1}{r}\right) + 2\eta\zeta \frac{\partial^2}{\partial y\, \partial z}\left(\frac{1}{r}\right)
$$
$$
\left. \left. + 2\zeta\xi \frac{\partial^2}{\partial z\, \partial x}\left(\frac{1}{r}\right) \right] + \cdots \right\} \qquad (6.65)
$$

The series in (6.65) converges provided $r > a$, where $a$ is the distance from the origin to the furthest element of source.

Putting (6.65) into (6.64) and integrating with respect to $\xi$, $\eta$, $\zeta$ (over the entire source distribution) permits the total potential to be expressed in the form

$$
\Phi_p = \frac{1}{4\pi\sigma} \int \frac{I_v\, dV}{R} \tag{6.66}
$$

$$
\Phi_p = \sum_{n=0}^{\infty} \Phi_n \tag{6.67}
$$

where $\Phi_n$ is the total contribution from terms whose radial dependence is $r^{-(n+1)}$.

The first term in (6.67), $\Phi_0$, is

$$
\Phi_0 = \frac{I^{(0)}}{4\pi\sigma r} \tag{6.68}
$$

where

$$
I^{(0)} = \int I_v(\xi,\eta,\zeta)\, dV
$$

and is the total algebraic sum of all the current sources. On physical grounds this must be zero for biological sources. (This is called the monopole term.)

The next term can be written as

$$
\Phi_1 = -\frac{1}{4\pi\sigma} \mathbf{I}^{(1)} \cdot \nabla\left(\frac{1}{r}\right) \tag{6.69}
$$

where

$$
I_x{}^{(1)} = \int I_v \xi\, dV \qquad I_y{}^{(1)} = \int I_v \eta\, dV \qquad I_z{}^{(1)} = \int I_v \zeta\, dV \tag{6.70}
$$

This is the dipole term, and $I_x{}^{(1)}$ is the $x$ component strength, $I_y{}^{(1)}$ is the $y$ component strength, and $I_z{}^{(1)}$ is the $z$ component strength. In a similar way

$$\Phi_2 = -\frac{1}{4\pi\sigma}\left[\frac{I_{xx}^{(2)}}{2}\frac{\partial^2}{\partial x^2}\left(\frac{1}{r}\right) + I_{xy}^{(2)}\frac{\partial^2}{\partial x\,\partial y}\left(\frac{1}{r}\right) + \frac{I_{yy}^{(2)}}{2}\frac{\partial^2}{\partial y^2}\left(\frac{1}{r}\right)\right.$$
$$\left. + I_{yz}^{(2)}\frac{\partial^2}{\partial y\,\partial z}\left(\frac{1}{r}\right) + \frac{I_{zz}^{(2)}}{2}\frac{\partial^2}{\partial z^2}\left(\frac{1}{r}\right) + I_{zx}^{(2)}\frac{\partial^2}{\partial x\,\partial z}\left(\frac{1}{r}\right)\right] \quad (6.71)$$

and

$$\begin{array}{lll} I_{xx}^{(2)} = \int I_v \xi^2\,dV & I_{yy}^{(2)} = \int I_v \eta^2\,dV & I_{zz}^{(2)} = \int I_v \zeta^2\,dV \\ I_{xy}^{(2)} = \int I_v \xi\eta\,dV & I_{yz}^{(2)} = \int I_v \eta\zeta\,dV & I_{zx}^{(2)} = \int I_v \zeta\xi\,dV \end{array}$$

$$(6.72)$$

The above terms evaluate second moments of the distribution in contrast to the first moments of (6.70). The pattern for the successive higher-order terms should be clear from the above details. It should also be observed that subsequent terms involve successively higher inverse powers of $r$, as noted earlier. A compact way of expressing the entire series is

$$\Phi_p = \frac{1}{4\pi\sigma}\sum_{n=0}^{\infty}\sum_{l=0}^{n}\sum_{k=0}^{n-l}\frac{(-1)^n}{l!k!(n-l-k)!}\left[\int_V \xi^l\eta^k\zeta^{n-l-k}I_v(\xi,\eta,\zeta)\,d\xi\,d\eta\,d\zeta\right]$$

$$\frac{\partial^n}{\partial x^l\,\partial y^k\,\partial z^{(n-l-k)}}\left(\frac{1}{r}\right) \quad (6.73)$$

and it is understood that $r > a$, as specified earlier, and also $n \geq l + k$ is required. The integer $n$ determines the order of the multipole, and we see that this corresponds to the condition noted earlier that $r$ have the exponent $-(n+1)$.

Equation (6.73) shows how the potential from an *arbitrary* source distribution $I_v$ can be thought of as arising from the sum of source terms, each of which is characterized by a different inverse power of $r$. The $r^{-(n+1)}$ term is called a multipole source of order $n$. Since bioelectric current sources have a zero algebraic sum, the zeroth multipole (monopole) term does not occur, and the leading term is $1/r^2$, which is the dipole term. (The component which depends on $r$ as $1/r^3$ is the quadrupole term, while $1/r^4$ is called the octopole.) When the field point is at a distance which is large compared with the overall source dimension, then the higher terms are ordinarily negligible. In this case, a multipole formulation is particularly advantageous since a few coefficients describe the resultant behavior of sources which may be fairly complicated.

To the extent that the electrical activity of the heart can be described by a dipole, then at any instant three parameters suffice to characterize this activity. If we recognize that at any instant the actual sources are distributed throughout the heart and that significantly more than three parameters would be required for their description, it is clear that a substantial loss of information is inherent in the dipole model. If possible, one would like sufficient surface information to reconstruct the source structure in some detail, thereby shedding light on the behavior of specific regions of the heart. We can see what is required mathematically by a consideration of (6.73). We first rewrite this expression as

$$\Phi_p = \sum_{n=0}^{\infty} \sum_{l=0}^{n} \sum_{k=0}^{n-l} \frac{(-1)^n}{l!k!(n-l-k)!} C_{nkl} \frac{\partial^n}{\partial x^l \, \partial y^k \, \partial z^{(n-l-k)}} \left(\frac{1}{r}\right) \quad (6.74)$$

where

$$C_{nkl} = \frac{1}{4\pi\sigma} \int_V I_v(\xi,\eta,\zeta) \xi^l \eta^k \zeta^{(n-l-k)} \, d\xi \, d\eta \, d\zeta \quad (6.75)$$

The coefficients $C_{nkl}$ are recognized as the joint source distribution moments. If we define the *characteristic function* of the source distribution as

$$M(v_1,v_2,v_3) = \int_{\infty}^{\infty} \int_{\infty}^{\infty} \int_{\infty}^{\infty} I_v(\xi,\eta,\zeta) \exp\left(jv_1\xi + jv_2\eta + jv_3\zeta\right) d\xi \, d\eta \, d\zeta \quad (6.76)$$

then the joint moments $C_{nkl}$ can be found from

$$C_{nkl} = (-j)^n \frac{\partial^n M(v_1,v_2,v_3)}{\partial v_1{}^l \, \partial v_2{}^k \, \partial v_3{}^{(n-l-k)}} \bigg|_{v_1=v_2=v_3=0} \quad (6.77)$$

Thus the joint moments serve to define the Taylor series expansion of the characteristic function about the origin ($v_1 = v_2 = v_3 = 0$). That is,

$$M(v_1,v_2,v_3) = \sum_{n=0}^{\infty} \sum_{l=0}^{n} \sum_{k=0}^{n=l} \frac{(j)^n}{l!k!(n-l-k)!} C_{nkl} v_1{}^l v_2{}^k v_3{}^{(n-l-k)} \quad (6.78)$$

If we assume that (6.78) converges sufficiently rapidly, then the source distribution may be found by taking an inverse transform of (6.76), namely,

$$I_v(\xi,\eta,\zeta) = \frac{1}{(2\pi)^3}$$
$$\int_{-\infty}^{\infty} \int_{-\infty}^{\infty} \int_{-\infty}^{\infty} M(v_1,v_2,v_3) \exp\left(-jv_1\xi - jv_2\eta - jv_3\zeta\right) dv_1 \, dv_2 \, dv_3 \quad (6.79)$$

Since the source function $I_v$ is confined to a finite region, it can be shown[1] that (6.78) must converge. Consequently, the joint moments $C_{nkl}$ uniquely determine the source distribution.

[1] M. G. Kendall, "Advanced Theory of Statistics," vol. I, Charles Griffin & Company, Ltd., London, 1948.

The electrocardiographic problem may be formulated as requiring the determination of $C_{nkl}$ from surface-potential measurements. A consideration of (6.74) shows a basic difficulty that arises in this process. For suppose the source consists entirely of components $C_{202}$, $C_{220}$, and $C_{200}$ and that

$$C_{200} = C_{220} = C_{202} \qquad (6.80)$$

Substituting these coefficients into (6.74) results in an expression for the potential field given by

$$\Phi_p = \tfrac{1}{2}C_{200}\left[\frac{\partial^2}{\partial x^2}\left(\frac{1}{r}\right) + \frac{\partial^2}{\partial y^2}\left(\frac{1}{r}\right) + \frac{\partial^2}{\partial z^2}\left(\frac{1}{r}\right)\right]$$

But the above expression is identically zero since $(1/r)$ is a solution of Laplace's equation when $r = 0$ is excluded, as is the case here. Since this distribution produces a null field over the bounding surface, it cannot be detected by potential measurements. Thus, quadrupole determination will be ambiguous since we can always add (6.80) (multiplied by an arbitrary coefficient) to any solution. If we examine the octopole terms, then we can verify that by choosing either of the following sets of source components,

$$C_{303}, C_{321}, C_{301} \qquad \text{where } C_{303} = C_{321} = C_{301} \qquad (6.81a)$$

$$C_{302}, C_{320}, C_{300} \qquad \text{where } C_{302} = C_{320} = C_{300} \qquad (6.81b)$$

$$C_{312}, C_{330}, C_{310} \qquad \text{where } C_{312} = C_{330} = C_{310} \qquad (6.81c)$$

a null field results. Thus octopole determinations are ambiguous also and involve three implicit constraining relations. Thus it is not possible to determine the coefficients $C_{nkl}$ from surface-potential measurements; hence the bioelectric sources themselves cannot be found.

The above remarks can be extended by noting that[1] for $r > r'$,

$$\frac{1}{R} = \sum_{n=0}^{\infty} \sum_{m=0}^{n} \epsilon_m \frac{r'^n}{r^{n+1}} \frac{(n-m)!}{(n+m)!}$$
$$P_n{}^m(\cos\theta')P_n{}^m(\cos\theta)\cos[m(\phi - \phi')] \quad (6.82)$$

where the source point has the spherical coordinates $(r',\theta',\phi')$, the field point is at $(r,\theta,\phi)$, $R$ is the distance between source and field point, and $\epsilon_m$ is the Neumann factor ($\epsilon_m = 2$, $m \neq 0$; $\epsilon_0 = 1$). Consequently, (6.66) may be written

$$\Phi_p = \sum_{n=1}^{\infty} \sum_{m=0}^{n} [A_{nm}Y_{nm}{}^e(\theta,\phi) + B_{nm}Y_{nm}{}^o(\theta,\phi)]\frac{1}{r^{n+1}} \qquad (6.83)$$

[1] P. M. Morse and H. Feshbach, "Methods of Theoretical Physics," p. 1274, McGraw-Hill Book Company, New York, 1953.

where

$$\begin{Bmatrix} A_{nm} \\ B_{nm} \end{Bmatrix} = \frac{\epsilon_m (n-m)!}{4\pi\sigma (n+m)!} \int_0^{2\pi} \begin{Bmatrix} \cos m\phi' \\ \sin m\phi' \end{Bmatrix} d\phi'$$
$$\int_0^{\pi} \sin \theta' \, P_n{}^m (\cos \theta') \, d\theta' \int_0^a I_v(r',\theta',\phi') r'^{n+2} \, dr' \quad (6.84)$$

and $r > a$. In (6.83) $Y_{nm}{}^e$ designates the even tesseral harmonic and $Y_{nm}{}^o$ the odd tesseral harmonic; that is,

$$Y_{nm}{}^e = P_n{}^m (\cos \theta) \cos m\phi \qquad\qquad\qquad (6.85a)$$
$$Y_{nm}{}^o = P_n{}^m (\cos \theta) \sin m\phi \qquad\qquad\qquad (6.85b)$$

Equation (6.84) is an expansion for the potential consisting of terms in inverse powers of $r$, just as is (6.74). Both expressions are multipole expansions, and the index $n$ yields inverse $(n+1)$ powers of $r$. The coefficients $A_{nm}$, $B_{nm}$ are obtained from the source function $I_v$ by the weighting functions specified in (6.84), which are analogous to, but not the same as, those in the expression for $C_{nkl}$ given by (6.75). What is of particular importance is that for a specific multipole order $n$ the number of distinct coefficients $A_{nm}$ equals $(n+1)$, while $B_{nm}$ contributes $n$ additional terms $(B_{n0} = 0)$, for a total of $(2n+1)$. On the other hand, we have $\frac{1}{2}(n+1)(n+2)$ different coefficients $C_{nkl}$ corresponding to the index $n$. If we contrast the two, we note that a redundancy in the $C_{nkl}$ coefficients exists and amounts, specifically, to $[\frac{1}{2}(n+1)(n+2) - (2n+1)]$. For second- and third-order multipoles the redundancies are one and three, respectively, and this is precisely the number of constraints enumerated in (6.80) and (6.81). For the first-order case (dipole) no redundancy exists and both representations are, in fact, identical.

The significance of the above remarks can be explained in the following way. Suppose the source to consist solely of an $n$th-order multipole. Then (6.83) makes clear that $(2n+1)$ independent potential measurements will suffice to determine the unknown coefficients $A_{nm}$ and $B_{nm}$; once determined, the potential at all other points is uniquely specified by (6.83). Thus, no matter how many potential measurements might be made, no more than $(2n+1)$ are independent. But the representation in (6.75) requires $\frac{1}{2}(n+1)(n+2)$ independent conditions for the evaluation of the $\frac{1}{2}(n+1)(n+2)C_{nkl}$'s. We see that this cannot be done since the unknowns exceed the number of equations that could be written by $[\frac{1}{2}(n+1)(n+2) - (2n+1)]$.

One might wonder whether redundancies exist in the representation (6.83). The answer is negative, for we shall show that the coefficients $A_{nm}$, $B_{nm}$ are uniquely determined by a specification of the potential over a closed surface. We shall verify this for an idealized model where the sources lie in a spherical conductor and shall defer to the next section consideration of a general shape. The potential field of an arbitrary source has the representation of (6.83) only when the medium is infinite in extent. The effect of a finite spherical boundary, at $r = R$, generates additional terms which are source-free, i.e., solutions of Laplace's equation. The total potential consists of a superposition of both expressions, and the result is

$$\Phi_p = \sum_{n=1}^{\infty} \sum_{m=0}^{n} \left( \frac{A_{nm}}{r^{n+1}} + r^n A'_{nm} \right) Y_{nm}{}^e(\theta,\phi) + \left( \frac{B_{nm}}{r^{n+1}} + r^n B'_{nm} \right)$$

$$\times Y_{nm}{}^o(\theta,\phi) \quad (6.86)$$

where the source-free terms are identified by the primed coefficients. By virtue of the boundary condition that $\partial\Phi/\partial r = 0$ at $r = R$, we require that

$$A'_{nm} = \frac{n+1}{n} \frac{A_{nm}}{R^{2n+1}} \qquad B'_{nm} = \frac{n+1}{n} \frac{B_{nm}}{R^{2n+1}} \qquad (6.87)$$

so that the surface potential is

$$\Phi(R,\theta,\phi) = \sum_{n=1}^{\infty} \sum_{m=0}^{n} \frac{2n+1}{n} R^{-(n+1)}[A_{nm}Y_{nm}{}^e(\theta,\phi) + B_{nm}Y_{nm}{}^o(\theta,\phi)]$$

$$(6.88)$$

Now if the left-hand side is known, then we see that $A_{nm}$ and $B_{nm}$ are readily found by utilizing the orthogonality of the tesseral harmonics. The result is

$$\begin{Bmatrix} A_{nm} \\ B_{nm} \end{Bmatrix} = R^{n+1} \frac{n\epsilon_m}{4\pi} \frac{(n-m)!}{(n+m)!}$$

$$\int_{S_0} \Phi(R,\theta,\phi) \, P_n{}^m (\cos\theta) \begin{Bmatrix} \cos m\phi \\ \sin m\phi \end{Bmatrix} \sin\theta \, d\theta \, d\phi \quad (6.89)$$

In the electrocardiographic problem the information available is the potential field over the torso surface. We have seen that this is inadequate to determine the underlying bioelectric sources. However, it is sufficient for the unique determination of multipole coefficients of the spherical-harmonic form given by (6.83); the greater the number of coefficients that can be determined, the more the aspects of the source that become known. The limitations in obtaining higher orders of multipole coefficients lie in the higher inverse power of $r$ that is involved and hence, in general, the smaller signal-to-noise ratio at the surface. From a practical standpoint the multipole representation permits a quantitative discussion of the nondipolar contribution to the surface-potential field. And even if some other physical model is ultimately to be employed, the mathematical formalism of the transformation of surface data to the multipole representation may be very useful.

The multipole formalism is clearly appropriate for a quantitative examination of the dipole hypothesis. By placing a beating turtle heart in a spherical electrolytic tank and measuring, the resultant potential field over the surface (6.89) can then be used to evaluate the multipole coefficients of the cardiac source. This procedure was followed by Heppner and Plonsey;[1] measurements were made at 60 surface points, and this permitted (6.89) to be evaluated numerically. Results of this work show that the quadrupole-to-dipole ratio increases (irregularly) to a peak and then decreases to zero through the QRS. The maximum rms quadrupole-to-dipole ratio, for the 1-in.-diameter heart in a 3-in.-diameter sphere, was approximately 0.2. This relative heart-to-"torso" ratio is in the order of magnitude of the human heart in the human torso.

## 6.14 INHOMOGENEITIES

In the previous section the simplistic dipole theory was abandoned for a formal approach which considers the bioelectric sources as distributed. However, the approximation of a uniform homogeneous medium was retained. We shall now consider a distributed source in a nonhomogeneous torso composed of regions each of which is assigned a uniform conductivity. This model will permit taking some account of the conductivity and shape of lungs, fat, muscle, bone, blood, etc. In addition to assuming each region to have a uniform conductivity, we shall also assume them to be isotropic. The latter assumption may not be as restrictive as might seem at first, as pointed out by Rush.[2] For simplicity, we shall also assume that the conductivity of surface tissue is uniform and of value $\sigma_s$.

On the basis of the above model, the surface potential at any arbitrary point $S$, using (5.174), is

$$\Phi(S) = \frac{1}{4\pi\sigma_s} \int_V \frac{I_v}{r}\, dV - \frac{1}{4\pi\sigma_s} \int_{S_0} \sigma_s \Phi\, \frac{\partial(1/r)}{\partial n}\, dS$$

$$+ \frac{1}{4\pi\sigma_s} \sum_j \int_{S_j} \Phi(\sigma_j'' - \sigma_j')\, \frac{\partial(1/r)}{\partial n}\, dS_j \quad (6.90)$$

[1] D. B. Heppner and R. Plonsey, Multipole Measurements of an Isolated Turtle Heart, *Proc. 19th Ann. Conf. Eng. Med. Biol.*, **8**:193 (1966). See also D. B. Heppner, Dipole and Quadrupole Measurements of an Isolated Turtle Heart, *IEEE Trans., Biomed. Eng.*, **BME15**:298 (1968).

[2] S. Rush, A Principle for Solving a Class of Anisotropic Current Flow Problems and Applications to Electrocardiography, *IEEE Trans.*, **BME14**:18 (1967).

where $I_v$ is the volume source density in the myocardium. The surfaces $S_j$ are those bounding the lungs, the ventricular blood mass, etc. Equation (6.90) can be interpreted, insofar as surface potentials are concerned, as relating that potential to the actual (primary) sources $I_v$ plus double-layer (secondary) sources on $S_j$, as if all existed within a homogeneous torso of conductivity $\sigma_s$ bounded by $S_0$. If no detailed information is available concerning the location, shape, and conductivity of all discrete conductive regions, then (6.90) makes clear that it will be impossible even to consider a resolution of the net effective source into primary and secondary components. Under these conditions one can only lump primary and secondary sources together to obtain a net equivalent source, which will be designated as a volume density[1] $I_v{}^e$.

Thus, the electrocardiographic problem of a distributed cardiac source in an inhomogeneous isotropic body can be replaced by one in which an equivalent source $I_v{}^e$ lies in a uniform homogeneous torso of conductivity $\sigma_s$. Thus the problem reduces to the one considered in the previous section except for the interpretation of the effective source $I_v{}^e$. We can, for example, characterize the equivalent source $I_v{}^e$ by its multipole expansion (in a medium of infinite extent), namely, (6.83) and (6.84). That is,

$$\Phi = \sum_{n=1}^{\infty} \sum_{m=0}^{n} [A_{nm} Y_{nm}{}^e(\theta,\phi) + B_{nm} Y_{nm}{}^o(\theta,\phi)] \frac{1}{r^{n+1}} \tag{6.91}$$

where

$$\begin{Bmatrix} A_{nm} \\ B_{nm} \end{Bmatrix} = \frac{\epsilon_m (n-m)!}{4\pi\sigma_s(n+m)!} \int_V I_v{}^e r^n \begin{Bmatrix} Y_{nm}{}^e \\ Y_{nm}{}^o \end{Bmatrix} dV \tag{6.92}$$

and (6.91) converges for $r > a$, where $a$ is the smallest radius that encloses all sources. In this case the radius $a$ must be drawn to include secondary as well as primary sources.

An alternative expression can be obtained by replacing $I_v{}^e$ by $-\nabla \cdot J_i{}^e$, which expresses all sources (primary and secondary) as dipole moments per unit volume. Then, by utilizing a standard vector identity, the integral of (6.92) can be written

$$\mathbf{J}_i{}^e \cdot \nabla r^n \begin{Bmatrix} Y_{nm}{}^e \\ Y_{nm}{}^o \end{Bmatrix} - \nabla \cdot \mathbf{J}_i{}^e r^n \begin{Bmatrix} Y_{nm}{}^e \\ Y_{nm}{}^o \end{Bmatrix}$$

The volume integral of the divergence term can be converted to a surface integral; and since $\mathbf{J}_i{}^e$ is zero on the boundary surface, this term is seen

---

[1] In the formalism used here, a surface source density is considered to be a degenerate case of a volume source density.

to be equal to zero.  Consequently,

$$\begin{Bmatrix} A_{nm} \\ B_{nm} \end{Bmatrix} = \frac{\epsilon_m (n-m)!}{4\pi\sigma_s (n+m)!} \int_V \mathbf{J}_i^e \cdot \mathbf{\nabla} \left[ r^n \begin{Bmatrix} Y_{nm} \\ Y_{nm}^o \end{Bmatrix} \right] dV \tag{6.93}$$

In Sec. 6.13 we showed how the multipole coefficients can be obtained from an appropriate surface integral when the body is homogeneous and spherical in shape.  We shall now generalize this result to a homogeneous medium of arbitrary shape, such as is now being considered. We make use of the reciprocity theorem of (6.51), namely,

$$\int_{S_0} K\Phi \, dS = \int_V I_v{}^e \psi \, dV \tag{6.94}$$

where $I_v{}^e$ sets up the potential field $\Phi$, and the applied surface current $K$ at $S_0$ sets up $\psi$ in the homogeneous torso of conductivity $\sigma_s$.  If we choose the current $K$ as

$$K \, dA = \sigma_s \mathbf{\nabla}\psi \cdot d\mathbf{S} \tag{6.95}$$

where $\psi$ is defined by

$$\psi = \begin{Bmatrix} \psi_{nm}{}^e \\ \psi_{nm}{}^o \end{Bmatrix} = \epsilon_m \frac{(n-m)!}{(n+m)!} r^n P_n{}^m (\cos\theta) \begin{Bmatrix} \cos m\phi \\ \sin m\phi \end{Bmatrix} \tag{6.96}$$

then $\psi$ is precisely the potential field which the surface current $K$, defined by (6.95), sets up.  (Note that $\psi$ is a solution to Laplace's equation, so that it is physically realizable.)  Under these conditions the right-hand side of (6.94) corresponds essentially to (6.92), and the left side must then satisfy

$$\begin{Bmatrix} A_{nm} \\ B_{nm} \end{Bmatrix} = \frac{1}{4\pi} \int_{S_0} \Phi\mathbf{\nabla} \begin{Bmatrix} \psi_{nm}{}^e \\ \psi_{nm}{}^o \end{Bmatrix} \cdot d\mathbf{S} \tag{6.97}$$

Thus the multipole coefficients of the net sources $I_v{}^e$ can be obtained from the surface integral of the actual measured potential over the bounding (torso) surface $S_0$.  Clearly, the multipole coefficients are uniquely determined by the surface distribution.  This result is of great importance and was first derived by Geselowitz,[1] although the development given here follows that due to Brody et al.[2]  Note that (6.97) reduces to (6.89) when the surface is spherical.

We have seen that the surface potentials uniquely determine the spherical harmonic multipole expansion of an arbitrarily distributed

[1] D. B. Geselowitz, Multipole Representation for an Equivalent Cardiac Generator, *Proc. IRE*, **48**:75 (1960).

[2] D. A. Brody, J. C. Bradshaw, and J. W. Evans, A Theoretical Basis for Determining Heart-lead Relationships of the Equivalent Cardiac Multipole, *IRE Trans. Biomed. Electron.*, **BME8**:139 (1961).

equivalent source. The multipole coefficients can, in turn, be inserted into (6.91) and the potential in the *infinite medium* obtained, provided $r > a$. If a sphere of radius $a$ lies completely within the torso, then (6.91) converges on $S_0$, and by straightforward procedure (to be discussed in Sec. 6.16) a solution for $\Phi$ on $S_0$ in an insulating medium can be obtained. If the spherical surface $r = a$ extends beyond the torso, then we may consider as our model the sources $I_v^e$ in a medium of infinite extent. Now if we apply (5.73), then $\Phi(S)$ must satisfy the following integral equation:

$$\frac{1}{4\pi} \int_{S_0} \Phi(S) \nabla \left(\frac{1}{r}\right) \cdot d\mathbf{S} = \Phi(P) \tag{6.98}$$

where $P$ lies on (say) a bounding spherical surface with radius $r > a$. In (6.98) $\Phi(P)$ is known since $P$ is in a region where (6.91) converges. Equation (6.98) can be solved by standard numerical techniques. Thus in either of the above cases the potential distribution on $S_0$ uniquely determines the multipole coefficients $\{A_{nm}, B_{nm}\}$, and, conversely, the multipole coefficients permit determination of the potential distribution on $S_0$. It is this characteristic of multipole theory that makes it an attractive tool in electrocardiographic research.

If the geometry of the inhomogeneities and their conductivities are known, then the multipole expansion of the heart sources alone, $I_v$, can be obtained (in principle). In this case one recognizes that

$$I_v^e = I_v + \sum_j I_{sj} \tag{6.99}$$

where $I_{sj}$ is a double layer on the $j$th surface of strength $(\sigma_j'' - \sigma_j')\Phi$. We can now evaluate the multipole coefficients of the secondary sources. By applying (6.94), this comes out[1]

$$\begin{Bmatrix} a_{nm} \\ b_{nm} \end{Bmatrix} = \frac{\epsilon_m (n-m)!}{4\pi \sigma_s (n+m)!} \sum_j \int_{S_j} (\sigma_j'' - \sigma_j') \Phi \, \nabla r^n \begin{Bmatrix} Y_{nm}^e \\ Y_{nm}^o \end{Bmatrix} \cdot d\mathbf{S}_j \tag{6.100}$$

Note that the geometry, conductivities, and the solution for the surface potential at all phase boundaries must be available in order to evaluate (6.100). The desired multipole expansion (of $I_v$ alone) is the difference between $\{A_{nm}, B_{nm}\}$ found from (6.92) and $\{a_{nm}, b_{nm}\}$ found from (6.100).

In principle, if the geometry and conductivity of each region were known, suitable values of $\mathbf{K}_{nm}^e$ and $\mathbf{K}_{nm}^o$ could be found such that the scalar field in the *heart region* (considered homogeneous) would equal that specified by (6.96). In this case the right-hand side of (6.51) would be

[1] See also D. B. Geselowitz, On Bioelectric Potentials in an Inhomogeneous Volume Conductor, *Biophys. J.*, **7**:1 (1967).

$4\pi A'_{nm}$ and $4\pi B'_{nm}$, which are the multipole coefficients of the heart source itself (that is, $I_v$). This procedure amounts to a generalization of (6.97) to

$$\begin{Bmatrix} A'_{nm} \\ B'_{nm} \end{Bmatrix} = \frac{1}{4\pi\sigma_s} \int_{S_0} \Phi \begin{Bmatrix} \mathbf{K}_{nm}{}^e \\ \mathbf{K}_{nm}{}^o \end{Bmatrix} \cdot d\mathbf{S}$$

Several investigations have been conducted to determine the importance of various inhomogeneities. The effect of an insulating sternum and a lung of four times body resistivity was considered for a two-dimensional problem by Nelson.[1] Results are plotted in Fig. 6.32 and show significant quantitative differences; however, qualitative similarities are also apparent.

An investigation of the effect of lung and blood inhomogeneities on the image surface for a two-dimensional frontal-plane model was performed by Lizzi and Grayzel.[2] The dipole source in their work was

[1] C. V. Nelson, Human Thorax Potentials, *N.Y. Acad. Sci.*, **65**:1014 (1957).

[2] F. Lizzi and J. Grayzel, EKG Boundary Potentials in Conductive Models of the Human Torso, *Proc. 18th Ann. Conf. Eng. Med. Biol.*, p. 96 (1965).

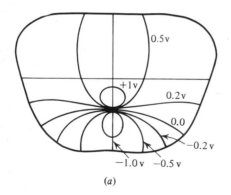

(a)

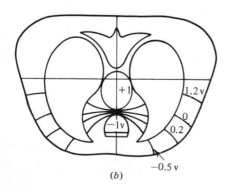

(b)

Fig. 6.32 Two-dimensional field plot for horizontal dipole. (*a*) Homogeneous torso section; (*b*) insulating sternum, "lung" resistivity four times "body" resistivity. [*C. V. Nelson, N.Y. Acad. Sci.*, **65**:1014 (1957).]

chosen at several different "heart" positions, and significant asymmetrical effects of the inhomogeneities result.

While the aforementioned studies are useful in estimating the qualitative effect on the field pattern of inhomogeneities, the shape of both bioelectric source distribution and the volume conductor deviate too greatly from two-dimensionality to permit quantitative generalizations from these results. Some studies of the effect of inhomogeneities on the lead vectors, utilizing three-dimensional torso models, by Frank[1] and Burger and Van Milaan[2] show relatively little perturbation. Some additional evidence arises from studies[3] where electrophysiological data were used to compute the potential distribution that would arise at the surface of a *homogeneous* torso. Since the results compare favorably with the surface potentials actually obtained, one infers that inhomogeneities do not enter in a major way.

On the other hand, a study of the inverse problem in electrocardiography performed by Barnard et al.[4] showed that inclusion of the lung and blood was necessary for their results to have physiological significance.

Of the several inhomogeneities the one considered most influential is that of the intraventricular blood. This is because the conductivity of blood is from two to five times that of the average body conductivity. A rough estimation of the effect is obtained from an idealized model in which the ventricle is considered to be spherical and surrounded by a concentric spherical (heart) muscle, as illustrated in Fig. 6.33. If the intracardiac blood is assumed to be a perfect conductor, then its effect on a dipole source in the muscle wall can be determined by utilizing image theory.[5] Thus, for a radially oriented dipole at the muscle-blood interface the image enhances the actual dipole, with the resultant being doubled in strength. On the other hand, tangential dipoles create an image in the opposite direction such that cancellation occurs. For a dipole located *within* the muscle wall its mirror image lies within the blood region; because of this spatial separation one can speak only

[1] E. Frank, A Direct Experimental Study of Three Systems of Spatial Vectorcardiography, *Circulation*, **10**:101 (1954).

[2] H. C. Burger and J. B. Van Milaan, Heart Vector and Leads. Parts I, II, III, *Brit. Heart J.*, **8**:157 (1946), **9**:154 (1947), **10**:229 (1948).

[3] R. H. Selvester, J. Soloman, and T. Gillespie, Mathematical Model of Total Body ECG with a Realistic External Torso Boundary, *Proc. Symp. Biomed. Eng.*, *Milwaukee, Wis.*, **1**:352 (1966).

[4] A. C. L. Barnard, M. S. Lynn, J. M. Holt, and L. T. Sheffield, A Proposed Method for the Inverse Problem in Electrocardiography, *Biophys. J.*, **7**:925 (1967).

[5] R. Plonsey, Current Dipole Images and Reference Potentials, *IEEE Trans. Biomed. Electron.*, **BME10**:3 (1963).

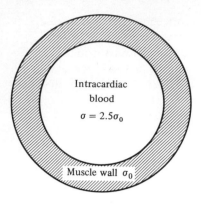

**Fig. 6.33** Idealized spherical ventricle with concentric spherical wall.

approximately of enhancement of radial and reduction of tangential dipole strengths.

The case of finite blood conductivity ten times that of the muscle and external medium was considered by Brody.[1] He showed that the effect could be approximated by utilizing images that arise for a perfect conductor but with a strength reduced by around 25 percent. The qualitative results stated above remain.

The effect of the high-conductivity blood can also be demonstrated by a lead-field approach. For a uniform applied field into which a spherical ventricular blood volume is placed, the resultant flow lines at the boundary will be approximately radial, as plotted in Fig. 6.34. The current density within the blood turns out to be uniform and enhanced over the applied current density.[2] Now an orthogonal lead system is designed precisely to set up such a uniform applied lead field in a homogeneous heart. Thus, with such a lead, subject to the approximation of a spherical geometry, the resultant lead-vector field will be radial. Consequently, the effect of intracardiac blood is a distortion of the lead such that it is sensitive to radial dipoles *only*. This effect might be expected to be strongest in the endocardium, in examining Fig. 6.34.

Specifically, for a representation of the heart muscle as a thin concentric spherical layer surrounding the spherical blood, the field can be approximated by that which would exist in the absence of the heart muscle. In particular, a uniform field within the blood will exist as illustrated in Fig. 6.34. Since normal components of current density are continuous across a conductive interface, the current density within the

[1] D. A. Brody, A Theoretical Analysis of Intracavitary Blood Mass Influence on the Heart-lead Relationship, *Circulation Res.*, **4**:731 (1956).

[2] W. K. H. Panofsky and M. Phillips, "Classical Electricity and Magnetism," p. 84, Addison-Wesley Publishing Company, Inc., Reading, Mass., 1962.

muscle at the "front" and "back" (i.e., points $A$ and $B$ in Fig. 6.34) is equal to that within the blood mass. Along the sides ($C$ and $D$ in Fig. 6.34) the continuity of the tangential electric field requires that the ratio of the current density in the heart muscle to that in the blood equal the respective conductivity ratio. Since the latter is, roughly, 0.4, an enhancement of radial to tangential dipole sources of 2.5 results. The effect is reduced when anisotropy is considered, according to McFee and Rush.[1]

If one assumes the combined right and left ventricle to be represented by a spherical conductor divided in half by a septum (dotted lines in Fig. 6.34), then to first order the direction of the flow lines within the blood is not affected by the septum, as can be seen in Fig. 6.34. Within a thin septum, as shown, the magnitude of the current density is the same as in the surrounding blood (continuity of normal current); consequently, the lead field in the septum exceeds that in the external tissue by the factor $3/(1 + 2a)$,[†] where $a$ is the ratio of tissue to blood conductivity. When $a = 0.4$, then a lead-vector enhancement of 1.7 occurs. On the other hand, if the direction of the applied lead-vector field is parallel to the (thin) septum, then the *electric field* in the blood and in the septum are approximately the same (continuity of tangential electric field); both are reduced from that in the external tissue by a factor of $3/(2 + 1/a)$, or 0.67 for $a = 0.4$. The ratio of the two is simply $1/a$, a result that can also be reasoned from the existence of a

[1] R. McFee and S. Rush, Qualitative Effects of Thoracic Resistivity Variations on the Interpretation of Electrocardiograms: The "Brody" Effect, *Am. Heart J.*, November, 1967.

[†] Panofsky and Phillips, *op. cit.*

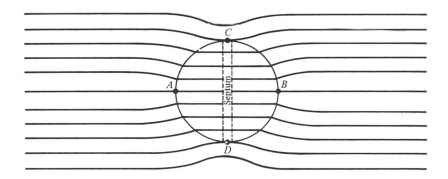

**Fig. 6.34**  Flow lines for a conducting sphere in a uniform field in a medium of lower conductivity. Also represents field structure for a thin concentric spherical heart muscle and thin septum.

uniform field within the blood and the condition that the normal components of current and tangential components of electric field are continuous across the (thin) septum. Thus the "Brody effect" may be said to extend to the septum since radial dipoles are enhanced while tangential dipoles are attenuated in their effects.

Another inhomogeneity of some importance is the muscle layer just beneath the skin. As discussed by McFee and Rush,[1] conductivity in a direction parallel to the surface can be considered as isotropic with a resistivity of 280 ohm-cm, while in the direction normal to the fibers (and torso surface) the resistivity is 2,300 ohm-cm. In their torso model the heart is surrounded by a "lung" medium, with a resistivity of 2,000 ohm-cm, bounded by the aforementioned anisotropic muscle layer; the heart is imbedded in the lung medium and of interest is the effect of the muscle layer on the lead field strength at the heart. A planar model of the muscle layer can be converted into an equivalent isotropic medium by a scale transformation of the type discussed in Sec. 7.3. In effect, a 1-cm anisotropic muscle layer with resistivities as described becomes equivalent to a 3-cm-thick isotropic layer with resistivity of 800 ohm-cm. (The stretching and resistance ratio factor is given by the square root of the high-to-low resistivities, that is, $\sqrt{2,300/280} = 2.9$.) The isotropic muscle at 800 ohm-cm can in turn be replaced by an equivalent layer of resistivity equal to that of the lung with no significant modification of the lead field in the lung, provided lead strengths are reduced by a factor of $\frac{2}{3}$. Thus, in summary, the actual 1-cm anisotropic muscle layer is equivalent to one 3 times thicker at lung resistivity but with the proviso that all lead voltages be reduced by $\frac{2}{3}$. The results predicted from the flat layer model are fairly well substantiated in a spherical model patterned on the human torso.

In the above model the lead field for a stratified planar conducting media is needed. This is found most easily by the application of image principles. Consider, for example, Fig. 6.35$a$ which shows a unit point current source in a region of resistivity $\rho_1$ which has a planar interface with a region of resistivity $\rho_2$. Then, for fields within region 1, the effect of the discontinuity can be accounted for by assuming all of space at $\rho_1$ and an image source located at the mirror image point of strength $(\rho_2 - \rho_1)/(\rho_2 + \rho_1)$. This condition is illustrated in Fig. 6.35$b$. For the fields within region 2, all of space may be considered to have a resistivity $\rho_2$ provided the actual source is replaced by an equivalent one at an amplitude $2\rho_1/(\rho_1 + \rho_2)$. This condition is illustrated in Fig. 6.35$c$.

[1] R. McFee and S. Rush, Qualitative Effects of Thoracic Resistivity Variations on the Interpretation of Electrocardiograms: The Low-resistance Surface Layer, *Am. Heart J.*, **76**:48 (1968).

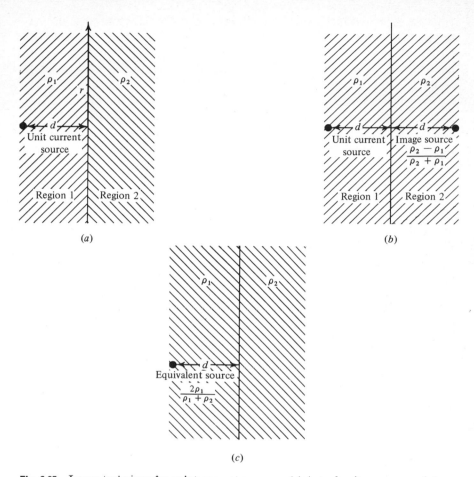

**Fig. 6.35** Image technique for point current source. (a) Actual unit source near interface of two different conducting media; (b) equivalent sources in medium of resistivity $\rho_1$, valid for region 1 only; (c) equivalent sources in medium of resistivity $\rho_2$, valid for region 2 only.

That these assertions are correct can be verified in the following way. First, the fields produced necessarily satisfy Laplace's equation and have the correct singularity at the actual source. Second, the boundary conditions at the interface, namely, continuity of normal current density and tangential electric field, are satisfied. Specifically, at a distance $r$ from the foot of the perpendicular from the source, located $d$ from the interface, the tangential $E$ field is

$$\frac{2\rho_1\rho_2}{4\pi(\rho_1 + \rho_2)} \frac{r}{(d^2 + r^2)^{\frac{3}{2}}}$$

when evaluated *either* by Fig. 6.35$b$ or Fig. 6.35$c$.  The normal current density is

$$\frac{2\rho_2}{4\pi(\rho_1 + \rho_2)} \frac{d}{(d^2 + r^2)^{\frac{3}{2}}}$$

as obtained *either* from Fig. 6.35$b$ or Fig. 6.35$c$.  Consequently, the boundary conditions are satisfied and the image solution confirmed.

## 6.15  MULTIPOLE PROPERTIES OF LEADS; CHOICE OF ORIGIN

For distributed source $\mathbf{J}_i$ (current dipole moment per unit volume) in a homogeneous volume conductor, as illustrated in Fig. 6.36, the potential difference created between two surface points $a$ and $b$ has been shown by (6.57) to be given by

$$V_{ab} = \int_V \boldsymbol{\nabla}\Phi_2 \cdot \mathbf{J}_i \, dV \tag{6.101}$$

where $\Phi_2$ is the lead field of $a$ and $b$.  In this section we shall develop an alternative expression which relates the lead voltage to the multipole expansion coefficients of the source along with factors that relate to the multipole sensitivity of the lead.

We begin by utilizing (6.51) and choose $I_v = -\boldsymbol{\nabla} \cdot \mathbf{J}_i$ and

$$\mathbf{K} = \delta(\mathbf{r}_a - \mathbf{r}) - \delta(\mathbf{r}_b - \mathbf{r}) \tag{6.102}$$

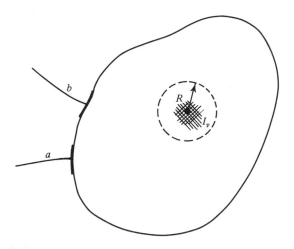

**Fig. 6.36**  Lead-source geometry for a homogeneous volume conductor.

Consequently,

$$\Phi_1(r_a) - \Phi_1(r_b) = V_{ab} = \int_V I_v \Phi_2 \, dV \tag{6.103}$$

where $\Phi_2$ is the lead field of $a$ and $b$ as before.   Since the lead field must satisfy Laplace's equation in $V$, it can be expanded in an infinite series of tesseral harmonics.   Thus

$$\Phi_2 = \sum_{n=1}^{\infty} \sum_{m=0}^{n} (s_{nm} \cos m\phi + t_{nm} \sin m\phi) r^n P_n{}^m(\cos \theta) \tag{6.104}$$

and

$$V_{ab} = \int_V I_v \sum_{n=1}^{\infty} \sum_{m=0}^{n} (s_{nm} \cos m\phi + t_{nm} \sin m\phi) r^n P_n(\cos \theta) \, dV$$

$$\tag{6.105}$$

Now if (6.84) is used, then (6.105) can be written as

$$V_{ab} = \sum_{n=1}^{\infty} \sum_{m=0}^{n} (S_{nm} A_{nm} + T_{nm} B_{nm}) \tag{6.106}$$

where

$$\begin{Bmatrix} S_{nm} \\ T_{nm} \end{Bmatrix} = \frac{4\pi\sigma(n+m)!}{\epsilon_m(n-m)!} \begin{Bmatrix} s_{nm} \\ t_{nm} \end{Bmatrix} \tag{6.107}$$

Equation (6.106) can be thought of as a multipole generalization of the simpler dipole relationship in (6.39).

The coefficients $\{S_{nm}, T_{nm}\}$ in (6.107) characterize the sensitivity of a particular lead to each multipole source at the selected origin.   That is, the lead voltage is the scalar product of a multipole lead tensor and the specific multipole source.   This characterization of multipole leads was described by Arzbaecher and Brody,[1] along with a suggested method for the experimental determination of $S_{nm}$ and $T_{nm}$.   This is done by setting up the lead field $\Phi_2$ in a torso electrolytic model.   This field is then explored over a spherical surface about the chosen origin, yielding, at $r = r_1$, values of $\Phi_2(r_1, \theta, \phi)$.   From (6.104) we have

$$\Phi_2(r_1, \theta, \phi) = \sum_{n=1}^{\infty} \sum_{m=0}^{n} (s_{nm} \cos m\phi + t_{nm} \sin m\phi) r_1{}^n P_n(\cos \theta)$$

$$\tag{6.108}$$

[1] R. C. Arzbaecher and D. A. Brody, Electrocardiographic Lead-tensor Measuring System, *IEEE Trans.*, **BME11**:29 (1964).

Using the orthogonality of the tesseral harmonics gives the desired result, namely,

$$\begin{Bmatrix} s_{nm} \\ t_{nm} \end{Bmatrix} = \frac{(2n+1)(n-m)!}{2\pi(n+m)!}$$

$$\int_0^{2\pi} \int_0^{\pi} \left(\frac{1}{r_1}\right)^{n-2} \Phi_2(r_1,\theta,\phi) P_n^m(\cos \phi) \begin{Bmatrix} \cos m\phi \\ \sin m\phi \end{Bmatrix} \sin \theta \, d\theta \, d\phi \quad (6.109)$$

Numerical-analysis techniques for evaluating (6.109) are discussed by Brody and Arzbaecher,[1] based on the sampling technique which they devised.

An alternative method is based on the conversion of (6.104) into a power series in rectangular coordinates. This permits determination of the coefficients $\{s_{nm}, t_{nm}\}$ in terms of appropriate derivatives of $\Phi_2(r,\theta,\phi)$ evaluated at the origin. Thus, enumerating the terms through quadrupole order, we get

$$\Phi_2(x,y,z) = s_{10}z + s_{11}x + t_{11}y + s_{20}\left(z^2 - \frac{x^2+y^2}{2}\right) + s_{21}(3xz)$$
$$+ t_{21}(3yz) + 3s_{22}(x^2-y^2) + 6t_{22}xy + \cdots$$

Thus, to illustrate for several selected coefficients, the following relationships hold:

$$t_{21} = \frac{1}{3}\frac{\partial^2}{\partial y\, \partial z}[\Phi_2(x,y,z)]\bigg|_{x=y=z=0}$$

$$s_{20} = \frac{\partial^2}{\partial z^2}[\Phi_2(x,y,z)]\bigg|_{x=y=z=0}$$

$$s_{22} = \frac{1}{3}\left\{\frac{\partial^2}{\partial x^2}[\Phi_2(x,y,z)]\bigg|_{x=y=z=0} - \frac{\partial^2}{\partial y^2}[\Phi_2(x,y,z)]\bigg|_{x=y=z=0}\right\}$$

$$s_{10} = \frac{\partial}{\partial z}[\Phi_2(x,y,z)]\bigg|_{x=y=z=0}$$

The extension to other terms should be evident. A practical measurement system can be devised to approximate the desired differentials by finite differences. This is the basis of the multipole probe suggested by Plonsey.[2] A disadvantage of this scheme is in the physical difficulties in maintaining the necessary mechanical tolerance in the measurement probe. On the other hand, the probe may be easily shifted from point to point, and can be implanted in an experimental animal.

The multipole lead-field coefficients, determined by either of the aforementioned methods (or by other procedures), may be used to esti-

---

[1] D. A. Brody and R. Arzbaecher, The Sampling and Analysis of Lead Fields in Electrocardiographic Torso Models, *IEEE Trans.*, **BME14**:22 (1967).

[2] R. Plonsey, Theoretical Consideration for a Multipole Probe in Electrocardiographic Studies, *IEEE Trans.*, **BME12**:105 (1965).

mate the multipole content of the human electrocardiogram. Two approximations are involved: one that the *actual* lead coefficients are essentially those determined from the torso measurement, and the second that only a finite number of multipole coefficients are required. Assuming only dipole plus quadrupole coefficients to be significant, then eight independent measurements of potential at a torso surface will reduce (6.106) to a set of eight simultaneous equations in eight ($A_{10}$, $A_{11}$, $B_{11}$, $A_{20}$, $A_{21}$, $A_{22}$, $B_{21}$, $B_{22}$) unknowns. By a matrix inversion the desired coefficients can be found. A more satisfactory procedure is to measure potentials at greater than eight points. In this case (6.106) is overdetermined, and one can solve for the $A$'s and $B$'s so that the system of equations is satisfied in a least-squares sense. Specifically, if $V_j$ represents the measured potentials, we seek coefficients $A_i$ (representing $A_{nm}$'s and $B_{nm}$'s) which minimize

$$\sum_{j=1}^{N} \left( V_j - \sum_{i=1}^{8} S_{ji}A_i \right)^2$$

where $S_{ji}$ represents the known $S_{nm}$'s and $T_{nm}$'s. The number of measured values $N$ is assumed to greatly exceed the eight hypothesized pertinent multipole coefficients. Taking derivatives with respect to $A_1$, $A_2$, . . . , $A_8$ and setting the result equal to zero gives eight equations of the form

$$\sum_{j=1}^{N} 2\left( V_j - \sum_{i=1}^{8} S_{ji}A_i \right)S_{jk} = 0 \qquad (k = 1, 2, \ldots, 8)$$

In matrix notation this becomes

$$[S^T][V] = [S^T][S][A]$$

where $[S]$ is the $N \times 8$ matrix of lead coefficients and $[S^T]$ its transpose; $[A]$ is the eight-dimensional column vector of multipole source strength, while $[V]$ is an $N$-dimensional column vector of measured potentials. The desired solution is

$$[A] = ([S^T][S])^{-1}[S^T][V]$$

Utilizing this approach results in coefficients that are much less sensitive to measurement "noise." The former procedure was followed by Hlavin and Plonsey,[1] while the least-squares technique was utilized by Schubert[2] and Arthur et al.[3]

[1] J. M. Hlavin and R. Plonsey, An Experimental Determination of a Multipole Representation of a Turtle Heart, *IEEE Trans. Biomed. Electron.*, **BME10**:98 (1963).

[2] R. Schubert, An Experimental Study of the Multipole Series That Represents the Human Electrocardiogram, *IEEE Trans. Biomed. Eng.*, **BME15**:303 (1968).

[3] M. Arthur, S. A. Briller, and D. B. Geselowitz, Representation of Human Electrocardiograms by Addition of Quadrupole to Dipole Terms, *Circulation*, **36**:II-57 (1967).

In evaluating the multipole lead coefficients $\{S_{nm}, T_{nm}\}$ as well as the multipole source coefficients $\{A_{nm}, B_{nm}\}$ an origin must be designated, and the aforementioned coefficients will depend on its location in general. Because of this dependence two distributions cannot be readily compared by means of their multipole expansions unless they are both generated with respect to an equivalently defined origin. However, even though the same source can be described by any number of multipole expansions that differ only in the choice of origin, all such representations are readily interconvertible. Appropriate transformations which depend only on the coordinates of the new origin with respect to the old are readily developed, and we proceed now to derive such relations for the dipole and quadrupole; expressions for higher terms can be determined by extending the procedure. The question of defining a suitable origin will be considered subsequently.

Starting with (6.93), we have the following expressions for the dipole coefficients:

$$A_{10} = \frac{1}{4\pi\sigma} \int_V \mathbf{J}_i^e \cdot \nabla (r \cos \theta) \, dV = \frac{1}{4\pi\sigma} \int_V \mathbf{J}_i^e \cdot \nabla z \, dV \qquad (6.110a)$$

$$A_{11} = \frac{1}{4\pi\sigma} \int_V \mathbf{J}_i^e \cdot \nabla (r \sin \theta \cos \phi) \, dV = \frac{1}{4\pi\sigma} \int_V \mathbf{J}_i^e \cdot \nabla x \, dV \qquad (6.110b)$$

$$B_{11} = \frac{1}{4\pi\sigma} \int_V \mathbf{J}_i^e \cdot \nabla (r \sin \theta \sin \phi) \, dV = \frac{1}{4\pi\sigma} \int_V \mathbf{J}_i^e \cdot \nabla y \, dV \qquad (6.110c)$$

An interpretation of (6.110) which is obtained from (6.101) is that $\{A_{10}, A_{11}, B_{11}\}$ corresponds to the $z$, $x$, and $y$ lead voltages of an ideal corrected orthogonal lead system. Further, $\{A_{10}, A_{11}, B_{11}\}$ are precisely the algebraic summation of the $z$, $x$, and $y$ dipole components of the distributed sources and are independent of the location of each source element $\mathbf{J}_i^e \, dV$. By the same token the dipole coefficients are independent of the location of the origin (and will be invariant under a displacement of the origin).

To see the effect of a change in origin on the quadrupole terms let us consider the term $A_{20}$. From (6.92) we have

$$A_{20} = \frac{1}{16\pi} \int_V I_v^e r^2 (3 \cos 2\theta + 1) \, dV \qquad (6.111)$$

$$A_{20} = \frac{1}{16\pi} \int_V I_v^e (4z^2 - 2x^2 - 2y^2) \, dV \qquad (6.112)$$

Now, if the new origin has the coordinates $(x_0, y_0, z_0)$ with respect to the old, then $x' = x - x_0$, $y' = y - y_0$, $z' = z_0$, where primes are used to designate the new system. Then

$$A'_{20} = \frac{1}{8\pi\sigma} \int_V I_v^e [2(z - z_0)^2 - (x - x_0)^2 - (y - y_0)^2] \, dV \qquad (6.113)$$

$$A'_{20} = \frac{1}{8\pi\sigma} \int_V I_v^e (2z^2 - x^2 - y^2) \, dV - \frac{2z_0}{4\pi\sigma} \int_V I_v^e z \, dV$$
$$+ \frac{x_0}{4\pi\sigma} \int_V I_v^e x \, dV + \frac{y_0}{4\pi\sigma} \int I_v^e y \, dV \qquad (6.114)$$

and remaining terms which involve only the volume integral of $I_v^e$ go to zero since there is no net current. By using (6.92) the terms in (6.114) can be identified, leading to

$$A'_{20} = A_{20} - 2z_0 A_{10} + x_0 A_{11} + y_0 B_{11} \qquad (6.115a)$$

and in a similar way we obtain

$$A'_{21} = A_{21} - z_0 A_{11} - x_0 A_{10} \qquad (6.115b)$$
$$B'_{21} = B_{21} - z_0 B_{11} - y_0 A_{10} \qquad (6.115c)$$
$$A'_{22} = A_{22} - \tfrac{1}{2} x_0 A_{11} + \tfrac{1}{2} y_0 B_{11} \qquad (6.115d)$$
$$B'_{22} = B_{22} - \tfrac{1}{2} y_0 A_{11} - \tfrac{1}{2} x_0 B_{11} \qquad (6.115e)$$

The previous result was utilized by Geselowitz[1] to locate an "optimum origin." This was taken to be that location for which the dipole field alone, as measured over a concentric spherical surface, differs from the actual field in the least-squares sense. For simplicity we choose the coordinate axes such that the $A_{11}$ and $B_{11}$ dipole terms go to zero.[2] Let the given source be described by multipole coefficients $\{A_{nm}, B_{nm}\}$ about the origin $(0,0,0)$ and denote the optimum origin for a dipole, with coefficients $a'_{10} = A_{10}$, $a'_{11} = b'_{11} = 0$, as $x_0$, $y_0$, $z_0$. The dipole and quadrupole expansion coefficients for the dipole may then be calculated with respect to the origin $(0,0,0)$ by utilizing (6.115). If we denote these with lowercase letters (unprimed), we have

$$a_{10} = a'_{10} \qquad (6.116a)$$
$$a_{11} = b_{11} = 0 \qquad (6.116b)$$
$$a_{20} = 2z_0 a'_{10} \qquad (6.116c)$$
$$a_{21} = x_0 a'_{10} \qquad (6.116d)$$
$$b_{21} = y_0 a'_{10} \qquad (6.116e)$$
$$a'_{22} = b'_{22} = 0 \qquad (6.116f)$$

[1] D. B. Geselowitz, Two Theorems Concerning the Quadrupole Applicable to Electrocardiography, *IEEE Trans.*, **BME12**:164 (1965).

[2] If the coordinate axes are oriented so that the $z$ axis is in the direction of the net dipole moment of the source, then from (6.110) we must have $A_{11} = B_{11} = 0$. The coefficient $A_{10}$, in this case, is proportional to the total dipole component strength.

We may now formulate the problem as requiring a choice of $(x_0, y_0, z_0)$ such that the following integral is minimized [note that matters have been arranged so that $\{A_{nm}, B_{nm}\}$ and $\{a_{nm}, b_{nm}\}$ are expanded about the same origin, namely $(0,0,0)$]:

$$\text{Mean-square error} = \int_0^{2\pi} \int_0^{\pi} \left[ \sum_{n=1}^{\infty} \sum_{m=0}^{n} \frac{A_{nm} \cos m\phi + B_{nm} \sin m\phi}{R^{n+1}} \right.$$

$$\left. P_n{}^m(\cos \theta) - \sum_{n=1}^{\infty} \sum_{m=0}^{n} \frac{a_{mm} \cos m\phi + b_{nm} \sin m\phi}{R^{n+1}} P_n{}^m(\cos \theta) \right]^2$$

$$\sin \theta \, d\theta \, d\phi \quad (6.117)$$

If all terms beyond the quadrupole can be neglected, then the result comes out

$$x_0 = \frac{A_{21}}{A_{10}} \tag{6.118}$$

$$y_0 = \frac{B_{21}}{A_{10}} \tag{6.119}$$

$$z_0 = \frac{A_{20}}{2A_{10}} \tag{6.120}$$

It is not hard to show that this result is independent of the original choice of origin $(0,0,0)$, except insofar as this affects the assumption that terms beyond the quadrupole are negligible.

If the actual source is a double layer with a planar rim, such as might be a reasonable approximation to the active heart sources at an instant of time, then the criteria given by (6.118) to (6.120) have an interesting physical interpretation. Since the $z$ axis has been chosen parallel to the dipole axis, then, in this example, it is oriented normal to the plane of the rim (the double-layer plane). From (6.110a) the magnitude of $A_{10}$ is proportional to the double-layer area and strength. Now if (6.93) is utilized to evaluate $A_{21}$, $B_{21}$, and $A_{20}$, one obtains (6.118) to (6.120), but with an identification of $z_0$ as the intercept of the double-layer plane with the $z$ axis and where $(x_0, y_0)$ locates its center of area. Details may be found in a paper by Brody and Bradshaw,[1] where it is shown that $A_{22} = B_{22} = 0$ for the planar rim.

A rational definition for the origin of the multipole expansion of a current-source distribution is suggested by analogy with the moment expansion in mechanics. For a body of variable (mass) density it is the zero moment (i.e., mass) that is independent of the choice in origin.

[1] D. A. Brody and J. C. Bradshaw, The Equivalent Generator Components of Uniform Double Layers, *Bull. Math. Biophys.*, **24**:183 (1962).

The first moment, which is a linear function of origin location, does not possess a minimum, but rather can be set equal to zero by a suitable choice of origin. The latter point is known as the *center of mass*. The second moments (principal moments of inertia) are defined with respect to this center, where they take on their minimum value.

In an analogous fashion, for a distribution of current dipole moments, the net dipole moment is invariant to translation of the origin since, in this case, there is no zero moment. There is no loss of generality in choosing the $z$ axis along the dipole direction, in which case $A_{11} = B_{11} = 0$. For displacement of the origin to the point $(x_0, y_0, z_0)$ the new quadrupole moments (primed) relative to the old (unprimed) are given by (6.115) and are

$$A'_{20} = A_{20} - 2z_0 A_{10}$$
$$A'_{21} = A_{21} - x_0 A_{10}$$
$$B'_{21} = B_{21} - y_0 A_{10}$$
$$A'_{22} = A_{22}$$
$$B'_{22} = B_{22}$$

Only three of the five quadrupole coefficients depend on the translation of origin; consequently, we can choose to make $A'_{20} = A'_{21} = B'_{21} = 0$. This condition requires the satisfaction, precisely, of (6.118) to (6.120) and therefore locates an origin which is analogous to the center of mass. We can designate it as the center of dipole distribution, or *dipole center*. This center would appear to be the appropriate origin for any cardiac-source distribution. A description of a cardiac source is thus given by

$$A_{10}, \ A_{11}, \ B_{11}, \ A_{22}, \ B_{22}$$

which are origin-independent, by the origin itself

$$x_0, \ y_0, \ z_0$$

and higher moments relative to this origin

$$A'_{20}, \ A'_{21}, \ A'_{22}, \ \ldots$$

## 6.16  THE FORWARD PROBLEM

The forward problem in electrocardiography considers the generation of potential fields within and on the torso resulting from the electrical activity of the heart. Specifically, it includes application of cardiac electrophysiology to the determination of a quantitative measure of the bioelectric generators and the determination of the current field set up in the torso due to these active sources, taking into account body inhomogeneities, anisotropy, and shape. The availability of high-speed computers to evaluate these potential fields has provided a powerful tool for the investigation of the forward problem. The algorithms are based on Eq. (5.174) or (5.182), either one being capable of taking account of inhomogeneities and body shape. Details may be found in the papers of Gelernter and Swihart[1] and Barr et al.[2] and are summarized in a paper by Geselowitz.[3] Some additional work on the equivalent fluid-flow problem is given by Hess and Smith.[4] The initial aim of forward-problem studies is to check the adequacy of available data and the model itself. Ultimately, model studies should greatly assist in understanding the relationship between various types of pathology and the electrocardiograms that they produce.

To illustrate the above procedure, assume for the moment that the torso region is homogeneous. In Eqs. (5.174) and (5.182), only the surface integral over $S_0$ then remains; Eq. (5.182), for example, becomes

$$\Phi = \int_V \frac{I_v}{4\pi\sigma r}\, dV + \int_{S_0} \frac{2\sigma E_0}{4\pi\sigma r}\, dS_0 \qquad (6.121)$$

The boundary condition on $S_0$, that $\partial\Phi/\partial n = 0$, yields the following equation derived from (6.121) by differentiating under the integral sign with respect to $n$:

$$\frac{1}{4\pi} \int_V \frac{I_v}{\sigma} \mathbf{\nabla}\left(\frac{1}{r}\right) \cdot \mathbf{n}\, dV + \frac{1}{4\pi} \int_{S_0} 2E_0 \mathbf{\nabla}\left(\frac{1}{r}\right) \cdot \mathbf{n}\, dS_0 = 0 \qquad (6.122)$$

where $\mathbf{n}$ is the outward normal over $S_0$. Equation (6.122) constitutes an integral equation for the desired unknown surface function $E_0$.

Consider a small element of $S_0$ denoted by $\Delta S_k$. For a field point which approaches the center of this area along a path that is normal to the surface,

$$\int_{\Delta S_k} 2E_0 \mathbf{\nabla}\left(\frac{1}{r}\right) \cdot \mathbf{n}\, dS_0 = -\int_{\Delta S_k} 2E_0\, d\Omega \rightarrow -4\pi E_0(k)$$

This result follows since a solid angle of $2\pi$ is subtended in the limit that the field point approaches $\Delta S_k$, while for $\Delta S_k$ sufficiently small $E_0 \approx E_0(k)$, where $E_0(k)$ is the essentially constant value of $E_0$ over the (small) area

[1] H. L. Gelernter and J. C. Swihart, A Mathematical-Physical Model of the Genesis of the Electrocardiogram, *Biophys. J.*, **4**:285 (1964).

[2] R. C. Barr, T. C. Pilkington, J. P. Boineau, and M. S. Spach, Determining Surface Potentials from Current Dipoles with Application to Electrocardiography, *IEEE Trans. Biomed. Eng.*, **BME13**:88 (1966).

[3] D. B. Geselowitz, On Bioelectric Potentials in an Inhomogeneous Volume Conductor, *Biophys. J.*, **7**:1 (1967).

[4] J. L. Hess and A. M. O. Smith, Calculation of Non-lifting Potential Flow about Arbitrary Three-dimensional Bodies, *J. Ship Res.*, p. 22 (1964).

$\Delta S_k$. Equation (6.122) may now be written

$$E_0(k) = \int_V \frac{I_v}{4\pi\sigma} \left[ \boldsymbol{\nabla}\left(\frac{1}{r}\right) \cdot \mathbf{n} \right]_k dV + \frac{1}{4\pi} \int_{S_0 - \Delta S_k} 2E_0 \left[ \boldsymbol{\nabla}\left(\frac{1}{r}\right) \cdot \mathbf{n} \right]_k dS_0$$
(6.123)

where $[\boldsymbol{\nabla}(1/r) \cdot \mathbf{n}]_k$ is to be evaluated at the field point $\Delta S_k$, and $(S_0 - \Delta S_k)$ is the surface $S_0$ with $\Delta S_k$ excluded.

The solution of (6.123) for the unknown distribution $E_0$ proceeds in the following manner. We divide $S_0$ into a finite number of segments $h = 1, 2, 3, \ldots, k, \ldots, N$ and make an initial guess for the values $E_0(h)$, perhaps the magnitude of the primary field normal to $S_0$. Then by starting with the first segment $h = 1$, the right-hand side of (6.123) is computed (the surface integral being approximated by numerical techniques), and this constitutes an improved guess for the value of $E(1)$ and represents the new value to be used (Gauss-Seidel iteration). One then proceeds to the element $h = 2$, obtaining an improved value of $E_0(2)$, and so on. The procedure can be shown to converge, the difference from the true answer depending on the fineness with which $S_0$ is subdivided. For the $k$th iteration, we have

$$E_0(k) = \frac{1}{4\pi} \int_V \frac{I_v}{\sigma} \left[ \boldsymbol{\nabla}\left(\frac{1}{r}\right) \cdot \mathbf{n} \right]_k dV + \frac{1}{4\pi} \sum_{\substack{j=1 \\ j \neq k}}^{N} 2E_0(j) \left[ \boldsymbol{\nabla}\left(\frac{1}{r}\right) \cdot \mathbf{n} \right]_k \Delta S_j$$
(6.124)

Some discussion of convergence and actual examples are given by Gelernter and Swihart,[1] Hess and Smith,[2] and Lynn and Timlake.[3]

An alternative procedure is to note that in the numerical evaluation of the surface integral in (6.124), one obtains an algebraic expression which is linear in the coefficients $E_0(h)$, $h = 1, 2, \ldots, N$ [excluding $E_0(k)$, which appears on the left-hand side]. A set of $N$ equations can be written corresponding to $k = 1, 2, \ldots, N$, and these may be solved simultaneously for $E_0(k)$ by gaussian elimination. In general, however, large matrices result so that some type of iteration procedure becomes necessary. Once a solution for $E_0(k)$ is obtained, the potential anywhere may be found from (6.121) to an approximation that depends on the degree of subdivision of $S_0$ (that is, the value $N$).

If, instead of the above, one starts with Eq. (5.174), then—subject

---

[1] *Loc. cit.*

[2] *Loc. cit.*

[3] M. S. Lynn and W. P. Timlake, The Numerical Solution of Singular Integral Equations of Potential Theory, *Numerische Math.*, **11**:77 (1968); The Use of Multiple Deflations in the Numerical Solution of Singular Systems of Equations with Applications to Potential Theory, *Siam. J. Numer. Anal.*, **5**:303 (1968).

face.   A consideration of Eq. (5.97) shows that the potential field produced by the double layer depends only on the solid angle subtended by
the field point; thus, measurements can determine only the solid angle
and, consequently, not the shape.   This example illustrates the fact that
surface measurements record integrated effects produced by the sources
and that there are basic difficulties in resolving the summation into its
components.

Another problem which affects uniqueness in a practical way is that
the intensity of the field falls off with the distance from the source.
Because of inevitable noise, signals that might be of importance in
resolving two different distributions could lie below the noise level.
Thus, two sources may look alike from a practical standpoint, given the
ambiguity in measurement.   Multipole theory gives some insight into
this question because it shows that higher-order source components
involve higher inverse powers of $r$ and hence will, in general, have decreasing contributions to the total surface potential.   As the multipole order
increases, the difficulty in detecting its contribution, in general, also
increases.   A discussion of these difficulties from an electromagnetic-
theory standpoint is given in Morse and Feshbach,[1] and some further
comments with respect to electrocardiography are given by Plonsey.[2]

It would be very useful if one could choose as the heart model the
multiple-dipole distribution mentioned earlier.   For such a model permits
direct physiological interpretations; i.e., the absence of excitation of a
dipole could be interpreted as an infarction in that region.   It is hoped
that the lack of uniqueness can be reduced or eliminated by adding
known physiological constraints.   Until these questions of uniqueness
are resolved, a conservative approach is to choose an equivalent heart
model that *is* uniquely defined by surface potentials as a fundamental
property.   Such a model is the spherical harmonic multipole model.

In addition to the fixed-dipole, multiple-dipole, and multipole
cardiac representations, other formulations have been proposed.   One
is the moving dipole; in this model, the dipole location is permitted to shift
so that some reflection of the activation region may occur.   The criterion
for origin location would be to achieve the best fit to the measured data.
This model has a disadvantage related to the uniqueness problem, namely,
that no unique *physical* interpretation of the significance of the shifting
origin can be given theoretically.   On the other hand, one could view this
model as increasing the heart vector from three to six dimensions, thereby
providing additional possibilities in specification of heart condition.

[1] P. M. Morse and H. Feshbach, "Methods of Theoretical Physics," McGraw-Hill
Book Company, New York, 1953.
[2] R. Plonsey, Limitations on the Equivalent Cardiac Generator, *Biophys. J.*, **6**:153
(1966).

Another approach[1] involves the application of (5.173) to a region external to the heart, but within the torso. The surfaces bounding the region in question are the torso boundary that we have denoted $S_0$, and a surface just enclosing the heart, $S_h$. If homogeneity is assumed, (5.173) reduces to the following expression for the potential in $V$ (where $I_v = 0$):

$$\Phi(P) = \frac{1}{4\pi} \left[ \int_{S_h} \frac{1}{r} \frac{\partial \Phi}{\partial n} \, dS - \int_{S_h} \Phi \frac{\partial (1/r)}{\partial n} \, dS - \int_{S_0} \Phi \frac{\partial (1/r)}{\partial n} \, dS \right]$$

(6.128)

Let the surface $S_0$ be divided into $L$ surface elements and $S_h$ into $M$ elements. Then, if the field point $P$ is chosen just inside $S_0$, it must equal the (known) surface potential, while if it approaches $S_h$ it must equal the (unknown) surface potential there. In this way $(L + M)$ simultaneous equations are generated in $2M$ unknowns ($\Phi$ and $\partial \Phi / \partial n$ at $M$ sites on $S_h$). If $L \geq M$, a solution can be obtained for the potential and normal derivative over the surface of the heart from data taken at the torso surface. The specification of $\Phi$ and $\partial \Phi / \partial n$ may be thought of as defining equivalent single- and double-layer sources of the heart at the surface of the heart.

While these equivalent sources appear to have a close relationship to the actual sources, it should be noted that they are, nevertheless, equivalent sources and, in general, reflect integrated effects similar to the surface electrocardiogram itself. In addition, since the basic data are those measured at the bounding surface, there is no way of restoring potential components that are lost because of insufficient amplitude for detection. (If the outer surface were so distant that only dipole components were measurable, then the equivalent sources over the heart's surface would reflect the absence of data associated with the higher-multipole components.) Nevertheless, equivalent sources of this kind may have advantages in pattern recognition (in diagnosis).

A useful intermediate step in inverse problem studies involves the transformation of measured surface potentials to infinite-medium potentials. The latter refer to potentials that would be measured over the torso surface were it to lie in a conducting medium of infinite extent having the conductivity of body tissue. By so doing the effect of the discontinuity in conductance at the torso surface is removed from further consideration.

The procedure will be illustrated for the homogeneous torso, in

[1] T. C. Pilkington, "Application of Green's Theorem to the Inverse Problem in Electrocardiography" (unpublished). T. C. Pilkington, W. C. Metz, and R. C. Barr, Calculation of Resistance of Three-dimensional Configurations, *Proc. IEEE (Letters)*, **54**:307 (1966).

which event (6.125) applies. Thus the measured surface potential at an arbitrary point $J$, $\Phi(J)$, is given by

$$\Phi(J) = \frac{1}{4\pi\sigma} \int_V \frac{I_v}{r}\,dV + \frac{1}{4\pi\sigma} \int_{S_0} \Phi\,d\Omega \qquad (6.129)$$

Now the first term on the right-hand side of (6.129) is precisely the infinite-medium potential at $J$, $\Phi_\infty(J)$; consequently,

$$\Phi_\infty(J) = \Phi(J) - \frac{1}{4\pi} \int_{S_0} \Phi\,d\Omega \qquad (6.130)$$

By following the method described in Sec. 6.16, the integral over $S_0$ is approximated by a finite summation over the component surface segments $\Delta S_k$. Letting the point $J$ be at the center of segment $\Delta S_j$ permits (6.130) to be written

$$\Phi_\infty(J) = \frac{\Phi(J)}{2} - \sum_{\substack{k=1 \\ k \ne j}}^{N} \Phi(k)\,\Delta\Omega_{jk} \qquad (6.131)$$

where $\Phi(k)$ is the measured potential at the $k$th point, while $\Delta\Omega_{jk}$ is the solid angle subtended at $j$ by the segment $\Delta S_k$. In principle, therefore, infinite-medium potentials may always be determined from the measured values and known torso geometry.

If a more realistic model of inhomogeneities is required, then additional terms enter (6.129) corresponding to each discontinuous conductive interface. In this way the effect of lungs, intraventricular blood, etc., could be taken into account. However, the determination of their geometry presents obvious difficulties.

The technique described above has been given the colorful description of "boundary-layer peeloff." It was described initially by Pilkington et al.[1]

### BIBLIOGRAPHY

Bayley, R. H.: "Biophysical Principles of Electrocardiography," Paul B. Hoeber, Inc., New York, 1958.
Berne, R. M., and M. N. Levy: "Cardiovascular Physiology," The C. V. Mosby Company, St. Louis, 1967.
Evans, J. R. (ed.): "Structure and Function of Heart Muscle," The American Heart Association, New York, 1964.
Fishman, A. P. (ed.): "The Myocardium: Its Biochemistry and Biophysics," New York Heart Association, 1961.

[1] T. C. Pilkington, R. C. Barr, J. P. Boineau, and M. S. Spach, The Use of Green's Theorem for Obtaining Infinite Media Surface Potential Maps, *Proc. Ann. 19th Conf. Eng. Med. Biol.*, **8**:132 (1966).

Hecht, H. H. (ed.): Electrophysiology of the Heart, *Ann. N.Y. Acad. Sci.*, vol. **65**, art. 6, August, 1957.

Hoffman, B. F., and P. F. Cranefield: "Electrophysiology of the Heart," McGraw-Hill Book Company, New York, 1960.

Hoffman, I. (ed.): "Vectorcardiography—1965," North Holland Publishing Company, Amsterdam, 1966.

Pozzi, L.: "Basic Principles in Electrocardiography," Charles C Thomas, Publisher, Springfield, Ill., 1961.

Wolff, L.: "Electrocardiography," W. B. Saunders Company, Philadelphia, 1962.

Woodbury, J. W., Cellular Electrophysiology of the Heart, in "Handbook of Physiology," sec. 2, American Physiological Society, Washington, D.C., 1962.

# 7
# Other Applications

## 7.1  INTRODUCTION

Chapter 6 consists of a fairly detailed account of the application of bioelectric principles to electrocardiography.  This chapter, in contrast,
contains only a relatively brief account of several additional topics to
which bioelectric principles apply.   The intention here is simply to show
the existence of a broader range of subjects to which the fundamentals
may be applied.

## 7.2  MEASUREMENT OF TISSUE RESISTANCE

For the analysis of current flow in a volume conductor, the electrical
impedance of the biological tissue must be known.   This information,
consequently, is important in the study of electroencephalography, electromyography, electrocardiography, and other such systems.   A simple
technique, readily applied to *in vivo* materials, involves the use of a four-
electrode device,[1] and its application will be discussed in this section.

---

[1] C. A. Heiland, "Geophysical Exploration," p. 29, Prentice-Hall, Inc., Englewood
Cliffs, N.J., 1940.

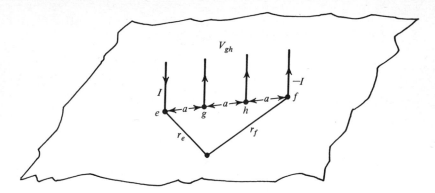

**Fig. 7.1** Four-electrode device for impedance measurement. Current $I$ is passed between outer electrodes $e$ and $f$, and potential is measured between inner electrodes $g$ and $h$.

As illustrated in Fig. 7.1, the four-electrode system consists of four equally spaced point electrodes. The outer elements carry the applied current $I$, while the inner electrodes are used to sense the voltage $V_{gh}$. When the device is placed on a tissue whose thickness $t$ is large compared with the electrode extent (that is, $3a \ll t$), the tissue may be considered semi-infinite. In this case, if we assume the tissue to be uniform and isotropic, the current spreads out from electrode $e$ (or $f$) along radial paths; equipotential surfaces are hemispheres. Under these conditions the current density at a point is the total current divided by the area of a hemisphere drawn through that point. Thus the current density due to electrodes $e$ and $f$ will be

$$J_e = \frac{I}{2\pi r_e{}^2} \tag{7.1}$$

and

$$J_f = \frac{I}{2\pi r_f{}^2}$$

where $r_e$ and $r_f$ are distances of the field point from each respective current source. The difference of potential between $g$ and $h$ is the superposition of the potential difference due to each current source acting separately. This is found by integrating the electric field (current density divided by conductivity) between $g$ and $h$, giving

$$(V_{gh})_e = \int_a^{2a} \frac{I}{2\pi\sigma r_e{}^2}\, dr_e = \frac{I}{4\pi\sigma a}$$

where $\sigma$ is the conductivity of the medium.  Since the contribution from electrode $f$ is the same, the total potential $V_{gh}$ is

$$V_{gh} = \frac{I}{2\pi\sigma a} \tag{7.2}$$

Solving for $\sigma$, we get

$$\sigma = \frac{I}{2\pi a V_{gh}} \tag{7.3}$$

Equation (7.3) is the desired formula for the tissue conductivity as a function of applied current $I$ and measured potential difference $V_{gh}$.

If the tissue is very thin compared with the electrode extent (that is, $3a \gg t$), then the current spread may be considered to be that in a thin lamina.  In this case the problem is two-dimensional, and the current density at a distance $r_e$ from electrode $e$ must be given by $I/2\pi r_e t$ (that is, $2\pi r_e t$ is the ring area at $r = r_e$).  The potential difference between $g$ and $h$ is determined, as before, by integrating the electric field from $g$ to $h$, and is

$$(V_{gh})_e = \int_a^{2a} \frac{I}{2\pi\sigma t r_e}\, dr_e = \frac{I}{2\pi\sigma t} \ln 2$$

The same contribution arises from electrode $f$; hence the total potential difference between $g$ and $h$ is

$$V_{gh} = \frac{I \ln 2}{\pi\sigma t} \tag{7.4}$$

The quantity $\sigma t$ is often considered as a separate entity and represents the conductance per unit area in a direction transverse to the lamina.  It is particularly appropriate for designating the "leakage" resistance of a membrane.  (Membranes are, of course, readily treated as two-dimensional entities.)  Designating $\sigma_m = \sigma t$, we have

$$\sigma_m = \frac{I \ln 2}{\pi V_{gh}} \tag{7.5}$$

as the value of conductance per unit area corresponding to measured values of $I$ and $V_{gh}$.  Note that the result does not depend on electrode spacing so long as they are equal and much greater than the tissue thickness.[1]

---

[1] In the analysis given here the extent of the sample is assumed to be very large, so that the effect of the edges (boundary) can be neglected.  For a circular sample of finite size a correction factor can be found in the paper by M. P. Albert and J. F. Combs [*IEEE Trans. Electron Devices*, **ED11**:148 (1964)].

## 7.3 ANISOTROPIC CONDUCTING MEDIA

Many biological tissues are electrically anisotropic; that is, the conductivity in different directions is not the same. This condition can be represented by three principal conductivities in three mutually perpendicular directions (principal axes) as determined by the structure of the material. If these directions are designated $x$, $y$, $z$, then

$$J_x = \sigma_x \frac{\partial \Phi}{\partial x} \qquad J_y = \sigma_y \frac{\partial \Phi}{\partial y} \qquad J_z = \sigma_z \frac{\partial \Phi}{\partial z} \tag{7.6}$$

and the conductivity coefficients are, in general, all different. For homogeneous anisotropic media (constant $\sigma_x$, $\sigma_y$, $\sigma_z$) the problem can be transformed to an equivalent one involving isotropic media.[1] The appropriate transformation is given by

$$x' = \frac{(\sigma_y \sigma_z)^{\frac{1}{2}}}{\sigma} x \qquad y' = \frac{(\sigma_x \sigma_z)^{\frac{1}{2}}}{\sigma} y \qquad z' = \frac{(\sigma_x \sigma_y)^{\frac{1}{2}}}{\sigma} z \tag{7.7}$$

where the new coordinates are designated by primes and where $\sigma$ is an arbitrary constant. Furthermore, we choose

$$\Phi'(x',y',z') = \Phi(x,y,z) \tag{7.8}$$

$$J_x' = \frac{\sigma^2}{\sigma_x(\sigma_y\sigma_z)^{\frac{1}{2}}} J_x \qquad J_y' = \frac{\sigma^2}{\sigma_y(\sigma_x\sigma_z)^{\frac{1}{2}}} J_y \qquad J_z' = \frac{\sigma^2}{\sigma_z(\sigma_x\sigma_y)^{\frac{1}{2}}} J_z \tag{7.9}$$

Now in the original medium we have

$$\boldsymbol{\nabla} \cdot \bar{\mathbf{J}} = 0 = \sigma_x \frac{\partial^2 \Phi}{\partial x^2} + \sigma_y \frac{\partial^2 \Phi}{\partial y^2} + \sigma_z \frac{\partial^2 \Phi}{\partial z^2} \tag{7.10}$$

By utilizing (7.7) and (7.8), Eq. (7.10) can be transformed to a relationship in the primed system, as follows:

$$\frac{\partial \Phi'}{\partial x'} = \frac{\partial \Phi}{\partial x} \frac{dx}{dx'} = \frac{\sigma}{(\sigma_y\sigma_z)^{\frac{1}{2}}} \frac{\partial \Phi}{\partial x} \tag{7.11}$$

$$\frac{\partial^2 \Phi'}{\partial x'^2} = \frac{\sigma^2}{\sigma_y\sigma_z} \frac{\partial^2 \Phi}{\partial x^2} \tag{7.12}$$

Similarly,

$$\frac{\partial^2 \Phi'}{\partial y'^2} = \frac{\sigma^2}{\sigma_x\sigma_z} \frac{\partial^2 \Phi}{\partial y^2} \qquad \frac{\partial^2 \Phi'}{\partial z'^2} = \frac{\sigma^2}{\sigma_x\sigma_y} \frac{\partial^2 \Phi}{\partial z^2} \tag{7.13}$$

Hence, substituting (7.12) and (7.13) into (7.10), we have

$$\left( \frac{\partial^2 \Phi'}{\partial x'^2} + \frac{\partial^2 \Phi'}{\partial y'^2} + \frac{\partial^2 \Phi'}{\partial z'^2} \right) \frac{\sigma_x\sigma_y\sigma_z}{\sigma^2} = 0 \tag{7.14}$$

[1] P. W. Nicholson, Experimental Models for Current Conduction in an Anisotropic Medium, *IEEE Trans. Biomed. Eng.*, **BME14**:55 (1967).

and thus Laplace's equation is satisfied in the transformed system. One can also show that the strength of a point current source is invariant under this transformation, although its position undergoes a change in accordance with (7.7). Constraining boundaries also change shape as required by (7.7).

The conductivity within the transformed region is found from (7.9) and (7.11). Thus for the $x$ component one obtains

$$\sigma_x' = \frac{J_x'}{\partial \Phi'/\partial x'} = \frac{\sigma^2}{\sigma_x(\sigma_y\sigma_z)^{\frac{1}{2}}} \frac{(\sigma_y\sigma_z)^{\frac{1}{2}}}{\sigma} \frac{J_x}{\partial \Phi/\partial x} = \sigma \tag{7.15}$$

It can be verified that

$$\sigma_x' = \sigma_y' = \sigma_z' = \sigma \tag{7.16}$$

that is, the transformed region is isotropic and its conductivity is the arbitrary constant in the transformation of (7.7).

The preceding discussion permits one to modify the four-electrode technique when applied to anisotropic media. If we assume that a principal plane (say the $xy$ plane) is parallel to the surface, then when the electrodes lie along $x$ their separation in the transformed region will be

$$a' = \frac{a(\sigma_y\sigma_z)^{\frac{1}{2}}}{\sigma}$$

Substituting $a'$ for $a$ in (7.2) and recalling that the $I$ and $V_{gh}$ have the same meaning after transformation gives

$$(V_{gh})_x = \frac{I}{2\pi a(\sigma_y\sigma_z)^{\frac{1}{2}}} \tag{7.17}$$

Hence

$$(\sigma_y\sigma_z)^{\frac{1}{2}} = \frac{I}{2\pi a(V_{gh})_x} \tag{7.18}$$

Similarly, one can show that

$$(\sigma_x\sigma_z)^{\frac{1}{2}} = \frac{I}{2\pi a(V_{gh})_y} \tag{7.19}$$

Thus the $x$ and $y$ component conductances are determined relative to the $z$ conductance.

In general, muscle will be found to be anisotropic with the conductivity along the fiber axis higher than in a crosswise direction. If the high conductivity $\sigma_h$ is defined as in the $y$ direction, then the low conductivity $\sigma_1$ will represent the $x$ and $z$ conductivities. In this case

(7.18) and (7.19) become

$$\sigma_1 = \frac{I}{2\pi a(V_{gh})_h} \tag{7.20}$$

$$\sigma_h = \left[\frac{(V_{gh})_h}{(V_{gh})_1}\right]^2 \sigma_1 \tag{7.21}$$

where $(V_{gh})_h$ and $(V_{gh})_1$ are the potentials measured with electrodes oriented along the high- and low-conductance directions, respectively. Using this technique, Rush et al.[1] measured $\sigma_1 = 0.0005$ mho/cm, $\sigma_h = 0.0067$ mho/cm for skeletal muscle; and $\sigma_1 = 0.0017$ mho/cm, $\sigma_h = 0.005$ mho/cm for heart muscle. Additional considerations in the measurement of conductivity parameters of anisotropic media are given by Rush.[2]

The torso muscle layer whose properties were considered in Sec. 6.14 can be reexamined here utilizing the above theory. If we let the $xy$ plane be parallel to the surface, we have $\rho_x = \rho_y = 280$ ohm-cm and $\rho_z = 2,300$ ohm-cm. By choosing $\sigma = (\sigma_y\sigma_z)^{\frac{1}{2}}$, (7.7) gives $x' = x$, $y' = y$, $z' = z\sqrt{\rho_z/\rho_x} = z\sqrt{2,300/280} = 2.9z$, as noted. Thus, there is stretching by a factor of 3 in the $z$ direction only. We furthermore confirm here that the resistivity of the transformed isotropic medium must be $(\rho_x\rho_z)^{\frac{1}{2}} = \sqrt{280 \times 2,300} \approx 800$ ohm-cm.

## 7.4 IMPEDANCE PLETHYSMOGRAPHY

Several studies have suggested that useful measures of blood flow in body segments can be achieved by monitoring the electrical impedance of that portion of the body. The early development of this technique was undertaken by Nyboer.[3] More recently Kinnen, Kubicek, and Patterson[4] have made detailed studies of the variation in thoracic impedance as a measure of cardiac output. We proceed now to a brief discussion of this application of electrical-impedance plethysmography, or EIP.

The basic instrumentation utilizes four circumferential electrodes,

[1] S. Rush, J. A. Abildskov, and R. McFee, Resistivity of Body Tissue at Low Frequencies, *Circulation Res.*, **12**:40 (1963).

[2] S. Rush, Methods of Measuring the Resistivities of Anisotropic Conducting Media *in situ*, *J. Res. Natl. Bur. Std.*, **66c**:217 (1962).

[3] J. Nyboer, "Electrical Impedance Plethysmography," Charles C Thomas, Publisher, Springfield, Ill., 1952.

[4] E. Kinnen, W. G. Kubicek, and R. Patterson, Thoracic Cage Impedance Measurements, Impedance Plethysmographic Determination of Cardiac Output, *SAM-TDR-64-15*, March, 1964.

two around the neck and two around the thorax below the xiphoid. The outermost electrodes carry an exciting current (around 5 ma at 100 kHz), while the inner pair are led to a bridge circuit for measurement of voltage. The electric torso model assumes that the electric current flows through two major parallel paths, the thorax and the pulmonary vascular bed. This is illustrated in Fig. 7.2. By assuming, further, that blood-volume change in the lungs is entirely responsible for the change in impedance, one can estimate the right ventricular output.[1] In particular, referring to Fig. 7.2, the total resistance is given by

$$R = \frac{R_b R_t}{R_b + R_t} \tag{7.22}$$

where $R_t$ is the (fixed) thorax resistance and $R_b$ the variable blood resistance. Utilizing the simplified cylindrical model shown,

$$R_b = \frac{\rho_b l}{A_b} \tag{7.23}$$

where $\rho_b$ is the specific resistivity of blood,[2] $l$ the length of the torso, and $A_b$ an effective (variable) cross-sectional area. The stroke volume $SV$ is is given by

$$SV = l \, \Delta A_b \tag{7.24}$$

where $\Delta A_b$ is the net change in cross-sectional area. The corresponding

[1] E. Kinnen, Estimation of Pulmonary Blood Flow with an Electrical Impedance Plethysmograph, *SAM-TR*-65-81, December, 1965.

[2] That $\rho_b$ is a constant throughout the cardiac cycle is not completely correct. The work of F. M. Liebman, J. Pearl, and S. Bagnol [The Electrical Conductance Properties of Blood in Motion, *Phys. Med. Biol.*, **7**:177 (1962)] shows that blood conductivity is a function of its velocity. This complicating factor has not been considered in current EIP work.

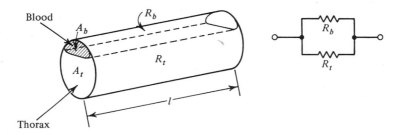

**Fig. 7.2** Simplified electrical-impedance model of torso. The thorax is considered as cylindrical with length $l$ and fixed cross section $A_t$. Blood has a variable cross section $A_b$ and length $l$. Blood and thorax resistances $R_b$ and $R_t$ are in parallel.

change in resistance is

$$\Delta R = R^2 \frac{\Delta R_b}{R_b{}^2} \tag{7.25}$$

where

$$\Delta R_b = -\frac{\rho_b l}{(A_b)^2}\Delta A_b \tag{7.26}$$

Consequently,

$$\Delta R = -\frac{\Delta A_b R^2}{\rho_b l} \tag{7.27}$$

and

$$SV = -\frac{\Delta R\,\rho_b l^2}{R^2} \tag{7.28}$$

Equation (7.28) was utilized by Kinnen[1] to calculate the stroke volume. These values were compared with that determined by the Fick technique as well as the indicator-dye-dilution method; the results are plotted in Fig. 7.3. The comparison is fairly good although the reliability is not completely satisfactory. (A typical plot of thoracic resistance and also an identification of $\Delta R$ are given in Fig. 7.4.)

In order to improve the basis for interpretation of the EIP, Kinnen has initiated a study of the current pathways in the inhomogeneous torso. The problem is similar to that considered in electrocardiography and corresponds to the determination of lead fields under reciprocal energization. The Barr-Pilkington or Gelernter-Swihart technique outlined in Chap. 6 could be applied here. However, because of the large number of phase inhomogeneities that must be identified in the present context, the difference-equation technique constitutes a more satisfactory approach.[2] This technique has been utilized by Kinnen to obtain a rough estimate (in preliminary computations) of current distribution.[3]

Efforts of a somewhat similar nature to that described above have been made to investigate blood flow in the brain. Known as *rheoencephalography*, its purpose, among others, is to determine pathological conditions that might interfere with normal flow (e.g., tumors). Considerable controversy exists over the reliability of this technique to evaluate hemodynamic factors. A review of the methodology is to be found in

---

[1] *Op. cit.*

[2] R. S. Varga, "Matrix Iterative Analysis," Prentice-Hall, Inc., Englewood Cliffs, N.J., 1962.

[3] E. Kinnen, Determining Electric Current Flow Patterns in the Thorax, *IEEE Reg. Six Conv. Record*, p. 379, April, 1966.

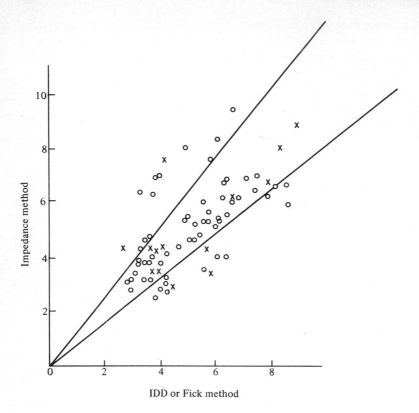

**Fig. 7.3** Comparison of cardiac output determined by electrical impedance with that found from indicator dye dilution (IDD designated by circles) and Fick method (designated by crosses). Patients with atypical impedance waveforms and with atrial fibrillation were excluded. Straight lines are at, roughly, 0.9 SD. (*From E. Kinnen, Estimation of Pulmonary Blood Flow with an Electrical Impedance Plethysmograph, SAM-TR-65-81, December, 1965.*)

papers of Lifshitz,[1] while a recent critical evaluation of the technique may be found in a paper by Laitinen.[2]

## 7.5  ELECTROENCEPHALOGRAPHY

Electrodes placed directly on the scalp are capable of detecting electric signals generated within the cortex. These surface potentials constitute

[1] K. Lifshitz, Rheoencephalography I and II, *J. Nervous Mental Disease*, **136**:388 (1963); **137**:285 (1963).
[2] L. V. Laitinen, A Comparative Study on Pulsatile Intracerebral Impedance and Rheoencephalography, *Electroencephalog. Clin. Neurophysiol.*, **25**:197 (1968).

the basis for the study known as *electroencephalography*. The magnitudes of the voltages obtained are in the low microvolt range, typically around 50 $\mu$v in man. Considerable effort has been expended to develop consistent interpretations of the electroencephalogram (EEG) and to understand its genesis.

The EEG signal persists even in the absence of obvious external stimuli. Although it would be characterized as basically stochastic, two principal rhythms have been identified, the alpha and the beta. The alpha activity is of low frequency (8 to 13 Hz) and high amplitude, while beta is of higher frequency (18 to 30 Hz) and lower amplitude. The alpha rhythm is associated with the resting condition, while the beta corresponds to an alert state.

In addition to the spontaneous activity the EEG shows a response to peripheral stimuli. Normally the background activity exceeds the

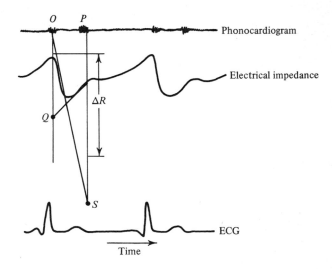

**Fig. 7.4** Measurement of thorax impedance (EIP) and both ECG and phonocardiogram simultaneously. The change in resistance, $\Delta R$, is due to pulmonary blood flow. It is evaluated from the record as follows: The negative slope during systole extrapolates to a minimum resistance (point $S$) at pulmonary-valve closure (associated with the second heart sound). The positive slope at this instant extrapolates to the first heart sound, determining the resistance $Q$. $\Delta R$ is the change from the peak resistance value (first heart sound) to the mean of $Q$ and $S$. (*E. Kinnen, Estimation of Pulmonary Blood Flow with an Electrical Impedance Plethysmograph, SAM-TR-65-81, December,* 1965.)

response, but through correlation techniques the signal can be separated from the "noise." The resultant wave is known as the *evoked response*.

As noted, the genesis of the EEG is not completely understood. If one considers the occurrence of an almost sinusoidal alpha activity with periodicity of some 100 msec against individual action potentials of 1 msec, it is clear that many cells must contribute to the resultant activity. Whatever the specific sources, the scalp potentials are a consequence of the resultant electric field within the volume conductor.

One model, due to Rush and Driscoll,[1] considers the cortex, skull, and scalp to constitute concentric spherical regions of uniform conductivity (as illustrated in Fig. 7.5). Comparison with a more realistic electrolytic model of the head shows the three-concentric-spheres model to be quite accurate except in those parts of the scalp region where the actual and model thickness differ most.

Utilizing such a model, Paicer, Larson, and Sances[2] showed that the potential field of evoked responses could be explained, roughly, by a single dipole source in the cortex. In their model the scalp and cortex had a conductivity of 0.0033 mho/cm, while the skull had a conductivity of 0.0002 mho/cm.

An interpretation of the relationship between source and lead can be made by utilizing reciprocity precisely as discussed in connection with the electrocardiographic lead field. Such a study, based on the three-layer model, was performed by Rush and Driscoll.[3]

[1] S. Rush and D. A. Driscoll, Current Distribution in the Brain from Surface Electrodes, *Anesthesia Analgesia*, **47**:717 (1968).

[2] P. L. Paicer, S. J. Larson, and A. Sances, Jr., Theoretical Evaluation of Cerebral Evoked Potentials, *Proc. 20th Ann. Conf. Eng. Med. Biol.*, **14**:1 (1967).

[3] S. Rush and D. A. Driscoll, EEG Electrode Sensitivity: An Application of Reciprocity, *IEEE Trans. Biomed. Eng.*, **BME16**:126 (1969).

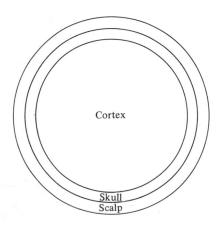

**Fig. 7.5**   Three-concentric-spheres model of the human head.

In Fig. 7.6 a lead field for a specific bipolar lead is shown.[1] From this field it is easy to visualize the relative lead sensitivity at different points. Some conclusions reached by Rush and Driscoll are that (1) an EEG electrode at the front and one at the back of the head constitute a pair that is sensitive to all sources within the brain almost equally well, (2) surface leads around 5 cm apart are approximately ten times more sensitive to proximal cortical sources than to sources in the center of the brain, and (3) electrode spacings closer than 5 cm are relatively inefficient because of the shunting effect of the scalp.

While the above is an extremely brief description of electroenceph- alography, it should be clear how the material of earlier chapters in

---

[1] The solution can be simplified by approximating the skull by a zero-thickness layer whose transverse resistance per unit area, $\rho_m$, equals that computed from the resistivity $\rho$ and thickness $t$, that is,

$$\rho_m = \rho t$$

This approximation is satisfactory because the skull thickness is small relative to scalp radius and because its resistivity is much greater than that of either the cortex or the scalp. A solution for this simplified model under the axially symmetric conditions of Fig. 7.6 requires that the potential in the cortex be of the form

$$\Phi_c = \sum_{n=1}^{\infty} a_n P_n(\cos \theta) r^n$$

while for the scalp

$$\Phi_s = \sum_{n=1}^{\infty} P_n(\cos \theta) \left( b_n r^n + \frac{c^n}{r^{n+1}} \right)$$

The three sets of unknown coefficients are found from the following three conditions:

1. The normal component of current density across the skull boundary must be continuous; i.e.,

$$\frac{1}{\rho_c} \frac{\partial \Phi_c}{\partial r} = \frac{1}{\rho_s} \frac{\partial \Phi_s}{\partial r} \qquad \text{at skull}$$

   where $\rho_c$ and $\rho_s$ are cortex and scalp resistivities respectively.
2. Ohm's law applied across the skull layer must be satisfied. Consequently,

$$\Phi_c - \Phi_s = -\rho_m \left( \frac{1}{\rho_c} \frac{\partial \Phi_c}{\partial r} \right)$$

3. The normal component of current density just within the surface of the scalp must equal the applied current density at the surface, $J(\theta)$. Consequently,

$$J(\theta) = \frac{1}{\rho_s} \frac{\partial \Phi_s}{\partial r} \qquad \text{at scalp surface}$$

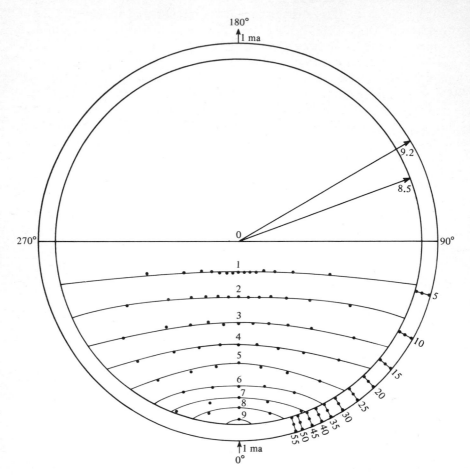

**Fig. 7.6** Plot of equipotentials resulting from application of a point current source and sink to a three-concentric-spheres conducting model. Source magnitudes are ±1 ma, resistivity of cortex and scalp is 220 ohm-cm, skull resistivity is 17,600 ohm-cm. Skull is approximated by a resistance layer of zero physical thickness but with transverse resistance equivalent to a 0.5-cm lamina. Cortex radius is 8.5 cm; scalp radius is 9.2 cm. The current source is at 0° and sink is at 180°. Potentials are in millivolts.

this book can be applied. In particular, the EEG bears striking similarities to the ECG. At this point similarities with electromyography, electroanesthesia, electrooculography, etc., should be apparent.

# Name Index

# Subject Index